The Fiction and Reality of Jan Struys

Also by Kees Boterbloem

LIFE AND DEATH UNDER STALIN

THE LIFE AND TIMES OF ANDREI ZHDANOV

The Fiction and Reality of Jan Struys

A Seventeenth-Century Dutch Globetrotter

Kees Boterbloem
Professor of History, Department of History, University of South Florida

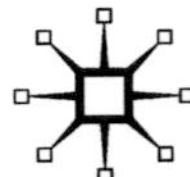

First published 2008 by
PALGRAVE MACMILLAN

Palgrave Macmillan in the UK is an imprint of Macmillan Publishers Limited, registered in England, company number 785998, of Houndmills, Basingstoke, Hampshire RG21 6XS.

Palgrave Macmillan in the US is a division of St Martin's Press LLC, 175 Fifth Avenue, New York, NY 10010.

Palgrave Macmillan is the global academic imprint of the above companies and has companies and representatives throughout the world.

Palgrave® and Macmillan® are registered trademarks in the United States, the United Kingdom, Europe and other countries.

ISBN-13: 978–0–230–55318–7 hardback
ISBN-10: 0–230–55318–4 hardback

This book is printed on paper suitable for recycling and made from fully managed and sustained forest sources. Logging, pulping and manufacturing processes are expected to conform to the environmental regulations of the country of origin.

A catalogue record for this book is available from the British Library.

Library of Congress Cataloging-in-Publication Data

Boterbloem, Kees, 1962–
The fiction and reality of Jan Struys : a seventeenth-century Dutch globetrotter / Kees Boterbloem.
p. cm.
Includes bibliographical references and index.
ISBN 978–0–230–55318–7
1. Struys, Jan Janszoon, d. 1694. 2. Struys, Jan Janszoon, d. 1694. Drie aanmerkelijke en seer rampspoedige Reysen. 3. Travelers–Netherlands–Biography. 4. Voyages and travels–Early works to 1800. 5. Russia–History–17th century. I. Title.

G236.S77B64 2008
910.92–dc22
[B] 2008021145

10 9 8 7 6 5 4 3 2 1
17 16 15 14 13 12 11 10 09 08

Printed and bound in Great Britain by
CPI Antony Rowe, Chippenham and Eastbourne

Contents

Acknowledgements

This book would not have been completed without the crucial help of a number of individuals and financial support from several institutions and organizations. To begin with the latter, the Social Sciences and Humanities Research Council of Canada, the Research Council of the University of South Florida, and USF's Department of History all deserve my thanks for funding parts of this project. The Struys project started thanks to a seminar conducted in 1989 by Philip Longworth, then at McGill University, who introduced me to 'Struys's' book, a copy of which is preserved in McGill's Rare Book collection. David Boruchoff of the same university helped to rekindle my interest in it a dozen years later. Crucial encouragement was given by Bruno Naarden of the University of Amsterdam and Daniel Kaiser of Grinnell College. For a variety of reasons, I am especially indebted to the following people who helped me in writing this book: Joyce Goggin and Karel Berkhoff of the University of Amsterdam, Ad Leerintveld of the Koninklijke Bibliotheek in The Hague, Igor Osipov of Syktyvkar (Komi Republic, Russian Federation), Arch Getty of UCLA and the various people of Praxis International in Russia, Marja Smolenaars of Leiden University, Barbara Berglund and Judy Drawdy at USF, David Schimmelpenninck van der Oye of Brock University, Thomas Lahusen of the University of Toronto, Tracy MacDonald of McMaster University, Steven High and Barbara Lorenzkowski of Concordia University, Anne Clendinning and Murat Tuncali of Nipissing University, Martha Gould in North Bay, Johanna Boterbloem-van Bockhoven, Klaas-Jan Boterbloem and Heleen van Ham in Haarlem, and Michael Strang, Christine Ranft and Ruth Ireland, as well as all their colleagues at Palgrave Macmillan.

Finally, this book would never have seen the light in this form without the continuous encouragement, love and support of my wife, Susan Mooney, an associate professor in USF's English Department, who read earlier versions of the text. She deserves my profound gratitude for weeding out a number of errors and for her crucial aid in clarifying many other points. Nevertheless, I am wholly responsible for any mistakes that remain in the following pages.

Introduction: From Sailmaker to Celebrity

> *De meeste reysigers, die derwaerts heen trecken, zijn gemeenlijck lieden, die het oogh meerder op eenigh gewin, als op een naeukeurige beschrijvingh van eenige plaets slaen ...*
> (Most travellers who travel there are commonly people who rather focus on personal profit than on a precise description of a place ...)
>
> *Jan Claesz ten Hoorn, Amsterdam bookseller, 1677*[1]

In 1676, under the name of Jan Janszoon Struys (*c.*1629–94), *Drie aanmerkelijke en seer rampspoedige Reysen* (*The perillous and most unhappy voyages of John Struys*), an autobiographical travelogue, appeared in Amsterdam.[2] Published by Jacob van Meurs (1617–80) and Johannes van Someren (d. 1678), *Reysen* became a bestseller and made out of Struys, a mere sailmaker at the time of its publication, a veritable celebrity.[3] The work went through a plethora of full and abridged editions in Dutch, French, German, English and Russian. Struys's book was known to Voltaire, and *Reysen* was mined by the Enlightenment naturalist Count Buffon. In 1879, Petr Iurchenko counted 22 printed versions of Struys's work in the Russian Imperial Public Library in St Petersburg, half of which were in French.[4] But by the time scholars such as Iurchenko 'discovered' the Struys text as a source for Russian history, the Western European reader had lost interest in crudely organized adventure tales such as Struys's.[5]

A historical Jan Struys indeed lived in Amsterdam and travelled the Seven Seas (and across much land as well) in the Golden Age of the Dutch Republic (*c.*1600–*c.*1700). A native of a region of the province of Holland where windmills had emptied lakes to create *polders* (stretches of drained land) just before his birth, this Jan Struys apprenticed with his father as a sailmaker in Wormer, his ancestral village, according to *Reysen*.[6] He took to sea to exercise his profession in 1647, travelling via Genoa to Sierra Leone, Madagascar, the Indonesian archipelago, Thailand, Taiwan and Japan. In 1651, he returned to Holland from Batavia, the capital of the Dutch East Indies. *Reysen* remains silent about Struys's whereabouts between 1651 and late 1655, when he likely made sails, ashore or on board, for the Dutch navy in the first Anglo-Dutch War (1652–54).

The book resumes with Struys's second major journey to Italy and the Eastern Mediterranean, when he served on Venetian ships fighting the Ottoman navy in 1656 and 1657. *Reysen* narrates Struys's captivity by the Turks and his subsequent escape from Ottoman (galley) slavery; at the end of 1657, he returned home safely from Italy. The book is silent again about his life between then and the summer of 1668. Archival documents show that during this period Jan Struys resided in Amsterdam, finding apparently sufficient employment ashore to provide for his wife and several children. But he buried some of these children and became a widower in late 1666.

Upon remarrying in 1668 he embarked on his third journey, with which *Reysen* resumes its account of Struys's movements. Accompanied by a number of other Dutch sailors, Struys enlisted in the tsar's service to sail on the first vessel of a Muscovite navy planned to command the Caspian Sea, but this ship, the *Oryol* (*Eagle*), never departed on her maiden voyage. Abandoning the *Oryol*, Struys and his mates narrowly escaped rampaging Cossacks at Astrakhan, only to be captured and enslaved by Dagestani in the Caucasus. With the support of the Dutch East India Company, the sailmaker regained his liberty and crossed Safavid Iran to the southern port of Bandar-e-'Abbas, from which he sailed in the summer of 1672 to Batavia. He finally reached the Republic, after a difficult return trip, in October 1673. At that point, *Reysen* ends, with Struys falling into the arms of his wife and children.

But Struys's life encountered further fascinating twists and turns, beginning with his agreement to have his odyssey published, somewhere in the first half of 1675. Jan Struys never actually wrote *Reysen*, for he was semi-literate, capable at best of merely some elementary reading.[7] But his experiences undoubtedly structure *Reysen*'s narrative, and his role in preparing *Reysen* was important enough to delay its publication until he returned from his second visit to Muscovy, in October 1676. In 1675, Struys travelled again to Moscow as a member of a Dutch embassy, since it offered him an opportunity to recover the arrears owed him by the tsar's government. When the book was published in late 1676, he was much better off than ever before in his life, fortified by the remainder of his Russian wages, payment for his service to the Dutch ambassador, and the publishers' fee he received for *Reysen*.

Although nearing the age of fifty, Struys set out in 1678 for Copenhagen, in an effort to build allegedly unsinkable warships on behalf of the Danish king. This experiment failed. Judging by the description of Struys's waterproofing methods, a version of which has come down to us in an elaborate tome about shipbuilding composed by the Amsterdam politican Nicolaas Witsen (1641–1717), this failure is unsurprising.[8] In the 1680s, Struys settled in the little town of Friedrichstadt in Schleswig. After making one final appearance in the Dutch Republic to defend *Reysen* against accusations of falsification, he returned to Friedrichstadt, where he died somewhere in the last decade of the seventeenth century.[9]

***

Besides presenting Jan Struys's biography on the basis of *Reysen*'s text and other evidence, the following pages reconstruct the creation of the book and ponder its value as an historical source for regions outside of Western Europe. *Reysen* will be compared and contrasted to the template of master narratives and other popular texts current in Western Europe about Muscovy, Persia, and other faraway regions during the Early Modern Age.[10] Influenced by structuralist analysis, some scholars ask whether the value of such epic travel accounts as *Reysen*, produced in an age unacquainted with the self-reflection of postmodernity, is restricted to helping us understand their authors and their worldview.[11] According to this viewpoint, these texts' descriptions are utterly unreliable since they refract other cultures through prejudiced prisms that distort them beyond salvage. Travel writers described mainly those things in foreign countries that had some sort of relevance for their own society.[12] Thus, in a survey of Western accounts of travelling in Russia, Françoise Liechtenhan notes how almost all Westerners were blind to the importance of icon worship in Muscovy because Orthodox churches were off limits for Westerners.[13] For a host of reasons, such as the residential segregation of Western Europeans, readers' expectations, travellers' habits of basing their accounts on authoratitive *Ur*-texts, and their lack of knowledge of the language, Westerners mechanically perpetuated stereotypes and clichés about non-Western cultures in their published accounts.[14] The blinkers of Western cultural superiority, especially infused by Catholic or Protestant religious prejudice, also affected such texts.[15] Furthermore, a traveller's social class, 'national' background, length of his sojourn as well as extent of his movements, and the parameters of the visit further affect the depiction of foreign regions in travel accounts to the point where, according to some scholars, they become virtually useless to the modern historian in learning about non-Western cultures.[16]

While such general objections against this type of source are already compelling, *Reysen*'s dubious authorship and the unequivocal presence in the text of plagiarized parts might lead to the conclusion that an investigation of the book and its alleged author might merely yield some insights into the process and context of producing books in the Dutch Republic and the mindset of its real author.[17] But M.G. Aune rightfully argues that '[a]s a record of cross-cultural encounters ... travel writing can reveal how [Western Europeans] regarded the cultural other, *and* how they regarded themselves [italics mine]'.[18] In addition, Edward Said suggests 'that you can be known by others in different ways than you know yourself, and ... valuable insights might be generated accordingly', although that does not mean that 'outsiders *ipso facto* have a better sense of you as an insider than you do of yourself'.[19] Historians of Safavid Iran, Muscovite Russia, the Mediterranean Sea and the Arabian Seas agree with such latter arguments and have therefore recently used *Reysen* as a source.[20]

Beyond archival documents attesting to the actuality of Struys's movements in *Reysen*, the book contains unique information about non-Western cultures in the seventeenth century.[21] For example, since Struys's movements in Muscovy were much less circumscribed than those of contemporary Western merchants or diplomats, his observations provide some rare insights about the tsar's realm.[22] Especially in the pages recounting the third journey *Reysen* supplies information hardly available otherwise about the history of Russia and Iran and the contested borderlands between them along the Caspian Sea's littoral.[23] Since source evidence beyond *Reysen* is particularly good for the third journey (constituting two-thirds of the book[24]), many of the following pages are dedicated to an analysis of this most authentic and interesting part of the book. This section also illustrates best how a clever combination of Struys's tales and other texts became *Reysen*'s story.

Reysen's singular quality is further reinforced by the information culled from other sources, sometimes no longer available to us, by Struys's masterful ghostwriter, in all likelihood Olfert Dapper (1639–89).[25] Dapper exhibited all the talents of an Early Modern scholar, capable of prodigious research and dismissing implausible tales. It is conceivable that *Reysen* occasionally surpasses the reliability of later writings, since Early Modern

Western-European writers did not espouse the same consistent sense of condescending superiority that typified those writing during the heyday of European imperialism.[26] *Reysen*, for example, does not yet present an ossified Orientalist image of Asia and Asians, as we will see in Chapters 9 and 10.[27]

Attesting to this open-minded curiosity about foreign others is the inclusion in *Reysen* of a section on Northern Italy. The publishers judged this area as sufficiently unknown or fascinating to the North-West European reader to warrant its inclusion.[28] Besides the Grand-Tour staples of Florence, Pisa, and Bologna, we can infer that two Italian cities were described in *Reysen* because of specific Dutch curiosity: Livorno, the main harbour for Dutch shipping in the Italian peninsula, and Venice, because her recently concluded long-term conflict with the Turks had seen many Dutchmen involved. Surely this section of *Reysen* was not included to express Dutch cultural superiority over the Italians, who remained the most admired producers of art and literature for most Western Europeans. Instead, it was offered for those contemplating the Grand Tour (or remembering it) and those curious about Mediterranean ports frequented by Dutch sailors.[29]

***

This book follows a modified chronological sequence. The first two chapters sketch the youth of the historical Jan Struys and ponder the veracity of some of the more striking episodes of his travels as recorded in the first two sections of *Reysen*. *Reysen*'s narrative and Struys's biography both provide insight into the modernizing society of the United Provinces and the collective mindset of the seventeenth-century Dutch at home and abroad. His character in the book and in real life were products of the Dutch Republic in its heyday, topic of the third chapter. Because the Dutch role in Muscovy has been rendered too modest in English-language historiography of the tsar's realm in the seventeenth century, the remarkable Dutch presence in Russia (leading to their key role in stimulating Peter the Great's naval project), crucial in understanding Struys's movements and *Reysen*'s text, will be outlined in the next two chapters.

The sixth chapter sketches on the basis of archival sources Struys's sojourn in the Republic between late 1657 and the summer of 1668, and explains why he set out on his third voyage, the outline of which is corroborated by a variety of other sources, as well as this journey's first stage. Chapter 7 describes Struys's sojourn in Moscow and his travelling along the Volga to the Caspian Sea and the significance of *Reysen*'s remarks about Russia and its inhabitants. Chapter 8 is dedicated to the *Oryol*'s crew harrowing experience at Astrakhan, Chapter 9 to Struys's enslavement, and Chapter 10 to his long journey through Iran and across the Indian and Atlantic Oceans in 1671–73. Especially in the latter two chapters the meaning of *Reysen*'s account beyond the issue of its possible veracity is gauged.

The three subsequent chapters analyse the story of creation of the book in 1675 and 1676 and its mixed genre, as well as the book's reception and popularity, first among a general readership and then, after 1800, among academics. The final chapter outlines Struys's later life, beyond the publication of *Reysen*, and the conclusion highlights some key insights deriving from the investigation of *Reysen* and Struys.

1
Struys's Youth and *Reysen*'s First Journey

Reysen opens with the young sailmaker Struys setting sail as a teenager, omitting like other seventeenth-century texts of its kind almost any information about its protagonist's youth.[1] Besides the few lines on Struys's birthplace, childhood, and adolescence in *Reysen*, no other sources are extant to fill in the blanks about his formative years, except that they confirm his birthyear as 1628 or 1629.[2] As in the subsequent two sections, in this first part *Reysen*'s charting of Struys's travels is interspersed with a series of *chorographies* or *cosmographies*.[3] These passages mix historical developments and ethnographic, geographic, and biological sketches of the exotic regions (culled from a variety of sources and only occasionally reflecting Struys's impressions) that had been visited by *Reysen*'s hero.

We can nevertheless sketch the contours of Struys's childhood. Given his profession as sailmaker, *Reysen*'s statement that Struys was born in Wormer, a village located twenty kilometres north-west of Amsterdam, appears plausible.[4] Wormer is situated in a region that is sometimes called the *Zaanstreek* (after the river *Zaan*), identified in the administrative argot of Holland as the *Noorderkwartier*. In 1566–67 at the outset of of the Dutch Revolt (1566–1648), as across the Netherlands, Protestants destroyed Catholic church interiors in Wormer before troops commanded by the Duke of Alba (1507–82) temporarily restored Habsburg rule in the Low Countries.[5] In 1572–73, when the revolt flared up again, the Zaanstreek saw fierce battling with Spanish troops, but the area fell permanently to Dutch Protestant rebels by 1574.[6] Despite the appearance of early radicalism in the iconoclasm of 1566–67, the region's Protestantization was slow, partially the result of inadequate numbers of Dutch Reformed ministers.[7] By 1630, Wormer and neighbouring villages had therefore become settlements in which Dutch Reformed, Catholics, and Anabaptists coexisted peacefully, which may have informed Struys's quasi-ecumenical Christian beliefs (if they do reflect the actual sailmaker's convictions) expressed in *Reysen*.[8] As Calvinists formed the majority in his village, it is most likely that Struys was raised as a nominal Calvinist, but neither *Reysen* nor archival records provide clear confirmation of his religion.[9] *Reysen*'s pages do not betray religious zealotry, and the few religious references may be oblique insertions by the book's ghostwriter.[10] Struys's final years in an originally Arminian enclave in Holstein suggests that the sailmaker felt most at home in a religiously tolerant environment.[11]

After 1580, behind the stabilizing Spanish-Dutch frontline that was located relatively far away, the northern Hollanders quickly recovered from the havoc of the previous fifteen years.[12] Operating from these tranquil parts of the United Provinces, refugee southern Netherlanders and native merchants expanded by the 1590s the radius of Dutch trade operations to the Mediterranean Sea, the White Sea, Africa, the Americas and Asia.[13] Although occasionally quarrelling among themselves, Dutch regents presided over a society easily absorbing annually thousands of immigrants from the Holy Roman Empire.[14] The booming Dutch economy strongly benefited from their labour. Benjamin Schmidt notes how after 1621 (when war with Spain resumed) the Dutch brimmed so strongly with confidence because of this economic success that their leadership followed a sort of global Grand Design in their war with the Spanish Habsburgs after 1621.[15] Capturing Spanish–Portuguese properties, the Dutch created a global mercantile empire of trading posts and conquered territories.

Dutch overseas activity built upon economic foundations and shipping experience predating their anti-Iberian turn, such as the crucial trade on the Baltic Sea (predominantly exchanging a variety of goods with the Polish Commonwealth for the grain that fed most of the Dutch), as well as the more recent, and more modest, trade on the Muscovite port of Arkhangel'sk.[16] The success of these ventures informed a willingness to engage and invest in a great variety of enterprises, from overseas joint-stock trading companies to domestic land-drainage projects.[17] Investment was sustained by the availability of credit (further facilitated by the creation of the Bank of Amsterdam in 1609). Technological invention further fortified an economy substantially geared toward processing imported raw materials and re-exporting them. This multifaceted economic development had Amsterdam replace Antwerp as the foremost staple market of Western Europe by 1600.

The Dutch capital and other frantically busy ports in Holland were in constant need of materials for their shipping, from ship's tack to sailcloth, ropes, and wooden planks, much of which was produced in the various villages of the Zaanstreek.[18] Extensive land-drainage projects occurred here in the early decades of the seventeenth century, which saw in 1626 the Wormer *polder* replace the Wormer lake. Struys's native village had a stereotypical Dutch look, with small houses with pointed tiled roofs, windmills and a church spire, in the midst of the flattest land.[19] But this exterior serenity was deceptive: By the time of Struys's birth, Wormer was situated in one of the most dynamic regions involved in the transformation of the Dutch economy into the most developed of Europe. Fishermen from Wormer and other nearby villages caught herring and eel at sea, a catch often shipped with other goods to German-speaking regions along the North Sea coast, east of the Dutch province of Groningen, including the Duchy of Holstein.[20] The men from Wormer sailed freight ships that returned from the Baltic Sea and Arkhangel'sk with grain, hemp, flax, wood and the like. Apart from working as sailors, the villagers of Wormer baked ship's tack, and made sailcloth.[21] The invention of sawmills in the 1590s also affected Struys's native region, for it produced the wooden planks for the mushrooming ships' wharves of Zaandam, later of such interest to Tsar Peter the Great.[22]

Wormer was therefore a dynamic and prosperous village in Jan Struys's youth. Probably before he was ten years old, he began to apprentice with his father to become a sailmaker. The Dutch historian A. van Deursen notes how '[in] a seafaring village [such as Wormer], men got accustomed from an early age to a strict discipline within a closed male society, in which physical force was one of the most important means to ensure the maintenance of one's own position. It seems almost inevitable that these long-standing habits also influenced manners and rearing.'[23] Instead of reading and writing, Jan Struys learned to rely on his wit, fists, and craftsmanship, like many fellow mariners.[24] Physical prowess and stamina certainly were inculcated in Jan Struys. While prone to get into fights when he was younger, he showed an ability in his later life to endure hardship for long periods of time.

According to *Reysen*, Struys took to sea after a thrashing from his father in December 1647, mere months before the Dutch conflict with Spain finally concluded. It may be, however, that unemployment loomed and a youthful desire to see the world beckoned. Work orders dropped off in anticipation of the peace, when the Dutch began to take their larger warships out of commission.[25] Although Struys at eighteen was slightly older than usual in taking to sea for the first time, he resembles many of his adventurous compatriots in joining the crew of a ship that was to sail from Amsterdam to Genoa.[26] Struys benefited from his training as he was given the job of under-sailmaker, a step up from a mere deckhand, the usual role of novices.[27]

Struys's first ship was called the *St John the Baptist* (*San Giovanni Batista*); its historical existence (as that of a companion Genoese ship, the *San Bernardo*) is confirmed by at least one other source.[28] *Reysen* tells us how the pair of ships encountered inclement weather and only arrived at Genoa on 25 February 1648, two months after setting sail.[29] Upon unloading the cargo, Struys did not have to seek long for his next job. The Doge of Genoa, who owned the two ships,[30] had them loaded with victuals and weapons to sustain the crews during the first stage of a three-year journey that was to establish a Genoese presence in the East Indies.[31] In joining this expedition, Struys was promoted to upper-sailmaker. The ships set sail in mid-April of 1648, passing through the Straits of Gibraltar six weeks later. The trip around the African Cape depicted in *Reysen* is enlivened by a tense stand-off with Barbary corsairs, a stopover to take in supplies at the Cape Verdian Islands, and a brush up with Sierra Leonians, the sort of incidents typical to befall Europeans sailing the Atlantic.[32] No mention is made about any landing at the southern Cape before the ships reached Madagascar in October 1648.[33]

Reysen suggests that the ships anchored in the Anton Gil Bay (located on the north-east side of the island), where armed Malagasy awaited them. Yet the locals desisted from an assault because their chief, Diembro, turned out not only to understand Dutch, but to have once been a slave of the ships' commander, Jan Maas.[34] This incredible story rather resembles the then famous encounter in 1664 between the Dutch Admiral Michiel de Ruyter (1607–1676) and the African friend of his youth, Jan Compagnie, on the island of *Goeree* (Gorée), located in the bay of the port of Dakar in Senegal, just off of the West-African coast.[35] Although an ex-slave, Compagnie served the Dutch West India Company as viceroy of the island and was said to have embraced his long-lost friend de Ruyter upon their reunion.

Maas and Diembro's reunion is certainly not the most plausible tale. Even if Jan Compagnie's story shows that some ex-slaves did not harbour vengeful feelings toward their European masters, it seems odd that a former slave would be as welcoming as Diembro was to a master from whom he had previously fled. This episode, wittingly or not, served to idealize and justify the burgeoning Dutch slave trade on Africa, presenting Africans as happy subjects of their righteous Dutch masters. Although the encounter with Diembro stretches our credulity, the visit to Madagascar may not have been merely an invention to present a chorography of the island.[36] Struys's actual sojourn there seems to be supported by *Reysen*'s fairly detailed rendition of a row that broke out among the expedition's officers ending with Maas shackling one of his lieutenants. The crew's stay at Madagascar lasted many months, for the monsoon season's disadvantageous winds prevented the ships from setting course eastward across the Indian Ocean.[37]

Reysen presents its sizable chorographic description of Madagascar as if based on the observations Struys made on the island during this long intermezzo, but it offered little original to the discerning Dutch reader.[38] Only a few years earlier, a much lengthier account of the island had been written by Olfert Dapper.[39] Given Dapper's likely involvement in producing *Reysen*, it does not surprise that it mainly summarizes or edits his work.[40] While *Reysen*'s subsequent chorographic descriptions all pay similar attention to geography, politics and history, resources, animals and vegetation of a series of regions, its description of Madgascar strongly emphasizes prevailing Malagasy female sexual lasciviousness, devil worship and infanticide.[41] In *Reysen*, Chief Diembro, acting as a superbly accommodating host, offered the crew the pleasure of the local women who apparently were 'very keen to breed from Dutch stock'.[42] This tale of utter subjugation and inhumane barbarity is reinforced by the story that other local women were kidnapped by some of the island's primates, raped, and torn to pieces, a commonplace tale in contemporary texts about the non-European world.[43] Alleged beastly Malagasy customs seem to be proffered here to affirm Western superiority over native populations, solidifying the expanding European slave trade and imperial rule. Undoubtedly, the salacious details were also included to capture readers' attention.

Readers familiar with Dapper's work may have been fooled in thinking *Reysen*'s Madagascar's description original because of the slightly different organization and wording in the 'Struysian' version.[44] But those who had read Flacourt's text, the template upon which much of both *Reysen*'s and Dapper's description of the islanders was based, may have been struck by strong similarities between all three works. Thus Flacourt's '[*d*]*e toutes les superstitions qui se pratiquent par les plus barbares nations du monde, celle par laquelle la nation de Madécasse rejette, délaisse et abondonne cruellement ses enfants, est la plus impie et abominable*' became in *Reysen* '[o]f all superstitions other peoples rival [them], but among them a certain horror and cruelty is common which I have not encountered in other parts but here, which is the throwing away and abandonment of one's own children'.[45] Likewise, *Reysen* faithfully copies Flacourt in noting how some Malagasy showed compassion and furtively ordered their female slaves to nurse and rear their rejected babies.[46]

After a half-year stay, the ships resumed their journey eastward in March 1649, reaching the Indonesian Archipelago near Sumatra in June.[47] Here the Genoese expedition captured two junks hailing from the Aceh sultanate on Sumatra. The Italians among the crew are said, in keeping with their 'horny and unbridled national character,' to have gang-raped the one woman they apprehended, while the Sumatran men were thrown overboard.[48] The violated woman was then put ashore, only to be mercilessly killed by her husband for her 'promiscuity'. In the vein of the negative portrayal of the Malagasy, this story contrasts Sumatrans' barbarian essence with the matrimonial fealty and exemplary Christian mercy as shown later in the text by Struys himself or by the Braks, faithful members of the Dutch Reformed Church.[49] The depiction of the lewd Italians conformed to the common European trope about degenerate Italians current in the age, showing how Christianity alone did not prevent sin.[50] The story (truthful or not) may have been included as a cautionary tale to the privileged reader about the danger of commoners ruled by an innate heart of darkness. Their wickedness emerged in foreign climes, when it was no longer bridled by Europeans laws, church morals, or the self-discipline that was the elite's desired norm for the European masses at home.

Reysen narrates how the Genoese ships continued to capture and loot Aceh junks for several weeks until in early July they encountered a flotilla of the Dutch East India Company, the VOC, sent out by its governor-general from Batavia. The Genoese vessels had imperilled the VOC's monopoly on Indonesian trade as well as the delicate peace in which the company preferred to conduct this trade.[51] Commander Maas, after consultation with his officers, decided to lower his sails and surrender, and the ships were convoyed to Batavia, the capital of the Dutch Asian empire.

The veracity of *Reysen*'s narrative of Struys's wanderings at this point seems confirmed by another contemporary source. In November 1649, a German servant of the VOC, Johann von der Behr, saw the two confiscated Genoese ships anchored at Batavia.[52] Von der Behr, whose memoir was published in German in 1668, recalled that the ships were called the *Ave Maria* and *Giovanni Batista* (in *Reysen*, Struys calls the ships *St Bernard* and *St John the Baptist*).[53] Because *Reysen*'s ghostwriter consulted sources in the VOC archives (see Chapter 12), it is possible that he merely blended data from VOC documents and von der Behr's book in creating the story of the first journey. But in examining various archives, the twentieth-century editor of von der Behr's text, S.P. L'Honoré Naber, uncovered a considerable amount of detail about the two ships, which the ghostwriter may not have had available to him. Most crucially, L'Honoré Naber discovered that the vessels had been built on the orders of Amsterdam and Genoese merchants, and that two-thirds of their crew was indeed Dutch.[54] *Reysen*'s story largely agrees not just with von der Behr's recollections, but also with L'Honoré Naber's investigations.[55] This suggests that Jan Struys was the source for the ghostwriter's narrative, and that he actually undertook a voyage to the East Indies in the late 1640s.

Reysen recounts how at Batavia VOC officers interrogated the crew of the Genoese ships, releasing all eventually without retribution.[56] Since Jan Maas died just after the ships had anchored at Batavia, VOC officials were spared from holding the commanding officer responsible for interloping.[57] This mercy to the crew is likely in the context of existing treaties of friendship between Genoa and the Dutch Republic; furthermore, the mild treatment of the Dutch contingent (although formally guilty of treasonous behaviour) derived from the chronic lack of personnel that plagued VOC.[58] Most of them joined the company, with the VOC chief even paying the arrears on the men's wages. *Reysen* suggests that Struys signed a three-year contract as sailmaker for eighteen guilders per month, the mandated VOC wage for his job, as other sources confirm.[59] It matched exactly the wage he had earned in Genoese service.

Reysen then explains how the sailmaker sailed from Batavia to Siam, about half a year later, in early 1650. In Thailand Struys supposedly spent once more enough time to learn of some of its cultural peculiarities.[60] But *Reysen*'s sizable description of Thailand seems to have been primarily included because of the fascination this mysterious country (rumoured to be awash in gold) exerted upon Western readers, as reflected, too, in the book's engravings. The publishers' desire to include a section on Siam was strengthened by the high cost of those copper plates (some of which they had in stock already for other publications; the print of Thailand's capital Ayutthaya-Iudia is a copy of an earlier painting).[61] *Reysen*'s detailed chorography (at times it resembles a chronicle) about the Siamese monarch, his subjects and their customs was almost fully plagiarized from a then unpublished manuscript written by the VOC officer Cornelis van Nijenrode.[62] Access to this fifty-year-old handwritten report was provided by one of the book's patrons, either Nicolaas Witsen, later a member of the VOC board, the *Heeren XVII*, or Koenraad van Klenk, whose brother Herman had been governor of Formosa.[63]

A few Thai passages in *Reysen* nevertheless seem to present Struys's genuine memories, although they are dated mistakenly. Even if *Reysen* errs in dating the shipping of elephants (in which Struys clumsily assisted) from Siam to Batavia, this transport actually occurred, albeit during the previous year, 1650.[64] A high number of other matches can be counted between names and events in *Reysen* and other historical records beyond von der Behr's treatise. The murder of a clergyman on Formosa, for example, seems an authentic recollection, even if his murderer cannot have been a man with a tail, as *Reysen* maintains in one of its most outrageous claims.[65] Struys or his editor here may have combined an actual memory with an invented part.[66] According to Lach and Van Kley, Struys additionally furnishes an observation unique among European writings from this era: he recalled mountain-dwelling indigenous men on Taiwan sporting tattoos.[67] *Reysen*'s fairly detailed description of Taiwan points nevertheless again at the use of one of Dapper's works as a crucial source: his 1670 description of Taiwan in *Gedenkwaerdig Bedryf*, a history of the VOC's overtures toward China.[68]

Much of *Reysen*'s chronology does not match dates for events the contours of which can be corroborated in archival or printed sources. The VOC's local chiefs' whereabouts around 1650 are thus rendered in error.[69] While these mistakes may have stemmed from Struys's faulty memory of events occurring a quarter century earlier or the ghostwriter's sloppy work, it is possible that the VOC's desire to protect its reputation and cover up company secrets played a role here. This secrecy may also have caused the sketchy quality of the book's engravings of the VOC houses in Isfahan and Bandar-e-'Abbas further down in the book.[70] In all three books published by van Meurs and van Someren in 1675 and 1676, details about VOC operations seem deliberately omitted or kept vague.[71] Throughout much of its existence, the VOC's Board of Directors (*Heeren XVII* ['Gentlemen Seventeen']) tried to suppress disclosure of key information about Asian countries and its operations there, as did the main and branch offices under its supervision.[72] It was in the publishers' interest to continue to abide by the rule of silence imposed by the mighty VOC on its employees, among whom had been Jan Struys, Frans van der Heiden, and Wouter Schouten.[73]

Neither could *Reysen*'s producers gain much advantage by blemishing the VOC's impeccable reputation through recalling in the book the scandals that had been rife in the Company around 1650. *Reysen* remains mum about the dismissal or forced resignation of the VOC officers Jan van Muiden, François Caron (1600–73), and Cornelis van der Lijn (1608–79) around 1650, all of whom Struys must have met.[74] Caron, especially, had become an odious figure by the 1670s, after he had set up the French *Compagnie des Indes* as a direct competitor for the VOC.[75] Instead, *Reysen* portrays VOC officials consistently in a most favourable light.

Altogether, the few concrete details about Struys's whereabouts given in this part of the book indicate that the editor and publishers were predominantly interested in finding a framework to present a lengthy account of ports of call frequented by Dutch sailors in Africa and Asia. If Struys never visited Asia around 1650, the editor may have based *Reysen*'s first voyage on documents in VOC archives, other printed or manuscripted works, and popular lore. Still, the case for Struys's presence in the East Indies around 1650 appears strong. Besides the VOC's apprehension of the Genoese ships, the cleric's murder and the habit of tattooing people on Formosa, Struys's wage, the elephantine episode, the correct naming of one Genoese ship and of various VOC officials, *Reysen* has Struys transfer on Formosa from the ship *de Zwarte Beer* to the *fluyt de Juffer*.[76] Ships with these two names sailed for the VOC in East Asia around this time, even if the *de Juffer* was wrecked near Sulawesi in March 1650 (thus Struys could not have sailed on it in May 1650, as the text maintains).

Finally, the details about his return ship roughly correspond with documentary evidence. Even though, according to his contract, Struys had another year left to serve the VOC, he was allowed to return to the Dutch Republic with the VOC fleet in early 1651, reaching the island of Goeree near Rotterdam in September 1651, according to *Reysen*.[77] *Reysen* mentions a return fleet of seven ships, the very number of vessels leaving Batavia for Holland in two groups on 11 December 1650 and 20 January 1651.[78] Struys remembered sailing on the *Zeelandia*: Other sources show that a ship called *Nieuw Zeelandia* in VOC service (under the administration of its Rotterdam Chamber and thus with that harbour as its destination) reached port in Holland on 24 June 1651.[79]

Reysen's depiction of Struys's movements in these years is sketchy, but it says even less about his mindset.[80] Assuming that he did sail to the East Indies, we can infer that his motivation to embark on a long-distance journey to the unknown derived from a mixture of Dutch *Wanderlust* and a wistful search for riches, in the footsteps of illustrious predecessors such as Jan Huyghen van Linschoten (1563–1611), Willem Barentsz (*c.*1550–97), or indeed Columbus and Marco Polo. Perhaps, more acutely, underlying his embarkation was the scarcity of work in Holland at the time of his departure. Although he did not find (or plunder) Eldorado, as a sailmaker he was not at the bottom of the ship's pecking order and earned a sufficient wage of which a young man without dependents could survive for a while. But the first 68 pages of *Reysen* that describe the first journey provide a mere glimpse of Struys's personality. Mainly, its hero appears as an impulsive and headstrong young man, albeit flexible enough to adjust to a variety of circumstances in his quest to make a living. Such qualities may have been genuinely those of the sailmaker Jan Struys, for they recur in the book and can be traced in other sources about him.

In its first section *Reysen* contained fairly detailed chorographic descriptions of Madagascar, Siam, and Formosa, three exotic areas that intrigued the seventeenth-century reader. Few Dutch descriptions of these areas were available around 1676, and some of them were only available in Dapper's expensive geographies. None of *Reysen*'s descriptions contained much original information. But for *Reysen*'s sketching of some actual movements of Struys, the text of the first voyage is predominantly derivative of other sources. The rather clumsy intersection of chorographic description with Struys's personal odyssey continued in the book's second and third parts. But the lengthy last section foregrounds the sailmaker's wanderings, making the book transcend a mere collection of chorographic passages framed by the travels of a cardboard-like protagonist. The first two journeys appear to have been included primarily as appetizers for the main course, the far more detailed narrative in which Jan Struys and his harrowing experiences in Muscovy and Iran come alive. The detail of *Reysen* regarding those two voyages also may show that Dapper had considerable time to conduct research awaiting Struys's second return from Moscow between July 1675 and October 1676 (see Chapter 12). The somewhat bland quality of the first two journeys concerned the publishers, who decided at the last moment to use a subterfuge.[81] They piqued readers' curiosity by placing the two letters about the dramatic events in Astrakhan in 1670 before the account of the first journey.

2
The Second Voyage

Struys's second journey in *Reysen* sees him traverse Italy and the eastern Mediterrean between December 1655 and December 1657.[1] *Reysen* states how Struys departed for Italy in late 1655 after a four-year hiatus ashore about which it is otherwise silent.[2] The book had no reason to bore the reader with the rather mundane (and possibly illicit, see below) work as a sailmaker in Amsterdam. Contemporary Dutch and Western European readers were uninterested in the daily lives of society's underclass, the rabble (*Jan Hagel, grauw*), unless depicted in a stylized form.[3]

Because most sailors whose home port was in Holland served on Dutch warships in the first Anglo-Dutch war (early 1652–mid-1654), however, it stands to reason that Struys was at sea for part of this period.[4] Dutch seamen were not impressed for naval duties, but the Republic's Estates General prohibited fishing, whaling, pirate, and merchant vessels from sailing out in wartime: Many sailors had no other option than join the navy to earn their keep.[5] Up to 25,000 men served in the navy during the wars of the second half of the seventeenth century, more than half of the total amount of seamen in the Republic.[6] A sign of Struys serving on warships appears to surface in his later life. In 1675 he departed in the retinue of the Dutch Ambassador Koenraad van Klenk for Moscow with the incongruous double title of palfryman and constable.[7] If there was any substance to his second office (both job assignments could have been a cover for a more important role as informer for Klenk), Struys may have learned skills in handling artillery pieces and ammunition (and leading the gunners) on a ship in the naval wars against the English.

Reysen is mum about Struys's whereabouts in the first two Anglo-Dutch Wars (1652–54 and 1664–67) or in the war with Sweden (1658–60), although the book ends with Struys's becoming entangled in the third Anglo-Dutch War (1672–74). If he was involved in any of these three wars, the book's silence seems odd in the light of the Dutch admiration for epic sea battles during the Golden Age, as the paintings by the van de Veldes, or the hagiography about some of the navigators and admirals, make amply clear.[8] The first two wars with England had, however, unfolded in a less than unequivocally triumphant manner for the Dutch, and had been led by the odious Johan de Witt (1627–72).[9] Nor was the war with Sweden an unmitigated triumph. The publishers may have wished to spare readers uncomfortable memories of those conflicts and of this disgraced leader.

But it seems most pertinent that the wars of the 1650s and 1660s bore little relation to *Reysen*'s unifying theme of charting Struys's three voyages to distant territories.[10] The Republic's third war with England intersected with Struys's third journey when he neared St Helena in 1673. That war's mention in *Reysen* aligned with the work's patriotic message of Dutch defiance of a dauntingly powerful alliance of enemies. By 1675, when *Reysen* began to be written, Holland only faced France (joined by Sweden in June of that year), after England and the prince-bishops of Münster and Cologne had made peace with the Republic. Even if the particular engagement at St Helena with an English squadron ended in a defeat, the war with England had been won. This victory was propagated as being in no small measure due to the resolute leadership by the country's 'legitimate' leadership of stadtholder William III (1650–1702; and, at lower levels, by the Amsterdam Orangists Witsen and van Klenk), rather than by the usurper whom he had succeeded, Johan de Witt.[11]

Of course, Struys was an artisan whose craft was in demand both ashore and at sea in the first two Anglo-Dutch wars. Dutch losses of warships and of merchant ships were heavy in both conflicts. As Jones remarks, most new warships during the first Anglo-Dutch War were built in Amsterdam and Zaandam.[12] Struys may thus have uninterruptedly worked on shore between 1651 and 1655, and again from 1657 to 1668.

Reysen resumes its account noting that in 1655 Struys resided near the capital in Durgerdam, a village just north of the river IJ, a short ferryride away from Amsterdam.[13] This place of residence seems plausible as at this location the sailmaker could easily make sail for the Zaan wharves to the north-west and the Amsterdam wharves to the south. In recounting his adventures to the ghostwriter, however, Jan Struys may have cunningly proffered Durgerdam as his domicile in the early 1650s to shield himself from unwanted attention to his labour past (and present).[14] It is likely that even before leaving on his second journey Struys resided at the Amsterdam address he supplied on his wedding licence in 1658, which was located in one of the newer quarters of the capital. The records of the Amsterdam sailmakers' guild do not name him, which may mean that (certainly after 1657) his work in the city as a sailmaker infringed on the guild's monopoly of the profession.[15] He thus was careful to identify himself as sailor ['*varensgezel*'] on his Amsterdam wedding licence in 1658, and pretended in *Reysen*'s pages that he lived in Durgerdam during the 1650s.

***

Further betraying its readership's lack of interest in the prosaic circumstances of life on board, *Reysen* divulges little to nothing about long-distance sailing aboard the various ships on which Struys sailed, omitting even the sort of sketchy details that had still been rendered in the popular work by Captain Willem Bontekoe (1587–*c*.1647), published a generation earlier.[16] Depicting navigational detail came to be seen in the course of the seventeenth century as dreary, and seafaring guides or portolans became part of a special niche of publications only of interest to specialists.[17] Readers' interest in the humdrum occupations of the common sailors was slight at best. Even Bontekoe's description of ship's life was far removed of the perspective of the average sailor. Cultivated seventeenth-century readers were more inclined to identify with the captain's point of view. *Reysen*'s publishers chose to concentrate on Struys's adventures in foreign lands and on the many cultures he allegedly observed there, rather than any offering any details about seafaring itself. The one exception to this is the fairly precise accounting of the trip across the Caspian Sea, but this was likely inserted in *Reysen* because it was an exotic 'sea' never before described by Dutchmen who had traversed it. Indeed, the Caspian had not yet been mapped properly, and 'Struys's' map follows the ancient Ptolemaic tradition in distorting its proportions.[18] Furthermore, the unusually rough Caspian crossing certainly stood out in Struys's memory, for the crew had made do in an improvised vessel, which was besieged by storms and pirates, and finally beached; this Caspian passage was also fairly recent and could thus be presented as even more authentic.[19]

Other sources supply us with accounts of the tremendous difficulties of life at sea as experienced by the Jan Struyses of the day, particularly during longer journeys such as those to the Far East. The VOC lawyer Pieter van Dam (1621–1706) explained their hardship:

> In general, people can sometimes bear the difficulties of travelling without replenishing for four months, but not for five months, as it

> particularly comes down to that last month. [It is] then [that] people begin to cave in because of lack of replenishments, and the scurvy begins to emerge, and [they] cannot respond and begin to expire.[20]

Most sailors preferred short trips over the long-distance journeys to the East or West Indies (which also paid comparatively poorly).[21] After his first trip to Indonesia, Struys, too, tried to sail in the vicinity of Europe, likely enjoying better conditions (that is having more food and less disease) on board than on ships sailing between Amsterdam and Batavia.[22] That he reached Batavia again in 1672 was due to circumstances beyond his control.

On the ships crossing the oceans seamen were plagued by malnutrition, illness, loneliness, and boredom, and exposed to mistreatment by superiors and fellow sailors.[23] Capture by pirates, drowning alone or with their mates, and sometimes dangerous foraging trips on land added to their challenges. A variety of authors point out that the chance of perishing in a wreck was not as high as might be supposed, but death was the sailors' constant companion.[24] Most men sailing from Holland to the East Indies never saw home again. And even during the relatively brief sorties of fleets during the Anglo-Dutch Wars, illness on the ships was rampant, especially brought on by poor-quality drinking water.[25]

VOC sailors, who may serve as examples, loaded and unloaded the ship's cargo, hoisted and lowered the sails, mopped the ship, caulked the cracks, cleaned the hull regularly from seaweed and shells, maintained the ropework, shifted and turned the gunpowder barrels, collected firewood and water ashore, and speedily and obediently executed commanders' orders, so as to avoid corporal punishment.[26] Some contemporary observers compared their fate to that of slaves. Higher commanders and politicians considered sailors, to use Jones's words, akin to 'expendable and consumable commodity like sails, cordage and planks'.[27] Most of the sailors on the vessels bound for the Indies were violent, rude and crude, boasting of few skills, rather different from the better skilled sailors on shorter hauls.[28] Pay was low on almost all ships, although it was normally a little higher than for comparable semi-skilled or unskilled work ashore.[29] VOC sailors, who for Dutch standards were paid comparatively low wages, enjoyed the privilege of being permitted to sell the contents of their chest (*kist*) upon return home, a habit that may have derived from Iberian practices.[30] Obviously, many tried to bring back in their trunks (or otherwise) more than what was allowed, and most engaged in petty trade at the various Asian and African ports of call.[31] Jan Struys must have brought back such a loaded trunk in 1651,[32] but in 1673 he missed out on this bonus because his goods had been confiscated by the English. If returning home safely, sailors might face prolonged unemployment, quickly depleting their purse from the proceeds of their trip.[33]

Somewhat surprising in a country that prided itself on its seafaring prowess, common mariners were looked upon with disdain in the Republic.[34] But the sailors did often express an independence of mind and a sense of solidarity, which hints at the development of an early sort of working-class consciousness.[35] They were also often incomparably tough, capable of facing many hardships. Moreover, the mariners were wont to exhibit a sense of opportunism and saw possibilities for upward social mobility others ignored or did not find worth the risk.[36] Struys's experiences attest to all of these qualities.

***

Although *Reysen* suggests that Struys sailed from Amsterdam to Livorno against his will, we may surmise from his determined movements after coming ashore in Italy in early 1656 that he planned to join the Venetian navy on its annual campaign fighting the Ottomans before he left Amsterdam.[37] The lead of compatriots may have informed Struys's decision: thousands of Dutchmen served the Most Serene Republic in this lengthy war (1645–69).[38] *Reysen* describes how Struys survived an eventful journey (he was robbed twice) on foot from Livorno to Venice.[39] In recounting this traversing of Italy, *Reysen* provides a sort of tour guide of the various Italian towns' sights, showcasing the investigative work of Struys's ghostwriter, and heightening the text's attraction for its target audience.[40] Did it stretch the credulity of contemporary readers to find their sailmaker suddenly behaving like a young Dutch patrician on the Grand Tour, sampling the various architectural delights through the prism of a cultured aesthetics?[41] Perhaps not. Western European readers' gullibility is manifest in their acceptance of Baronesse d'Aulnoy's (1650/1–1705) two-volume memoirs as authentic soon after *Reysen*'s publication.[42] Rather than travelling herself to the foreign countries described in her recollections, Madame d'Aulnoy's work was entirely a compilation of various travel accounts by others and sundry sources. No one questioned the authenticity of her voyages, however, and her books were reprinted as much as Struys's tales.

In Venice, which *Reysen* has him reach in March 1656, Struys found temporary lodgings with a Dutch sea officer, and was hired as sailmaker on the ship *Vergulde Haan*.[43] The Venetians were in the midst of equipping their fleet for its annual maritime campaign against the

Turks.[44] By the mid-1650s, this conflict had primarily become a struggle over possession of the island of Crete and the last Venetian stronghold there, its capital *Candia* (Herakleion). By annually blocking the Turkish navy from sailing out of the narrow waterways linking Istanbul with the Aegean Sea, the Venetian fleets attempted to stop significant Ottoman reinforcements from reaching the island, and grain and other supplies from reaching Istanbul.[45]

Led by the flagship *(Groote) St Joris* (*San Giorgio grande*) under the command of the renowned Danish-Norwegian commander Cort Sivertsen Adeler (1622–75), Struys's naval detachment left Venice on 10 April 1656.[46] A series of harrowing tales follows in *Reysen*, interspersed with the description of some of the Aegean coastline and islands, which its Dutch readers recognized, since they were usually well versed in their Classics.[47] Struys's alleged personal recollections provided once again the thread connecting the various sea battles, chorographies, and classical references.[48] *Reysen*'s outline of the progress of the war in these years is confirmed by other sources, which do not, however, identify minor participants such as sailmakers.

In the spring of 1656, the Venetian fleet camped out in the Dardanelles, near what *Reysen* suggests were the ruins of Troy.[49] Apart from some minor skirmishes, both fleets refused to engage in a major battle, and scurvy began to spread among the Venetian crews because of a dwindling supply of fresh food.[50] Struys was dispatched with seven other crew members to collect fresh water on the Asian coast, but was captured by Turkish soldiers.[51] He was brought to a galley, where he had to strip to his loincloth and his head was shaved. The use of galley ships was in decline by this time in the Mediterranean Sea, as they began to lose the competition with sailships in battle and trade: this had been one reason the Venetians hired many Dutch ships and sailors to fight the Turks.[52] But the transition from rowing to sailing was gradual. In the Venetian-Turkish war galleys, mainly rowed by captives, still fought in most battles. Struys's future looked bleak: Chained to each other, galley slaves suffered a 'living death' of insufficient sleep, hunger, thirst, lashings, unbelievably hard labour pulling oars, filth, and an early death; despair was enhanced by the knowledge that the Turks never released captives who had served the Venetians.[53]

Reysen informs the reader how Struys was seated at an oar with five other men, and was chained to one of his neighbours.[54] This was a Russian (*Rus*) who claimed to have been a galley slave for twenty-four years and who had lost his ears and nose as punishment for his failed escape attempts. If the story has substance, the two likely communicated in the original *lingua franca*, which Struys by then mastered sufficiently because of his close contact with Italians during his first journey and his travels across northern Italy.[55] *Reysen* recounts how the Russian had hidden a file bought on one of his previous visits on shore and had been awaiting an opportunity to use it in an escape attempt. When the two were assigned to fetch some water ashore, Struys and the Russian fled, eventually making it back to a Venetian ship of Dutch heritage, the *Abrahams Offerande*.[56]

Whereas the general outline of Struys's movements in Venetian service acquires at times the appearance of being historical, this episode of capture and escape seems more than likely a concoction by the ghostwriter. For someone who had endured twenty-four years in Ottoman captivity, the Russian's continued defiance is surprising.[57] The identification of Struys's oarsmate as Russian may have been a deliberate play on Dutch readers' lingering memory of Eastern Slavic renegades' role in the Venetian-Turkish war. The Ottoman Fleet in 1656 was commanded by a 'Russian' convert to Islam, Chinam Pasha, a fact possibly inspiring the invention of the Russian oarsmate.[58]

Reysen proceeds to supply a list of the Venetian commanders who led their navy to victory the battle of the Dardanelles of June 1656.[59] The officers' last names agree with many of those given by the Dutch historian de Jonge for this engagement, but it seems impossible that an illiterate sailor could remember the detailed list of Italian commanders as presented in *Reysen*.[60] The sources (such as the popular monthly journal *Hollandtsche Mercurius*) that modern historians used to depict the battle may have been the very same sources mined by *Reysen*'s ghostwriter in 1675–76.[61] The exploits of the Venetian fleet and especially the Dutch in Venetian service (many of the ships, although nominally under the command of a Venetian, were in fact navigated by Dutchmen[62]) had been celebrated in a poem by the greatest of Dutch poets of the Golden Age, Joost van den Vondel (1587–1679). Vondel's *Op den Zeetriomf der Heerschappije van Venetie* (1656) burned in the Dutch collective memory the heroic lore of the Dutch feats against the infidels at the Dardanelles.[63] Thus there are good reasons to suspect the Aegean episode as a grandiose forgery. But a firm rejection of any participation by Struys in the Venetian-Ottoman battling is complicated by *Reysen*'s rather wide array of people and incidents confirmed by sources beyond those mentioned.[64]

Having defeated the Ottoman adversaries, the Venetian fleet began to cruise across the Aegean Sea, forcing various islands to remit tribute [*brandtschatten*] in exchange for Venetian 'protection'.[65] On the shores of Samos, however, Struys, together with a bunch of companions, was once more captured by Turks. They were transported on a Turkish brigantine to Rhodes in order to be sold as slaves. An outbreak of pestilence had driven commodity prices down on Rhodes, and Struys and company were shipped to Scio where the Turks hoped to fetch better prices.[66] But prices there also fell below expectations, which allowed the captured Christian sailors to persuade the Turkish captain to sell them back to their own Venetian captain. For, *Reysen* tells its readers, his ship was still anchored near Samos.

After this second brief enslavement, Struys's ship set course home to Venice since it had sprung several leaks and was in dire need of repairs.[67] Its dilapidated state forced it to pause on the way, during which stop Struys proved instrumental in plugging some of the holes. If *Reysen* is at all reliable, Struys showed here some early signs of his prowess as a shipwright. Later, in Denmark, he would use his skills handily, earning him a place in the standard seventeenth-century work on Dutch shipbuilding by Nicolaas Witsen.[68]

After Struys's return to Venice, he left port again in the spring of 1657 to serve on another Venetian ship for a modest eighteen guilders a month.[69] His ship, which carried a Venetian *proveditor* (an official who inspected army and navy readiness and facilities), toured the Venetian islands in the Ionian Sea on the way to Crete and beyond.[70] This tour served as a happy pretext to supply a sort of geographic summary of seas and shores in *Reysen*'s pages. Struys's vessel eventually set course again for the Dardanelles to resume battle with a resurgent Turkish navy and army that tried to reconquer Tenedos. Several minor skirmishes in the early phase of this campaign went the Venetian way: during one of them a Barbary ship commanded by a Dutch renegade captain surrendered to the Venetians. This captain was one of several who fought with the Turkish squadrons against Dutch-made Venetian ships manned by Dutch crews.[71] Native Dutch sailors who had converted to Islam served in the Turkish navy in some number. An entire Dutch mercenary unit fighting on behalf of Venice against the Turks, including their commander, Charles van Dijck, converted to Islam after surrendering in the 1640s.[72]

Christian conversion to Islam in the Early Modern Period constitutes a complex topic.[73] These conversions raise questions about the extent and intensity of seventeenth-century 'confessionalization': Under the impact of reforming parsons, ministers, nuns, monks, consistories, and theologians, historians suggest, Western Europe's population intensified its Christianity by discarding lingering remnants of traditional folk religion, pagan survivals, and superstitions.[74] Among Dutch Calvinists, confessionalization received a boost from the Synod of Dordt (1618–19) and the Bible translation (completed in the 1630s), but long thereafter religious views remained often poorly supervised and defied the orthodoxy desired by the consistories.[75] Even in the province of Holland half of the population never joined the Dutch Reformed Church. The fairly high number of Dutch converts among Ottoman and Barbary sailors confirms that a deepening and purification of Christian beliefs neither happened overnight nor was everywhere thoroughly inculcated.

Because of the nature of their trade, meanwhile, Christian European mariners maintained tenuous ties to their home ports. When after capture by Barbary corsairs or Turkish ships their uncertain religious views were challenged by the tangible advantages – legal, material, cultural, social – offered by conversion to Islam, the temptation to convert was strong. For many, to become a Muslim was not such a stretch as one might surmise from the sinister image of the infidel painted by the contemporary iconography and literature of Christian Europe.[76]

Reysen tells us how, in late July 1657, the Venetians inflicted another defeat upon the Ottoman fleet near Tenedos.[77] Although the Venetians won several battles against the Turks, two of which are depicted in *Reysen*, the tide in the war definitively turned against them in subsequent months (about which *Reysen* is silent). The architect of the Ottoman counteroffensive was Mehmet Köprülü, the new Grand Vizir, who began the Ottoman recovery by landing troops on Tenedos and Lemnos.[78] The Venetian loss of these islands opened the route to Crete once again for Ottoman vessels. When much of the Venetian fleet was shipwrecked in a storm in 1658, the fall of isolated Candia became a matter of time.[79] The Turkish victory was finally confirmed in a peace treaty in 1669.

In *Reysen*, in another educational interlude, a stop at Delos leads Struys's ghostwriter to describe various ruins of the Greek temples, most of which were dedicated to Apollo.[80] These were further incongruous observations for an unlettered seaman. By contrast, his ship's visit to the island of Milo perhaps constitutes a more genuine recollection of Struys himself. As usual for islands under Venetian rule in the Aegean or Ionian Sea, Milo had both Orthodox and Catholic inhabitants. The reader is told how Italian rather than Greek was the language of the majority of the islanders. *Reysen*'s protagonist narrates how this was the consequence of the Venetian fleet's frequent visits, when not just supplies were loaded, but the commanders allowed their crews to have sex with the local women. The latter thus took the opportunity to learn Italian '*met byslapen*' (by having sex), by which the book happily returned to the trope of Italian lasciviousness.[81]

After reaching Venice again, Struys joined, unwittingly, a pirate corsair (*kaper*) with Livorno as its first destination.[82] Before any raid had even been attempted, however, the ship was confiscated at Livorno on the orders of the Grand-Duke of Tuscany: According to *Reysen*, Struys was spared punishment for his participation in piracy, and found a ship that brought him back to Holland.

***

Circumstantial evidence attests to the veracity of Struys's first two adventures. But the issue whether or not Struys truly completed these journeys seems to have been almost immaterial from the publishers' perspective. Most of the text rather reflects the research conducted by his ghostwriter. Seventeenth-century readers were willing to overlook the book's imperfect form, and, as the example of Baronesse d'Aulnoy's work shows, inclined to believe the details of Struys's odyssey, even if much of it was derivative. They perused the sections that engaged in history and physical and human geography with keen interest.

Dutch culture after 1600 featured an unrivalled openness to, and curiosity about, other worlds because its society and economy were intertwined with seafaring, overseas trade, and the encounter with other cultures near and far. But while this predeliction may make this erudite and curious Dutch readership appear 'modern', this was not a very 'open' perception and reception of the outside world. The Dutch elite's thinking and worldview were heavily grafted on the texts of Antiquity in which they had been immersed at their schools following a humanist curriculum inspired by Erasmus. The ghostwriter made a concerted effort to appeal to this education as can be gauged from the modelling of *Reysen*'s chorographic sections on Herodotus's work. Dutch interest in the outside world was not simply the expression of a curiosity that was wholly empirical or objective in its recording of facts. It was infused by a sense of their superiority based on their identification with this classical heritage (adding to their identification with a Biblical chosen people).[83]

Struys, his editor, publishers, and readers looked at the world through the prism of fundamental concepts developed by Greek philosophers and Christianity's morality and eschatology.[84] Informed by ideals such as piety, sobriety, and the importance of hard work that structured its readers' lives, *Reysen* concurred with the prevailing metanarrative about the foreign regions the protagonist-narrator Struys visited. Throughout the Early Modern Age, when observing other societies, travellers rarely transcended the spiritual parameters of their own culture, nor did the readers of their tales.[85] Only slowly did the transformative quality of travelling gain ground, moving toward acceptance of both 'Copernican and Columbian worlds'.[86]

But *Reysen* furnishes enough valuable material for the historian.[87] Some historical phenomena are rendered with an amount of detail greater than available otherwise in the meagre evidence extant about them; and the chorographic parts are more than mere fantasy even if we cannot always separate facts from fiction.[88] Following the example of books earlier published by van Meurs, *Reysen* supplies much information about foreign cultures and represents the contemporary state of knowledge about them, based on what was considered solid research by the era's standards.[89]

What is clearly a handicap for anyone using the book as a source for the *histoire événementielle* is its almost continuous supply of mistaken dates of events. Their error meanwhile adds strength to the case for the authenticity of Struys's recollections about them. Would an otherwise meticulous researcher such as *Reysen*'s ghostwriter deliberately distort the dating of events that he encountered in his sources? Struys's tales often seem to have indeed substituted for written sources. For if Struys participated in the events described in the tale of the first two journeys, it stands to reason that he could not remember a quarter century later when things had happened, especially because he did not write down notes (even though such a habit is ingenuously suggested at the end of the book!).[90] A case can be made for the greater accuracy of an illiterate's reminiscing compared to that of a literate person, but the illiterate will be only dimly aware of the organization of the calendar and the sequence of years.[91] When *Reysen* does supply very specific details, as in naming the Venetian fleet commanders, its text's authenticity actually diminishes.

In Struys's native region, sailors were wont to tell villagers tales of their adventures at sea and in faraway lands.[92] Undoubtedly, they embellished many a story, and they wove their own adventures into stories they had heard from others. It is conceivable that in Struys's narration of his first and second travels he appropriated tales of others' adventures in East Asia or the Mediterranean. Consulting private libraries or other print and document collections, his ghostwriter then tried to improve the anecdotes' verisimilitude by unearthing or correcting names of ships, VOC personnel or Venetian commanders whom Struys should have encountered. Throughout the book most dates of the events to which Struys was witness (or about which he heard) have been printed in Latin script, hinting that they were added after an effort at fact-checking or merely based on calculating guesswork (and the close reader might have nevertheless noticed that some of those dates could not be accurate). But incontrovertible proof that Jan Struys travelled to East-Asia in the 1640s and in the Mediterranean in the 1650s as rendered in *Reysen* is missing, and the earliest outside confirmation of his historical existence is his 1658 wedding licence.[93]

3
The Dutch Republic

Although *Reysen*'s title page announces that it features 26 years of Struys's travels, it describes in fact no more than the ten years that the sailmaker spent in faraway regions. By the time the book was published its alleged author had lived for four-fifths of his almost fifty years in the Dutch Republic. Its society's habitus and economic ethos informed the behaviour of the historical Jan Struys and determined the depiction of him in *Reysen*'s pages. In understanding Struys and *Reysen*'s most fascinating and revealing voyage to Muscovy and Iran (as well as *Reysen*'s creation and Struys's later years), the context of this anomalous seventeenth-century country needs to be taken into account.

The Dutch seventeenth-century advance toward capitalism and modernity has fascinated thinkers and scholars from Adam Smith and Karl Marx to Jan de Vries and Ad van der Woude.[1] If we adopt a slightly altered version of Geraldine Phipps's definition of modernization, the Dutch Republic after 1580 forged ahead by '[introducing] technological, procedural, and practical innovations in important areas' of the economy, government, and military, which led to 'changes in significant areas of society, as in social and cultural mores, political and religious ideology and practices, law, language, and national character'.[2] '[Importing] directly foreign techniques, [producing] native versions of foreign models, and [creating] new national institutions and technology', the Dutch created an economy that outstripped the value of every one of their European rivals.[3] Foreign visitors marvelled at the prosperity and power of the United Provinces.[4] Across Europe, elements of the Dutch military, naval, technological, entrepreneurial, and cultural model were admired and emulated. Seventeenth-century Dutch merchants, businessmen, soldiers, sailors, and artisans helped to modernize the economy, military, and navy of many countries, especially in Northern Europe, from the Duchy of Holstein to the Tsardom of Moscow.[5] Exploring and exploiting the various possibilities this burgeoning global economy and modernizing society offered, the historical Jan Struys increasingly asserted his abandonment of the traditional Christian-European worldview in which birthright, rather than merit, determined one's fate.[6] Telling his tales to his ghostwriter and lending his name to *Reysen* was one telling sign of his enterprising confidence, but as Chapter 14 will show, he explored a host of other opportunities to enrich himself after the return to Holland in 1673 with which *Reysen*'s pages end.

That the Republic presented an early version of capitalist modernity can be argued on a great number of grounds, beginning with the dimensions of its economy. In 1648, when *Reysen* has Struys departing Amsterdam for the first time, the Republic gained official independence at the Peace of Münster, and the Dutch economy entered its 'zenith' as the most developed capitalist economy of its day.[7] This thriving economy had a labour force predominantly employed outside of agriculture, while its own agriculture was specialized to a considerable degree, and most of the grain the Dutch consumed was imported (and a great amount of luxury goods such as colonial goods were consumed by much of the population as well).[8] The Republic was a popular destiny for immigrants, beckoned by an exceptionally high standard of living and seemingly abundant employment. A country geographically as vast as contemporary Russia could merely sustain six times the number of the Republic's inhabitants (c. 11.2 million versus c. 1.9 million), another sign of Dutch prosperity.[9] The Republic had a moneyed economy with a stock market, insurance, and banking system.[10] A legal system facilitated business and protected its contracts and proceeds, through property rights, patent law, insurance, and other guarantees.[11] The Republic boasted a substantial manufacturing sector even if its capacity was constrained by its reliance on the use of inefficient organic energy sources as its fuel.[12] The printing and the publishing industry that produced *Reysen* is a telling example of the early use of sophisticated technology and an elaborate, modern-type, division of labour (see Chapters 11 and 12).[13] Its production reached phenomenal dimensions, with Dutch presses responsible for half of the total number of books produced in Europe during the seventeenth century.[14]

The sophisticated mechanization of the major drainage projects in northern Holland around 1600 was another proto-industrial technological feat, while the reclaimed land itself became the location of further semi-mechanized production. Immediately after completion of the *polders*, sawmills and ships' wharves mushroomed in Struys's native region.[15] In 1670, the Dutch merchant marine had more than 2,100 ships, manned by tens of thousands of sailors.[16] Ships' construction was a complicated and costly matter that also involved a highly developed division of labour.[17] Such complex ventures could be executed in the Republic with ease thanks to the far-reaching commercialization of its economy.

Not just the political and economic elite behaved according to a modern ethos in the United Provinces. The Dutch rate of literacy neared levels only reached two hundred years later elsewhere in most of Europe.[18] Many of humbler birth, such as Jan Struys, explored the opportunities that were presented for potentially lucrative employment, and engaged in modern-style mass consumption of luxuries such as imported tobacco, coffee, tea, and sugar.[19] The very act of selling his stories to *Reysen*'s publishers attests to the sailmaker's precocious 'economic rationalism'.[20] That a sailmaker could credibly albeit mistakenly appear as a writer of a book of almost four-hundred pages is testimony to these literacy levels. In participating as a subject in capitalism's developing international labour market in the seventeenth century, Jan Struys behaved like other Dutch sailors, soldiers, and artisans seeking employment abroad, who were anything but 'lumpenized' objects.[21] Following J.A. Hobson's ideas, Lenin proposed around 1900 that the working class of the global hegemonic powers 'became bourgeois', because it benefited from the profits of the imperialist policies its ruling class executed in other parts of the world.[22] This *embourgeoisement* manifests itself early in the Dutch Republic, where Struys and his fellow sailors reached far higher living standards than the underclass in much of the rest of the contemporary world.

But life for Dutch workers was fraught with dangers. The choice for a life at sea represented a risky gamble, since a few made a fortune, but the great majority of Dutch sailors lived a short and brutish existence, particularly when engaged in long-distance sailing.[23] The rewards were tempting enough, however, as sailing skills were in demand far beyond the Republic itself.[24] Even though exposed to periodic epidemics, the vagaries of economic crises and working in a particularly dangerous occupation, some enjoyed an exceptional career, as the cases of the admirals Maarten Harpertszoon Tromp (1598–1653) and Michiel Adriaanszoon de Ruyter (1609–76) show.[25]

From 1672 onwards, a relative decline of the Dutch position in certain sectors of European trade was offset by a growing volume of trade on Spain and the Spanish Empire, as well as Africa and East Asia.[26] Nevertheless, the rise of England and France as competitors coincided with the passing of the great age in the Republic of the dynamic entrepreneur or merchant-captain, who had been the engines of the earlier Dutch capitalist boom.[27] Key financial investors now took the lead in keeping the economy humming, usually university-educated descendants of merchants and entrepreneurs who had made their fortune in an earlier age; by the second half of the seventeenth century, they also usually doubled as the political elite of the Republic. This transition again is apparent from the history of Jan Struys and of *Reysen*: Struys's book's patrons Koenraad van Klenk and Nicolaas Witsen were scions of the new patriciate.[28] Their peers retired to landed estates, sometimes brandishing quasi-noble titles and increasingly taking on the contours of a wealthy oligarchy.[29] By 1670, the Republic was thus becoming a country led by *rentiers* preferring the status quo which allowed them to live comfortably on their investments. For a while, such complacency went unpunished, as the Dutch economy only began to enter a phase of genuine (rather than relative) decline after 1715.[30]

***

The Republic's elite stood at the apex of a complicated, male-dominated, society composed of a layering of discernible groups that were primarily distinguished by wealth, political power, and social status.[31] Peter Burke has discovered the first modern use of the term 'class' (distinguishing people by property and wealth) in the seventeenth-century Republic, and it appears apt to perceive Dutch society as divided into gradations of classes based on men's professions.[32] Struys belonged for much of his life to the working class of sailors, soldiers, agricultural workers, and fishermen, for whom regular underemployment and unemployment shaped an uncertain existence.[33] They sometimes sank into the *lumpenproletariat* of day labourers and beggars, but as Struys's life shows, they could also reach middle-class affluence and respectability.[34] Social categories easily blended, and moving up and down the scale was common during many people's lives before 1700. Many poorer people discerned chances to rise above their station into the broad middle class. It is especially noteworthy for Struys's case that 'Dutch seafaring communities were ... much less resigned to their hard lot than were the more submissive urban workers and agricultural labourers.'[35] Dutch society stratified after 1650, when the political and social turmoil of the sixteenth century and the *Wirtschaftswunder* of the first half of seventeenth century gave way to a comparative socio-political calm and economic slowdown.[36]

The strength of patronage networks partially explains the remarkable stability of the Republic's social order, even in times of recession. The Republic's elite monopolized political bodies at town, province, and 'federal' level, distributing patronage and favours to those of lesser station; the upper crust enjoyed a wealth (including its 'symbolic capital') that steadily grew while the income of the groups below it usually stagnated or even declined after the middle of the century.[37] Increasingly after 1670, bourgeois patricians emulated the behaviour of French courtiers, even adopting French as their language of mutual communication in the final decades of the seventeenth century.[38] They all belonged to informal clans (among the elite) and were patrons of a retinue of clients (with society's elite subduing or controlling subalterns by selectively distributing favours).[39] An Amsterdam burgomaster of Struys's day, Joan Huydecoper (1625–1704), 'kept complete records of all the services he rendered as well as all the favours granted him. Only in this way could he control the network [...] that was one of his greatest social assets.'[40] Struys's life, the genesis of *Reysen*, and the activities of the Dutch community in seventeenth-century Russia and Asia testify to the operation of Dutch patronage and kinship networks at home and abroad.[41] VOC board members doubled as Amsterdam city councillors and dispensed favours in both capacities. The Amsterdam-centred networks had ties to others across the Republic and were linked with substantial Dutch interests abroad, the Dutch evidently realizing that '[r]eal capitalism is crony capitalism'.[42]

The financial backers of *Reysen*, Witsen and Klenk, had business interests in the East Indies and Muscovy, while serving as leading members of the Amsterdam city council and envoys of the Republic's Estates-General.[43] In Muscovy a closely connected network of foreign entrepreneurs and traders existed linked to the Dutch Republic.[44] Holland's ways and mindset supplied the frame of reference according to which this Western European network operated.[45] Kinship and patronage networks sustained the dominance of the Dutch trading 'houses' in the Russian-Western European import and export trade. The Dutch merchant houses and their agents waylaid the strong Muscovite xenophobic streak by their continual presence in Russia through several generations, consistently proving their value and loyalty to the tsar's government.[46]

Informal patronage networks and kinship networks also forged a sense of belonging to the abstract collective of a nation.[47] Struys was beholden to a series of patrons who occupied leading positions in the Dutch economy at home and abroad, as well as in Dutch politics, and he reaped the benefits of showing proper deference to his patrons. The praise *Reysen* lavishes upon VOC officials and the two patrons of the book show an editorial strategy that solicits the patriarchal benevolence the elite exercised toward its clients.

Social stability was further bolstered by the physical proximity between the population and the governmental institutions of the provincial estates and town councils, the prime *loci* of formal power in the Republic. To avoid major social unrest, the wealthy and powerful further ensured a constant availability of grain reserves that prevented famine from breaking out, while they funded charitable institutions to relieve searing poverty.[48] The political authority of the estates and councils originated in medieval customs, charters and privileges.[49] Governmental practices and institutions only evolved piecemeal in response to new political challenges, usually in times of crisis. This intermittent and cautious evolution ground to a halt after 1700, so that an increasingly ossified ruling class monopolized all positions of power throughout the eighteenth century.[50]

Despite its awkward and ostentatiously decentralized governmental structure, the seventeenth-century Republic was a strong state, effectively organizing its military and finances.[51] In Struys's day, a state's power was measured in military capability. The Dutch Republic was exceptionally capable of mobilizing resources for war, allowing it to survive a series of conflicts with mighty opponents on land and at sea.[52] Dutch army drill was so admired, that a manual for military exercise, Jakob de Gheyn's *Wapenhandelinghe* (1607), became the first secular book printed (in 1,200 copies) in Russia in 1647.[53] For the young Peter the Great (1672–1725) the Republic was still the 'first country on earth' with its seapower, wealth, science, and military prowess.[54] The United Provinces have recently been described as a prime example of a successful fiscal-military state, collecting enough revenue in the 1670s to field and pay a land army nearing 100,000 troops *and* equip and man a navy of warships and converted merchantmen that well-nigh matched the British Royal Navy.[55] To draw a blunt comparison, the tsar's government fielded a *barely larger* army mainly composed of warriors who, conditional upon their service, had been given land and peasants serfs to equip themselves and some retainers.[56] Partially because of high cost, the Muscovite regime had to abandon the deployment of its *single* warship in the late 1660s.

This Dutch mobilizing capacity was accompanied by considerable diplomatic savvy and an ability to supply lavish subsidies to a variety of countries. As a consequence, the Republic was a popular partner in a number of coalitions and alliances, playing a leading role in European diplomacy, which is reflected by the signing of major peace treaties at Breda (1667), Nijmegen (1678), Rijswijk (1697) and Utrecht (1713). In Asia, the VOC was powerful enough to be treated as a sort of sovereign state by a variety of rulers, including the shah of Iran.[57] The Dutch ruled regions even more distant from Amsterdam than the tsar's peripheral lands were from Moscow (obviously, control over outlying possessions was often nominal for both governments, as was Spain's in Spanish

America[58]).[59] De Vries and van der Woude conclude that the Dutch state appears 'unambiguously modern', being 'a well-ordered government long capable of protecting the security of its citizens, nurturing the economic interests of its merchants and fishermen, establishing vigorous institutions to advance its colonial ambitions, and maintaining domestic tranquillity.'[60]

The Dutch strongly adhered to the rule of law, holding even their regents accountable to it.[61] The Republic's legal practices had significantly advanced toward the darker version of disciplinary modernity emphasized by Michel Foucault: carefully mapped plans for the outlay of towns such as Haarlem and Amsterdam in the 1660s testify to Dutch authorities' desitre for emuneration, while they confined unproductive and delinquent (or unwanted and uncooperative) subjects to houses of detention.[62] In 1663, Olfert Dapper suggested that '[t]hieves have to be terrified away from crime by the prospect of hard labor, because a corporal penalty did not cause enough fear in them'.[63] Judicial authorities did not sentence people to prison out of humanitarian concern. The English diplomat Sir William Temple joined other visitors to the Republic (including later Peter the Great) in marvelling at its well-kept and orderly 'Hospitals'.[64] These institutions were part of a *modern* 'internalized colonialist project' that aimed at greater control over (or more disciplined behaviour by) subaltern groups, implemented by an elite interested in creating a better ordered, modern-type, society.[65]

On several occasions in *Reysen*, Jan Struys implies that the governments of the countries he visits were nowhere near the standard of orderliness, impartiality, and justice of his native country.[66] The '*slaafachtigh*[*e*]', or slave-like, nature of the Russians, for example, surprised him as a Dutchman who proudly worked for the highest bidder rather than being coerced.[67] But he does not seem to have been unduly disturbed by the brutal penal practices he encountered on his path.[68]

A Dutch 'Calvinist mindset' (which even affected the many Dutch Catholics) played its role in maintaining this well-ordered society. Sobriety was encouraged, ostentatiously sumptuous living discouraged. *Reysen* betrays occasionally this Calvinistic streak. For most Europeans in Struys's day, religion remained usually a more important identity marker than country of birth (let alone mother tongue). Confronted with foreign temptations that could have led him astray, *Reysen*'s Struys remains unwavering in his loyalty to his country and church (although this church's doctrine is exceedingly ill-defined in *Reysen*'s pages). The text demonstrates the righteous worldview of the Dutch, God's chosen people.[69] As a personalized religion, Calvinism seems linked to the a growing modern emphasis on self-reflection, self-discipline, and individualism in the Republic, even if the strengthening of such a mindset and attitude are discernible across seventeenth-century Europe.[70] Individualism was most pronounced among the economic and cultural (and literate) elite, but even in his increasing use of a personalized last name, Jan Janszoon Struys seems to have been affected by this greater emphasis on personal identity. Traditionally, most of his low-born peers used merely a first name and patronymic.[71] Especially from 1668 onwards, Jan Struys began to consider himself more than a mere cypher, as we will see in the next chapters.

The Dutch sense of a unique common destiny had historical roots going back to the Middle Ages, as Simon Schama has cogently argued.[72] Strong feelings of duty and loyalty to their country infused most Dutch, and their most powerful economic and political leaders exhibited paternalistic responsibility for their poorer or weaker compatriots. Identification with the Republic and Dutch culture was fortified when abroad and confronted with hostile others.[73] Considering that one in ten adolescent and adult men in the Republic may have sailed the Seven Seas, this acute awareness of being Dutch was rather widely diffused.[74] In Struys's encounters with Muscovite-Orthodox or Iranian-Muslim 'Others,' which are the most elaborately detailed of such confrontations in *Reysen*, the contours of a specific Dutch national identity can be traced even beyond the ghostwriter's embellished patriotism of the protagonist. This Dutch identification abroad presented a contrast to Holland itself, where often regional differences rather than supraregional commonalities were stressed.[75] Such 'discrimination' was nevertheless muted by the fact that the Republic was really an immigration country, whose cities harboured many foreign-born inhabitants.

In this respect, the Dutch entrepreneur A.D. Vinius's petition to become a subject of the tsar is instructive. It demonstrates the development among the Dutch of a more clearly defined national identity, associated with the impersonal collective of the state rather than the individual of the monarch. In 1648, Vinius explained to Tsar Aleksei (r. 1645–76) that his *podanstvo* [status of subject] of the Estates of Holland had lapsed during his thirteen-year sojourn in Russia and that therefore no one protected him '*ot sil'nykh*' [from strong ones] in Muscovy.[76] He implored the tsar to accept him as a subject. The petition shows how Vinius recognized that he was supposed to be legally subject to a sovereign government.[77] Such political or legal identification reflected in much of Europe a sense of dynastic rather than any national allegiance, but by 1648 in Vinius's native country no monarch ruled: he was therefore subject of an impersonal state. In Russia all subjects ruled by the tsar called themselves slaves [*kholopy*] of the tsar when addressing him, but Vinius speaks in his petition about wanting to switch his formal subjection [*pod(d)anstvo*] from the Dutch Estates-General to the tsar, indicating a different, and more modern, sensibility in this matter.[78]

Nationalism defines itself against inferior others. Seventeenth-century Dutchmen thought about those others in hierarchies derived from the idea of a 'Great Chain of Being', at the apex of which stood the Dutch.[79] As *Reysen* shows in its pages, Netherlanders ranked above the English, who ranked higher than the Italians, followed by Muscovites, Persians, and Turks, with Tatars or Malagasy bringing up the rear, considered barely more 'civilized' than apes.[80] In religious terms for the Dutch, Protestantism (Calvinism before Lutheranism and Anabaptism) was superior to idolatrous Catholicism, which was nevertheless preferable over (Russian) Orthodoxy, while all of Christianity was superior to Islam and Judaism: the Jews of Caucasian Derbent, whom *Reysen* contemptuously calls '*Schacchers*', are depicted as inveterate traders in stolen goods.[81] In turn, Judaism's and Islam's kinship with Christianity made those faiths superior to Hinduism or Buddhism, which were equated with backward and blasphemous paganism. Struys thus describes one group of Volga Tatars as 'the stupidest pagans that I have ever seen, neither having churches nor priests'.[82]

Racism needed another two centuries to develop into a fully coherent system of thought, but the Dutch called almost all people with darker skin pigmentation than those living along the Mediterranean shores '*zwart*' (black). This implied not just a different skin colour but an inferior ugliness: *Reysen* contemptuously notes that 'these Kalmyks are such terribly ugly people that the Hottentots and the most ugly Moors have rather decent mugs compared'.[83] Most others (including Italians) in *Reysen* are depicted as depraved, especially plagued by an overly robust sex drive leading to deplorable promiscuity.[84] This hierarchical and racist grouping of people combined with a perspective tending to detect binary oppositions, implicitly and explicitly, as is evident from *Reysen*.[85] Thus in judging the world, the Dutch served as the golden standard as well-dressed, sexually prudish, monogamous, and frugal Christians. Others, particularly non-Western-Europeans, were their alter egos (or evil twins), and in varying degrees defective, from improvident, poorly dressed, sexually orgiastic and promiscuous, to non-Christian. This mindset soothed the Dutch conscience when they forced non-European slaves to tend to the nutmeg trees on the Spice Islands, transported 'human commodities' across the ocean, or exploited African slaves in the Americas.[86]

And yet this view of humanity is occasionally challenged in *Reysen*. Certain individuals in *Reysen*'s third part, such as the Franciscan friars and Haji Biram Ali, are depicted as human beings rather than crude caricatures.[87] Curiosity and humanistic tolerance could offset wholesale rejection of those adhering to other religions or speaking alien tongues.[88]

Adopting Benedict Anderson's analysis of modern nationalism confirms how far the transition from a universal Christian to a specific Dutch consciousness had progressed in the Republic.[89] As a result of the Dutch Reformation, Latin had lost its hallowed tradition; monarchs no longer ruled the country; and history and cosmology had become separate in the Dutch discourse about the past. Schama emphatically argues that the Dutch exhibited a sense of loyalty to their fatherland in the seventeenth century resembling a modern civic identity.[90] The independence war against the Spaniards (1566–1648), the Calvinist dogma of the elect, historical tracts such as P.C. Hooft's *Histoorien*, or the cult of the *Batavieren* (the Batavians of Antiquity)[91] – all combined to forge a specific Dutch identity shared by Calvinists and non-Calvinists alike, as did the patronage system mentioned earlier.[92] Nationalism emphasizes national unity over class solidarity, and its propagation in the seventeenth-century Republic thus also 'played a vital role in maintaining social stability'.[93]

By its portrayal of sailors' heroics, *Reysen* affirmed key elements of seventeenth-century Dutch nationalism. As Vondel wrote in his eulogy of 1655 '[n]ext to God, the Great Blesser, the Batavian has acquitted himself with council and deed, on the altar of the sea'.[94] The historical Jan Struys, too, behaved abroad as a person with a Dutch identity and was recognized as Dutch by others. The VOC's favourable treatment of the Dutch contingent among the crew of the interloping Genoese ships in the 1640s, the assistance Struys received during his Mediterranean adventures during the next decade, the aid he and his comrades received from Dutch expatriates in Muscovy and VOC agents in Persia around 1670, and the mutual support among the *Oryol* sailors all indicate an acute awareness of belonging to a specifically Dutch nation. Judging by their actions between Riga and Batavia, Struys and his mates believed in, and benefited from, their membership of this nation.

In literal and figurative senses, national boundaries were nevertheless often far from clearly defined, even in seventeenth-century Holland. A word as *Duytsch* (from which English derives Dutch) signified in seventeenth-century Holland (as they do in Struys's work) both Dutch and German. In contemporary Russian *nemets* (which originally meant 'dumb one') was a generic rubric under which the Dutch were grouped, because the Muscovites could often not distinguish them from other Western Europeans. An unwavering sense of a national identity was not felt by all Dutchmen. Its strength did not overcome certain temptations, as the Dutch 'renegades' among the Barbary pirates or Ottoman fleet make evident.[95] Similarly, Vinius's petition to the tsar demonstrates that individual patriotism was trumped by material motives.[96] The necessity of providing for one's daily bread overrode sentimental attachment to the fatherland, as is evident from Struys's enlistment, as a middle-aged husband and father, in the service of the Russian tsar in the late 1660s.[97]

In other areas of culture,[98] such as its populace's high level of literacy and widespread education, the Republic reached a unique level of development for its day. But in the second half of the seventeenth century declining economic buoyancy was accompanied by waning cultural vibrancy. In literature and painting, the most original works belong to the era before 1670.[99] The Dutch Republic's history in the course of the seventeenth century seems a fine demonstration of the historical 'law of the disadvantageous lead', developed by the Dutch historian Jan Romein (1892–1962).[100] It holds that one society's advance (technological, economic, cultural) over other societies 'tends to act as a brake which hinders further progress'.[101]

But until the 1720s the Republic's culture continued to show dynamic sides. In the Republic as much as elsewhere, the increasing sway of Bacon's empiricism became an important impulse among the learned for travelling, collecting data, and writing or reading about exotic locales.[102] In 1647, the Holsteiner scholar-diplomat Adam Olearius (1603–71) laid down an empiricist template for those exploring Russia and Persia in his *Offt begehrte Beschreibung*.[103] Olearius had meticulously recorded his impressions travelling as secretary of two embassies sent by Grand-Duke Frederick of Holstein-Gottorp (1597–1659) to the courts of the Muscovite tsar and the Persian shah.[104] After their first publication, Olearius's publications served as a standard of scientific precision for chorographies in much of continental Europe for half a century.[105] Originally published in German, the Holsteiner's work appeared in a host of translations from the 1640s onward; a number of them were into Dutch.[106] Olearius was the most important source for *Reysen*'s ghostwriter, and *Reysen* seems to model itself after the Holsteiner's treatise in an attempt to appeal to the empiricist vogue that swept the Dutch Republic after 1650.

Scientific hunting and gathering were reinforced by the founding of the Royal Academies of France and England in the 1660s, with which various Dutch scientists (including Witsen) corresponded.[107] In the age of Newton and Leibniz, the Republic was an important centre of the burgeoning international community of scientists and scholars, some of whom sought refuge there.[108] Around the time *Reysen* was published, an early form of modern scientific inquiry had become the norm among Dutch biologists and physicists, and Jonathan Israel presents a persuasive case that early roots of the Enlightenment can be traced to the the Republic during the 1660s and 1670s.[109]

Before 1700, North-Western European medical doctors and surgeons, despite practising what are now often considered questionable healing methods, were in demand among the elite of several non-Western societies for their supposed superior expertise.[110] Cossack *hetman* Stenka Razin respected the reputation of Dutch medical skill enough to spare the life of the Frisian surgeon Jan [van] Termund[t] at Astrakhan in 1670.[111] Freed from the daily challenge to eke out a living, a number of talented Dutchmen found time to dedicate to the exploration of the parameters of existence, whether in philosophy, theology, or science. But Dutch scientists at home were far from isolated from the society in which they lived. It is possible that Struys himself, after his final return from the Far East, befriended Jan Swammerdam (1637–80), a globally renowned biologist.[112] In the Republic scientific curiosity was indeed not limited to the elite alone.[113]

Famous in his day for his works on shipbuilding and on Siberia's peoples and nature, the omnivorous scientist-scholar Nicolaas Witsen was a prime example of this generation.[114] Witsen's *Noord en Oost-Tartarye* is a sublime expression of this age of herbaria, curiosity cabinets, and the desire to finally fill in the remaining blanks spots on the world's map.[115] Thus Dutch creative energy that seemed to have been expended on painting, architecture, or literature in the first two-thirds of the seventeenth century now found different outlets.[116]

Even if only a handful of Dutch men and women were unhesitatingly agnostic or atheist in the seventeenth century, an increasingly secular and sceptical mindset was linked to this inquisitive curiosity.[117] Secularism's growth underscored the Republic's remarkable tolerance for religious dissenters, for which it was unique in Christendom.[118] Thanks to the persistent confessionalization counteroffensive by various Calvinist ministers, religion became only fitfully more of a private and individual matter toward 1700, but secularization and rationalism did gain ground.[119] Thus Balthasar Bekker ridiculed witchcraft trials as signs of obscurantist superstition in the 1690s.[120] Max Weber's *Entzauberung* (disenchantment) of society is reflected in the Hollanders' and Zeelanders' scepticism about witchcraft, of which Bekker's work was a learned expression.[121]

The absence of superstitious allusions on its pages seems to reflect *Reysen*'s ghostwriter's rejection of the folk tradition and aberrant beliefs strongly held in the seafaring communities from which the historical Jan Struys hailed, and to some of which the sailmaker must have adhered.[122] This scepsis about superstition on the part of the ghostwriter[123] seems a consequence of his secularizing and precocious scientific worldview, rather than deriving from an inclination to present his protagonist as a strict adherent of a new religious orthodoxy innocent of pagan beliefs that might be linked to the age's confessionalization. *Reysen* was of course not presented as a sort of agnostic work, but when it mentions Struys's religion it seems to engage in an obligatory exercise rather than in a rousing declaration of faith. *Reysen*'s references to Protestant Christianity almost seem consciously provided in an effort not to offend the religious sensibilities of its readership.

Reysen denies the influence of supernatural (rather than divine) forces on daily life, and depicts Russian Orthodoxy as a somewhat reprehensible eclectic mixture of superstition and popery. The book especially ridicules the Muscovite fetishist worship of icons, which Russians covered with a piece of cloth if they were having sexual intercourse.[124] But this condemnation lacks the pedantic references to the superior truths of the Dutch Reformed Church in the manner of Wouter Schouten's parallel text, in which one can readily discern the influence of the confessionalization project.[125] There are some tantalizing tidbits hinting at some sort of link between van Meurs and van Someren and religiously unorthodox groups, but there is insufficient evidence to propose that around 1676 they maintained connections with some of the radicals of their day (who then influenced *Reysen*'s text).[126]

This concludes the overview of the specific Dutch parameters that informed Struys's movements and *Reysen*'s text. From 1668 onward, in life and in *Reysen*, Struys's fate was also determined by the context of the state of Muscovy and the Dutch community there, to which we will now turn.

4
The Dutch in Muscovy

The third and overwhelmingly longest part of *Reysen* encompasses the tale of Jan Struys's years in Russia and Iran between September 1668 and October 1673. Other sources confirm how the sailmaker arrived in Russia in the autumn of 1668 enticed by a golden opportunity, joining a party of his compatriots who, for extremely high wages, were to sail a ship, called the *Oryol* (Eagle), for the Muscovite Tsar Aleksei Mikhailovich (r.1645–76). The choice for a Dutch team to supervise the ship's construction and to man the vessel was the consequence of the extraordinary presence of Dutch merchants, entrepreneurs, and artisans in Russia in the 1600s. Their role in modernizing Muscovy will be explored in this chapter. The *Oryol*'s construction on behalf of the tsar was not only an early example of the transfer of modern technology from the developed Republic to a developing Muscovy, but its main organizer, Jan van Sweeden (d. 1668), appears the equivalent of a modern-day entrepreneur capitalizing on the economic possibilities offered by a Third-World country.[1]

Before 1700, Dutch natives were numerous among the Western Europeans in Muscovy and several Dutch natives (such as A.D. Vinius, A.A. Vinius, van Sweeden, and Frans Timmerman) residing in Russia rose to such prominence that they influenced tsarist policy. Throughout the seventeenth century, Dutch ships carried most Muscovite goods leaving the country at Arkhangel'sk. Tsars Mikhail (r. 1613–45) and Aleksei used Dutch fortification builders to strengthen their peripheral towns, and after 1630 hired numerous Dutch military instructors and officers to train and command the Muscovite army. As head of various chancelleries dealing with foreign recruitment, Il'ia Danilovich Miloslavskii (d. 1668), Aleksei's father-in-law and one of his closest confidants until 1665, enlisted scores of Dutchmen as military and civilian specialists.[2] Adding to the familiarity with Dutch visitors and settlers was the towering Dutch reputation as sailors and shipwrights. All of this made Tsar Aleksei and his close advisor A.N. Ordin-Nashchokin (d. 1680) select Dutchmen for the *Oryol*.

Dutch trade with Muscovite Arkhangel'sk may have developed as a spin-off of the crucial shipping of grain and of naval stores from Baltic ports to the Netherlands.[3] Regular commerce with Russia began in earnest in the late sixteenth century, when Dutch merchants began to compete with the pioneering English Muscovy Company at Kholmogory and Arkhangel'sk (the latter port built in 1583–85) on the White Sea.[4] After Olivier Brunel and Jan van de Walle reached the mouth of the Dvina in the 1570s (concomitantly, Ivan IV [r.1533–84] tried to hire Dutch shipwrights and sailors),[5] it was a second generation of Dutch merchants who established a permanent presence in Muscovy, especially after the Time of Troubles (*c*.1598–1618).[6] Among them were Isaac Massa (1586–1643), Marcus de Vogelaer sr. (1564–1610), and Georg van Klenk (*c*.1580–*c*.1643). Also involved in the early stages of this exchange was the de Moucheron family, Calvinist refugees from the Southern Netherlands.[7] David Ruts (1595–1659), Jan van Sweeden's father-in-law, was married to a de Moucheron. Most of these individuals belonged to the Republic's new elite, engaging in a variety of overseas business ventures, and some were among the early directors of the Dutch East India Company. Meanwhile, Georg van Klenk's father had been a retainer of the Dutch stadtholder William of Orange's brother, Germans who became Dutch in the Netherlands' war of independence.[8]

Dutch merchants and entrepreneurs involved in foreign ventures were often linked by kinship ties, as the Butler-van Sweeden-Ruts-de Moucheron link in Muscovy illustrates.[9] Jealous of their riches and status, wealthy families intermarried, and in Muscovy they vigorously defended each other's business interests before the tsarist authorities. Families' collaboration often lasted for several generations, and involved careful grooming of successors.[10] Women assisted their spouses and relatives often with some panache, as in the case of van Sweeden's widow Maria Ruts, who continued her husband's business after his death.[11] Her self-confident behaviour in Muscovy is remarkable given that Russian elite women were barred from Muscovite public life. Friendship ties played sometimes an equally important role. The famous Scottish mercenary officer in Russian service Patrick Gordon (d. 1699), a key advisor to Peter the Great in the 1680s and 1690s, was both friend of Jan van Sweeden, the Dutch initiator of the *Oryol* project, and related by marriage to Cornelis van Bockhoven (d. 1678), the Dutch supervisor of the wharf where the ship was built.[12] All three played an important role in Muscovy in the second half of the seventeenth century.

Strong ties of amity and kinship among the Dutch merchants may have substituted for organization in a monopoly company. As individual traders, they formally enjoyed less favourable trading conditions than the English Muscovy Company, but the volume and value of Dutch commerce on Arkhangel'sk far outstripped English merchants long before the latter were banished from the tsar's realms in 1649.[13] Between 1613 and 1698, almost half of Western merchants and their representatives trading with Muscovy were Dutch (644 out of 1,361).[14] Dutch traders shipped to the Republic Russian raw or semi-raw materials, such as naval stores (tar, hemp, timber), *iuft* (soft leather, in Dutch *juchtleer*) and potash.[15] These goods were then processed in the Netherlands, after which the finished products were sometimes sold back to the Muscovites. Dutch traders intermittently purchased grain (rye mainly) in Muscovy, but its volume never rivalled that of the Dutch 'mother trade' with the Baltic ports.[16] In exchange for Muscovite resources, Dutch merchants brought bullion, luxury goods (such as wines), ammunition, and weapons, more than 'all other nations together' enriching the tsar.[17] Several dozen Dutch ships annually arrived at Arkhangel'sk, always forming the majority of bottoms anchoring there.[18] The Netherlandish reach in the international economy is underlined by the Muscovite re-export of Dutch woollen cloth via Astrakhan to Persia (with Russian merchants shipping it from Arkhangel'sk to Astrakhan).[19]

The *kooplieden* fared well as Frans Hals's portrait of Massa and his wife and Rembrandt's portrait of David Ruts's father in Russian dress reflect.[20] Many Dutch merchants prospered in the Russia trade, even when they sold their wares at profit margins lower than their foreign competitors.[21] The Dutch traders flourished for a variety of reasons. Their superior freight ships needed fewer hands to sail. Amsterdam functioned as a staple that could store many goods until prices reached a profitable level; the city linked trade on Muscovy with a host of other countries.[22] As elsewhere, experience was enhanced by a well-developed system of information gathering and a competent network of agents. Ventures were financed at lower cost than available to other European traders on Muscovy. Competition among the Dutch themselves was occasionally fierce, but they frequently closed ranks if challenged by English competition or native Russian merchants.[23]

Dutch economic clout, pronounced as much in Arkhangel'sk as along the shores of the Baltic Sea, also found cultural expression along the shores of the Baltic and White Seas. Peter the Great had sufficient command of Dutch to converse in this language without an interpreter with the Prussian-Brandenburger Elector Frederick III (r. 1688–1713) in 1697.[24] In the seventeenth century, Dutch and Middle Low German, two languages then so closely related that they seem almost interchangeable, served as the *lingua franca* of the Baltic Sea, as Italian did in the Mediterranean, an expression of respectively Dutch and Italian commercial omnipresence in each respective sea.[25] Janis Kreslins notes that a Hollander could communicate with a Dane or Swede (as well as with any German speaker he encountered in this region) without using a third language, using a variety of techniques that had evolved to avoid any misunderstandings.[26] Peter Burke suggests that 'Dutch was probably the Western language that a Russian of this time was most likely to know.'[27]

We can get a sense of the Dutch dominance over the Russian export trade through the venomous reflections of Tsar Aleksei's former personal physician, the Englishman Samuel Collins. The doctor condemned the Russians as 'false, Truce-breakers, subtile Foxes, and ravenous Wolves', but their innate duplicitous nature had much worsened from 'their traffick with the Hollander, by whom they have much improv'd themselves in villany and deceit'.[28] Collins further likened the Dutch to 'Locusts', who slandered the English by distributing 'lying pictures, and libelling pamphlets [making] the Russian think us a ruined Nation', bribing 'the Nobility with gifts', and 'eat[ing the] bread out of the English-mens mouths'.[29]

As part of the extensive Dutch economic activity in Northern Europe along the Baltic shores and in Muscovy, some of the bolder Dutch merchants became capitalist entrepreneurs who branched out beyond mere commodity exchange. Rapidly expanding in territory and population, by the 1660s Muscovy was a regional Power in Northern Europe with a growing demand for arms, uniforms, and other military stores, as well as luxury goods. The Dutch tried to satisfy this demand, not only by shipping goods to Arkhangel'sk, but also by starting a variety of businesses within Muscovy.[30] These were initiated by ambitious Dutch up-and-comers who decide to pursue riches abroad when toward 1650 a dominant elite began to monopolize the most lucrative economic pursuits in the Dutch Republic.[31] Among these upstarts were A.D. Vinius and Jan van Sweeden.[32] In 1627, when Dutch demand for Russian grain suddenly experienced an upsurge, Andries Denijsz. Vinius first visited Arkhangel'sk to buy grain, which he may have exchanged for arms.[33] In 1634 Vinius definitively left Holland 'to trade and seek his fortune in Moscow', setting up an arms manufacturing plant near Tula.[34] Vinius became officially a subject of the tsar in March 1648.[35] By 1655, his fourteen-year-old son Andrei, who was to rise to the very top of the Muscovite bureaucracy, was baptized in the Orthodox Church.[36] Successful businessmen such as Vinius came with formidable business savvy and were part of an extensive network of fellow merchants trading all over the world.[37]

In their endeavours, Dutch businessmen abroad, motivated much like multinational companies today, often prioritized profit over *patria* in the manner of Vinius. Dutch 'interlopers' resentful of the VOC persuaded the Danish king to charter a Danish East India Company and thus break the VOC's monopoly.[38] Admiral Cort Adeler, by then in Danish service, was a business partner of Koenraad van Klenk.[39] Sweden benefited from the activities of Louis de Geer (1587–1652) and Elias Trip (1569–1636), and Russia from those of Peter Marselis (1602–72), Vinius, van Sweeden, and Julius Coyett.[40] Riches could be reaped, as by 1670 the decades-old iron and armament works of Marselis, Ackema, and Vinius, and the equally old glass-blowing enterprise of the Coyett family, proved.[41]

Besides Tsar Aleksei and the Boyar Ordin-Nashchokin (whose roles will be investigated in the next chapter), the driving force behind the plan for a Muscovite sailing fleet on the Caspian Sea was Jan van Sweeden, an outstanding example of a seventeenth-century Dutch merchant-entrepreneur in Moscow. Van Sweeden followed the footsteps of people like Vinius in moving from the commodity trade along the Amsterdam-Arkhangel'sk route to the pursuit of business ventures inside Russia.[42] Van Sweeden, like many other Dutch businessmen abroad, had begun as a factotum of an Amsterdam trading house in Russia.[43] In 1646, David Ruts hired van Sweeden in Amsterdam to work for two years as wine-cooper in Moscow, for the meagre wage of 150 guilders per year (although food and drink were free for the duration of his contract).[44] David Ruts's wife was a de Moucheron, a daughter of one of the most powerful families in the Dutch Arkhangel'sk trade. Eventually, van Sweeden impressed David Ruts well enough to gain his permission to marry his own daughter Maria.

After completing his first contract for Ruts, van Sweeden set up a vineyard for the tsar near Astrakhan in 1648–9, and was delivering arms to the tsar by 1659.[45] Van Sweeden curried favour with the tsar in other ways. In 1659, for instance, he sold Aleksei Mikhailovich an exquisitely crafted baroque bed made by Dutch or German artisans, which the tsar believed to be magnificent enough to present as a gift to the Persian shah some years later.[46] In the early 1660s, van Sweeden developed a series of initiatives aimed at innovating Russia's economy, ranging from the setting up a regular postal system connecting Muscovy with the rest of Europe to the foundation of textile and paper mills and a glassblowing enterprise.[47] At the moment of his sudden death in late 1668, Jan van Sweeden seemed poised to become the most successful foreign entrepreneur ever to have been active in Muscovy.[48] The success of his efforts was partially due to the felicitous timing of his ventures, since they coincided with the Russian government's move toward economic protectionism and import substitution (see the next chapter).[49] Van Sweeden would not have been able to pursue his plans without active support from the tsar's government, and especially by the tsar's favourite at this time, Afanasii Nikitich Ordin-Nashchokin.[50]

As a result of his variegated activities in Muscovy over more than two decades, van Sweeden became a veritable connoisseur of tsardom's human and physical geography. He proved an invaluable source for the geographical and ethnographical information that the young Amsterdam patrician Nicolaas Witsen collected on his trip to Muscovy with Dutch ambassador Jacob Boreel in 1665, much of which found its way into Witsen's magnum opus *Noord en Oost Tartarye*, an encyclopedic cosmography of Siberia.[51] When Witsen visited Moscow as part of Boreel's retinue, van Sweeden provided Witsen with information about the Caspian Sea and its surrounding regions.[52] Not long after this conversation, van Sweeden began to work toward the establishment of a Dutch-operated Muscovite fleet on this sea. The first signs of van Sweeden's involvement in the *Oryol* project can at least be traced to 1666.

While Witsen took down notes during his conversation with van Sweeden, these were never published in his lifetime, and even both printed editions of Witsen's work on Siberia (for which he used some of these notes) were only made available to a handful of people after 1692. Van Sweeden may have urged upon Witsen not to divulge this information, jealously guarding the rich promise of the trans-Caspian trade on behalf of the tsar (and probably himself). Even if these plans had not yet hatched in van Sweeden's mind, the Dutch expatriate business community in Muscovy mirrored VOC practice in keeping most of its operations under wraps (indeed, some of them were concomitantly among the VOC's directors). This strategy happily accorded with the tsar's government's anxiety about leaking state secrets abroad.

Politically, the Dutch Republic and Muscovy had little reason to quarrel, and Dutch traders on Russia doubled as agents for both governments, who usually both rewarded them well for their services: Georg van Klenk provided the *Posol'skii prikaz* (foreign office) at least once with a written memorandum about the European political situation, even if he was a loyal client of the Princes of Orange, the Dutch stadtholders.[53] Together with acquaintances, such as Georg's son Koenraad van Klenk, Jan van

Sweeden sold arms from Holland to the Russians. Likewise, A.D. Vinius and Tieleman Lus Ackema did not hesitate to set up arms manufactories in Russia.[54] The Dutch government, run by and for its wealthy citizens raised few objections to this sort of business.

Europeans from all social backgrounds shifted both national and religious allegiances for personal gain. Despite the appearance of great religious zeal in this era, many obeyed governments faithful to other religions without objecting. Mercenary soldiers fought for pay in religiously fueled wars. The Catholic Patrick Gordon fought Catholic Poles on behalf of Muscovy, as long as the tsar paid him.[55] Whereas of the major Dutch tycoons in Muscovy only A.D. Vinius converted to Orthodoxy, all behaved loyally toward the tsar.[56] Legally considered subject to the tsar's government, law, and courts, Dutch merchants residing in Muscovy were spurred on to 'Russify'. Since Dutch traders' interests were not defended by a monopolistic company, they only received temporary relief from paying import and export duties on an individual basis.[57] Profits could rise by becoming *de facto* subjects of the tsar; after residing in Muscovy for a while, they were considered *Moskovskie (torgovye) inozemtsy* ('Moscow (trading) foreigners'), despite its appearance a title indicating favoured status.[58] Thus, similar to their attitude in North America, these early Western-European 'imperialists' in Muscovy often met the native population and government on a 'middle ground', adapting to local rules and customs in their pursuit of profit.[59]

Western European expatriate communities across Muscovy shared scarce information about trading conditions and business opportunities, and mutual solidarity offered a measure of protection against whimsical manoeuvres by the Russian government. Demkin notes how the tsar was accustomed to receive an annual delegation of 'Western European' merchants in the Kremlin: already in 1614, Georg van Klenk served as the representative of such traders in presenting the tsar with a variety of gifts.[60] Usually, such compacts included various native Hamburgers, Bremers, Swedes, Danes, and even English operating in Muscovy (ironically mirroring the Muscovite use of the generic term for Westerners as *nemtsy*).[61] The Westerners' sense of common interest was shared by artisans and mercenary officers, and was enhanced by their forced segregation in the *sloboda* (foreigners' suburb) in the 1650s.

Dutchmen in Muscovy, however, like other Western Europeans (Gordon and John Hebdon sr. both acted as agents for the Stuarts), were not devoid of affection for their native land (the Vinius case is rather unique). Their allegiance to their fatherland was reinforced because it yielded tangible material benefits.[62] *Dutch* mutual support abroad can be recognized elsewhere as well. It explains why the VOC in Asia exerted efforts to shield its personnel, extending its assistance to Dutch natives who were not in its pay, as Struys was to find out to his luck.[63] For most seventeenth-century Dutch expatriates religion or profit usually trumped patriotism, but an appeal to Dutch identity might offer support or even salvation in times of need.

Vinius's petition of 1648 to the tsar provides evidence of the development of a sense of national identity among the Dutch even before the war with Spain came to an end in 1648. Still, Vinius's shift in allegiance underlines again that the loyalty of most Dutchmen (and women) was first and foremost to their economic well-being: They were inclined to be most loyal to whoever paid or allowed for a profit to be gathered. Many Dutch natives (and even those born elsewhere who had initially migrated to Holland for economic or political reasons), like Vinius, moved from the United Provinces to the various countries surrounding the Baltic Sea and then gradually began to identify themselves as Hamburgers, Swedes, or Danes.

Muscovy and the United Provinces thus interacted intensively in the seventeenth century, marking '[t]he most important period [...] of Dutch-Russian relations'.[64] Strong commercial relations only gradually were translated into more regular political parlaying, and diplomatic discussions were always laced with economic topics. Before the Jacob Boreel embassy of 1664–5, the Dutch sent a mere three (full) embassies to Moscow, primarily to discuss trade issues.[65] The diplomatic ties between the two countries became closer in the 1660s and 1670s. Boreel's visit was followed by embassies led by Nicolaas Heins(ius) in 1669–70 and Koenraad van Klenk (Georg's son) in 1675–6.[66] Van Klenk left behind the first Dutch permanent resident to the tsarist court, Johan Willem van Keller, who was the Republic's envoy in Moscow until his death in 1698.[67] From then onward, the Netherlands retained a permanent representative at the tsar's court.

In Holland, the merchants trading with Moscow often doubled as politicians, in both Amsterdam and The Hague. It is no surprise, then, that the Republic's embassies regularly championed Dutch merchants before the tsar.[68] In 1676, van Klenk behaved rather as the advocate of the Dutch (and Western European) business interest in Muscovy than as a diplomat placing a premium on a military alliance. Recognizing him as an authoritative voice representing their interests, van Klenk was lavishly feted in Moscow by a consortium of various Dutch, Hamburg, and Danish merchants.[69] For the Dutch in the seventeenth century, the economic and the political could never be disentangled.[70]

The leading entrepreneurs and merchants were far from the only Dutch people earning their keep in seventeenth-century Russia. In Muscovy, foreign-managed companies continuously employed specialized Western-European craftsmen, many of whom hailed from the Republic.[71] Dutch artisans had been hired by the Russian government since the late sixteenth century.[72] Many veterans of the incessant wars of Western and Central Europe were recruited to modernize the Muscovite army.[73] One upsurge in the Muscovite hiring of foreign specialists accompanied a broad army and administrative reform programme implemented during the early years of Aleksei's reign (1645–54).[74] Aleksei repeatedly dispatched agents and envoys to the Republic to recruit Dutch officers and craftsmen, a sign of their contemporary reputation as highly proficient experts.[75] The Miloslavskii Embassy departed for Holland in 1646 with the brief to hire both military officers and artisans, which it did rather successfully (even if it left behind a trail of unpaid bills and damages).[76] At exactly this time, the engravings from Jacob de Gheyn's *Wapenhandelinghe*, portraying the various moments of military drill, appeared within the first secular book ever printed in Muscovy.[77] Around 1660, tsarist agents such as John Hebdon, Andries Vinius, and Jan van Sweeden did not just purchase arms for the tsar in the Dutch Republic but also hired soldiers, artisans, doctors, and apothecaries.[78] In 1660 in Amsterdam, Hendrik Swellengrebel, a brother-in-law of van Sweeden, hired Struys's later acquaintance Lodewijk Faber (Ludwig Fabricius, 1648–1729) together with his stepfather Paul Rudolf Beem (as a colonel in the artillery) to serve the tsar as military specialists.[79] Faber was to play a crucial role in Struys's life.

While proportionally fewer Dutchmen could be counted among Western mercenaries than among Western merchants and their agents operating in Muscovy, their numbers were still considerable and their importance disproportionate, due to the Dutch reputation for military innovation.[80] Thus, alongside Dutch merchants, commissioners, and entrepreneurs, a Dutch contingent of military specialists lived in Muscovy in the 1660s.[81] Along Muscovy's southern borders that faced the Crimean Tatars, Dutch engineers, too, aided the construction of a fortification line which was linked in the middle of the century.[82] And it was again Dutch 'industrialists' (Vinius and Ackema) who set up the first Muscovite firearms works at Tula in 1632, using some of their compatriots in managing the plant and training Russians in manufacturing the weapons.[83] Seventeenth-century Dutch experts in Muscovy included doctors and surgeons, such as Jan Termundt, whose path would also cross with Struys's.[84] Other Dutch speakers should be added to this expatriate community, such as diamond cutters, carpenters, church ministers, glassblowers, and paper makers. Many of these men were accompanied by their family members when moving to Russia. The Dutch contingent even harboured a few early examples of exchange students, people who claimed to want to learn the Russian language.[85] Between 1668 and 1671, and then again during his second stay in 1675 and 1676, Struys frequently encountered Dutch residents from all of these categories in Muscovy.

Precise numbers for this Muscovite Dutch contingent remain elusive.[86] Russian-language records are highly inconsistent in the spelling of names (for example, the usually Dutch 'van' often became the German 'von' (*fon*]), and the Russians often russified Dutch names. Jan Struys's name varies in one document from *Ian' Iansen* and *Ian' Iagans* to *Ian Iansen Struf*.[87] Russians used terms such as *galandets* ('Hollander'), *go(a)l(l)an(d)skii* ('Hollandish') or *go(a)landia* ('Holland'), but simultaneously referred to all Germanic-language speakers as *nemtsy* [singular *nemets*], reflecting how Northern and Western European national identity remained a hazy concept for Muscovites. Russians interpreted the European political constellation in a quasi-traditional, Byzantine-Orthodox manner. All *nemtsy* were ultimately subjects of the Western emperor (and, according to many in the West, the Dutch had legally been such subjects until 1648). Tsarist documents often do not distinguish whether someone identified as *nemets* is from the Holy Roman Empire or from the Dutch Republic (or even from Denmark, Sweden, England, Scotland, or Ireland). In Moscow, the foreigners' suburb was called *nemetskaia sloboda*, often translated as 'German suburb', although 'German*ic* suburb' might be a better rendering.[88]

Only a tentative estimate can be made for the number of Dutch people temporarily or permanently residing in the tsar's realm in the fall of 1668 and winter of 1669. In 1665 the foreigners' suburb near Moscow counted 206 'courts' (*dvory*), possibly housing some 1,200 people, of whom approximately one thousand were Western Europeans.[89] There were also foreigners' quarters in Tula, Arkhangel'sk, Nizhnii Novgorod, and Kazan, while smaller Western European communities survived elsewhere.[90] Extrapolating and combining these and other data, it is safe to assume that between five hundred and one thousand men and women considering themselves 'Netherlanders' then lived across Russia.[91] The size of this group of Dutch residents in Muscovy was similar to that of the contemporary Dutch overseas emigrant settlements on Manhattan and in Nieuw Nederland in North America, at the African Cape, and in Batavia.[92]

This conspicuous Dutch presence and the disproportionate importance and scope of Dutch residents in Muscovy explain why the tsar and his advisors decided to place the *Oryol* shipbuilding project in Dutch hands.[93] The tsar's court, too, had sufficient information about the Republic to deem the Dutch as the most economically, militarily, and technologically advanced in Europe. In an apt phrase, Kotilaine has dubbed the Western merchants who predominantly hailed from Holland as 'the midwives of [the Russian] empire' in the seventeenth century.[94] And it was not just their trading skills that played a key part in the emergence of Muscovy as a burgeoning European Power.

5
Muscovy

The tsar's decision to develop a Caspian fleet was linked to the 1667 New Commercial Code, the terms of which betrayed Muscovite xenophobia even if it was primarily rooted in the mercantilist ideas that dominated European economic thought. Mercantilist-type convictions informed the advice of people such as van Sweeden and Ordin-Nashchokin. Such sentiments found further expression in the concomitant Russo-Iranian trade treaty, which aimed at diverting much of the Persian silk export through Russia. But there were other than purely economic motives behind the building of the *Oryol* (and hence Struys's movements and *Reysen*'s text): the military-political context of 1660s Muscovy and the triumphant mood sweeping the court in 1667.

Throughout much of Aleksei's rule, the tsar was torn between two mutually incompatible goals for his country.[1] He tried to maintain his realm as the haven of uncorrupted Orthodoxy, but was forced to succumb to the necessity of fighting foreign foes – especially to the west and south – at huge expense. The resulting fiscal-military realities dictated reforms that led to a 'transformatory mindset' among the Muscovite elite.[2] Without the steady influx of Western military equipment and mercenaries, traders, entrepreneurs, and various artisans, Russia's fortunes would have shown worse against long-standing foreign enemies as well as against restless quasi-subjects such as the Don Cossacks.[3] The trauma of the Time of Troubles had made the government acutely aware that survival in contemptuous Orthodox isolation was not an option: to preempt any surprises, the *Posol'skii prikaz* closely monitored Western European developments by debriefing Russian ambassadors after their return from abroad, questioning foreign officers and merchants, analysing statements from foreign diplomats, maintaining an intensive correspondence with border areas and foreign ports and capitals, and reading Western books and newspapers.[4]

Armed with such information, technology and expertise, by 1667 Muscovite armies had triumphed over Polish foes (albeit torn asunder by domestic strife). But the tsar's military might still bowed before Swedish forces, and never challenged the Ottoman-Tartar foe in the south. In addition to the tsar's middling success in foreign wars, Aleksei was hardly master in his own house yet. Similar to his monarchical counterparts across Europe and Asia, he faced his share of violent domestic unrest. The fickle Cossacks, who sometimes fought for the tsar in his foreign wars, were never reliable allies, even if they let themselves be paid by the tsar.[5]

Facing the same foes as his predecessors, Aleksei followed their lead in implementing another round of 'Westernizing' reforms.[6] During the 1550s, a first comprehensive military reform had modernized the armed forces to match the military standards of the enemies in the west and south. This round of modernization was followed by several others, following Western and Central European changes in the conduct of war that began to resemble a genuine revolution.[7] This military revolution in its fullest (Western European) incarnation involved the development of special (sailing) warships and tactics in naval battles, the use of artillery at sieges and on the battlefield, a change in defensive fortifications, use of mining, a growing importance of infantry relative to that of cavalry, the spread of hand-held firearms, and an increasingly effective use of firepower in battle.[8] These technologically driven innovations made warfare ever more expensive, but their (partial) adoption was imperative to maintain Muscovy's military power and prestige.[9] The government extracted therefore every drop of the meagre surplus generated by Muscovy's economy.[10]

Tsar Mikhail Romanov and his son Aleksei sought to redress the humiliations and territorial losses Muscovy suffered at the hands of the Polish and Swedish armies in the Time of Troubles. Part of their programme was to expand their empire to the shores of the Baltic Sea. Once the terms of the Peace of Kardis of 1661 left this littoral firmly in Swedish hands, Aleksei's attention shifted to the Caspian Sea.[11] Ivan IV's capture of the Tatar strongholds of Kazan and Astrakhan during the 1550s had given the Muscovite heartland a direct waterway to the Caspian Sea through nominally Russian-held territory by way of the Volga. In the south and south-east, however, the steppe grasslands lacked the easily defensible boundaries of natural obstacles. Thus, from the sixteenth-century onwards a gradually linking fortification line was erected that moved in stages southward during the next century, which in its easternmost reaches it ended just south of Simbirsk on the Volga.[12]

The economic benefits that after the 1550s ensued from intensified trade with Iran and Central Asia did not at first outweigh the costs of defending a territory which had been overrun by nomadic peoples for millenia. Until the late seventeenth century at least, Russian rule along the lower parts of the Volga remained nominal, with Russian garrisons merely observing from the relatively safe ramparts of fortified urban citadels the trek across the steppe of Bashkirs, Crimean and Nogay Tatars, Kalmyks, Cossacks, and others.[13] Volga flotillas carrying merchandise had to be convoyed by armed detachments. Such feeble control let these nomads see themselves as Muscovite partners rather than the tsar's subjects, a view underlined by the regular payments made to them by the tsar's government.[14]

But the Tatar threat to the Muscovite heartland in the course of the seventeenth century subsided, thanks to the fortified border and persistent Cossack activity diverting Tatar attention. Cossack loyalty, however, was bought at a high price in money and kind. The mounted steppe warriors' allegiance to Moscow proved unreliable after the Time of Troubles, as is evident from the series of rebellions they unleashed under Vaska Us' and Stepan Razin (1667–71), Bulavin (1706–08), Mazeppa (1708–09), and Pugachev (1773–75), and many localized acts of brigandage. Cossacks did not just live by pillage and plunder on land, but were also the most audacious seamen among the Orthodox believers of Eastern Europe. They raided coastal areas along the Black Sea and Caspian Sea, and entered merchant ships on those seas.[15] One remarkable feat testifying to their seafaring prowess was Razin's Cossack fleet's victory over the shah's navy before the northern shore of Iran in 1669.

This Cossack victory, too, explains why Tsar Aleksei was interested in the creation of a modern navy. Besides Cossack piracy, the Caspian's dangers also included raids by Muslim marauders based in the Caucasus range. Evidently, if a fleet of rag-tag freebooters could defeat a naval flotilla of the mighty shah of Iran, seafaring on the Caspian Sea was technologically underdeveloped.[16] Muscovite sea patrols seemed to promise suppression of pirate activity and greater control over a peripheral region. Such pacification would be conducive to trade, increasing the tsar's income.

Given the size of the Muscovite empire and the enormous cost of its land-based defence, establishing a permanent seafaring navy and merchant fleet harboured in faraway imperial outposts seems nevertheless frivolous.[17] It would not only be expensive to build and maintain such a fleet, but safeguarding the security of its seaports would come at a high cost. Because of the vast distances, it was cumbersome and costly enough to ship goods between Moscow and the two main Russian seaports, Arkhangel'sk and Astrakhan.

Although Russians fished in the open sea, they avoided long-distance sailing in large ships in western directions from Arkhangel'sk.[18] They were hindered by navigational obstacles (most were unaware of how to use sail against the wind), by the unprofitability of competing with the Western Europeans, and by their awareness of the failure of a few Muscovite daredevils, who had encountered resolute trade boycotts when they had ventured to Western Europe.[19] Instead, in or on the way to Arkhangel'sk (or the Baltic littoral) the Russians sold their goods to British, Dutch, and northern German merchants who carried them to Europe.[20]

From Astrakhan to Iran or Central Asia, mainly Russian-produced goods (hides especially) were usually taken across the sea by small boats, sometimes navigated by Russian skippers, but more often by subjects of the Persian shah or even the Indian mughal.[21] Muscovy received in Astrakhan various goods from Asia, among which Iranian-produced raw silk predominated.[22] Before the 1630s, economic grounds made the tsars reluctant to police the Caspian Sea and stimulate the long-distance trade in bulk goods across it.[23] But once Tsar Mikhail tried to modernize his land army and strengthened and solidified the fortification lines that moved further southward, he also began to probe the possibilities at sea. In a sort of joint venture with the Holsteiner Grand-Duke Friedrich III (r. 1616–59), Mikhail allowed the construction of the ship *Friedrich*, built and sailed by foreigners, which however sank on its maiden voyage on the Caspian Sea during the 1630s.[24]

***

In 1667 several conflicts were settled to the advantage of Tsar Aleksei Mikhailovich, triggering a self-congratulatory mood at the court. Paul Bushkovitch has noted how in the heady atmosphere of that year the tutor of the tsar's children, Semyon Polotskii, composed a poem called 'Russian Eagle' [*Oryol Rossiiskii*] dedicated to the formal elevation of Aleksei's eldest son as his heir.[25] *Oryol Rossiiskii* painted Russia as the emerging hegemon of Eastern Europe. The ship-building project can be linked to this optimism, which was generated by the victory over Poland and the truculent Orthodox Patriarch Nikon, and the absence of major domestic uprisings in recent years.[26]

After a decade of bickering between Nikon, traditionalists, and the tsar over church reforms and the place of the patriarch within the

Muscovite political system, a church council under the auspices of senior Orthodox patriarchs settled matters in Aleksei's favour, deposing Nikon and unwittingly affirming a breach in the church's unity.[27] This schism would never be repaired, but in 1667 the tsar could hardly have foreseen this. Instead, order and unity seemed restored to a church now clearly subordinate to the tsar. The conclusion of the church council coincided with the truce of Andrusovo with the *Rzeczpospolita*, yielding Muscovy what is now Ukraine east of the Dnipro river, including the capital of medieval *Rus'*, Kyiv.[28] The tsar's realms thus expanded to include many more adherents of the 'true faith' [*pravoslavie*, the literal translation of the Russian word for Orthodoxy], and two decades of violent turmoil appeared finally at an end.

In the spring of 1667, the tsar (b. 1629) was still young enough in spirit to consider some bold plans to strengthen Russia's strategic position further. The war with Poland had yielded some important gains, but the Russian attempt to capture the Baltic harbour of Riga had been thwarted by the Swedes in a brief conflict (1656–58) in the midst of the Thirteen Years' War.[29] The Peace of Kardis had confirmed the Swedish-Russian status quo in 1661. It was thus evident that the Scandinavian state, like the Ottoman Empire, was too formidable an opponent yet for the Muscovites to challenge. Muscovite energy needed to find another outlet.

Around this time, chancellery secretary Gerasim Semenovich Dokhturov (d. 1676), a key player in the *Oryol* project, and Tsar Aleksei himself inspected the documents available in Muscovy regarding the failed search for the North-East Passage by the Dutch navigators Willem Barentsz and Jacob van Heemskerk in 1596–97.[30] The tsar apparently pondered exploring a possible searoute from Arkhangel'sk along northern Siberia to China, but soon realized that any expeditions probing east of Novaia Zemlia would be blocked by ice. Russia's western, north–eastern, and southern seaborne outlets thus cut off, the tsar turned to the north-south axis from Astrakhan to Arkhangel'sk. At Arkhangel'sk, foreign merchants could hardly be dislodged from their control over commerce with Western Europe, but they could be made to pay more for the privilege. At Astrakhan, Aleksei saw an opportunity to actively increase the trade in which his country engaged.

Muscovy benefited from the many luxury and strategic goods it received from Western Europeans. The tsar personally enjoyed many of these exotic goods: Vinius imported parrots and parakeets for Aleksei's pleasure in August 1654.[31] Nevertheless, Aleksei, in consultation with his more influential advisors such as Ordin-Nashchokin, decided in early 1667 that he could risk undermining the role of Western European merchants in his empire's trade by increasing the fees levied on their business, and limiting their activities to the ports. The New Commercial Code of 7 May 1667 increased the tariffs on foreigners' commercial transactions and reduced foreign involvement in domestic Russian trade.[32] The statute did not exclude foreigners from a share in the Muscovite import and export trade, but aimed to raise more revenue for the government while curtailing the movements of Westerners in Muscovy.[33] A comprehensive ban on foreigners carrying Russian goods overseas was inconceivable, given the absence of native Russian (Orthodox) long-distance overseas traders and of an ocean-faring merchant navy.[34] Terms of trade could therefore not be made so onerous that they would chase away the Western Europeans from the Arkhangel'sk trade. In practice, as Lahana points out, trading volume hardly declined.[35] In the course of the seventeenth century, the volume of trade conducted by Western European merchants in Russia grew four times in size in an almost uninterrupted trend.[36]

The architect of this Code was Ordin-Nashchokin.[37] He seems to have taken a calculated risk, convincing himself that the Code would not likely be met by Western European military, political or economic reprisals. The English were grateful for Alexis's stalwart support of the Stuart cause between 1649 and 1660, while the volume of their trade on Muscovy was too modest to ponder retaliation by a naval attack. A Dutch naval assault on Arkhangel'sk might damage its modest port facilities, but Dutch ships were unlikely to intimidate the tsar by it.[38] The Dutch and the English were meanwhile too distracted by their own conflict to launch any meaningful political protest against this unilateral Russian decision. Swedish efforts to relocate the Arkhangel'sk trade through Estonian ports also seemed a 'pipe dream', even if substantial trade was conducted with the Baltic littoral.[39]

While mercantilism was of course never a precisely defined theory of economic policy, most European governments believed that 'wealth was finite, that one nation could become wealthier only by depriving its rivals of their share of international trade, and that it was by state action that this should be achieved, by protective tariffs, prohibitions and subsidies and by naval warfare'.[40] After 1640, crucial tsarist advisors (Boris Morozov, Miloslavskii, Ordin, and Artamon Matveev alike) generally supported a coherent economic policy along mercantilist lines that was to benefit the Muscovite state and increase its wealth, including its treasury's specie.[41] To stimulate domestic trade and to integrate the domestic market, the government introduced a uniform sales tax in 1653, replacing the myriad tolls and fees previously collected.[42] Tsarist economic policy strove for protectionism of Muscovite traders and domestic trade, and for an increase of the state's bullion reserves. An early Muscovite effort at import substitution and establishing self-sufficiency in one key manufacturing branch dated from the 1630s: the arms industry run by a series of Dutch natives, emulating their compatriots' successful establishment of an armament industry in Sweden.[43]

Given the well-developed state of their intelligence gathering, the tsar and his closest advisors were vividly aware of the causes and consequences of the Anglo-Dutch military conflict that had begun in 1651.[44] The adoption by the English of their Navigation Acts in that year appeared to stimulate British overseas trade and domestic manufacturing.[45] The protection offered by the acts allowed the British to substitute their own products for many foreign-made goods previously imported by Dutch bottoms, and replaced the Dutch as carriers of many goods brought to England, thus stimulating the shipbuilding trades as well.[46] The Navigation Acts were a result of the widespread conviction among British politicians, economic thinkers, and merchants that the world's wealth was fixed and their country could only become richer at the expense of the wealth of others. Thus an economic war (twice becoming outright war) had been declared upon the Dutch. By 1667, Britain's economic growth was taking off, and appeared linked to the Navigation Acts and the muscle of the Royal Navy.[47] England's flourishing seemed to demonstrate the advantages of economic protectionism. The Russian government, in which in 1667 the imaginative Ordin-Nashchokin was at his most influential, decided to follow her example.[48]

The Russian New Commercial Code expressed the tsarist court's desire to capture greater revenue from foreign trade and protect and stimulate a fledgling manufacturing sector producing goods to replace foreign imports. Imperative economic motives aimed at reducing foreigners' roles combined with more half-hearted cultural ones. The pious, cautious, and traditional Tsar Aleksei was suspicious about traders moving unencumbered throughout his realm, even if he was impressed with foreigners' economic and technological prowess.[49] Xenophobia was an old affliction shared by Russia's temporal and spiritual elites,[50] resulting from the realm's isolation during the centuries of Mongolian rule and its unique position as sole independent Orthodox country after 1453 (or 1480).[51] Xenophobia had been reinforced in the Time of Troubles, when assorted foreigners, and especially the Catholic Poles, had almost succeeded in destroying Muscovy. After 1613, however, Tsar Mikhail was forced to turn to foreign help in resurrecting his devastated country.[52] Mikhail had not been in the position to be too picky about circumscribing foreigners' privileges. His son Aleksei, however, presided over a country that had regained its self-confidence and military prowess.

Thus in 1652, prompted by complaints from Orthodox priests and monks, Russian authorities moved Western European (non-Orthodox) residents of Moscow into the *sloboda* outside of the capital.[53] In other Russian towns, too, foreigners' offices and residences were closed during the 1650s. In 1653, Western foreigners who had received manorial estates as part of their pay for their service to the tsar lost their property title.[54] Yet in 1675 Aleksei issued an *ukaz* (decree) that prohibited Russians from adopting Western dress and from shaving their beards.[55] This religiously driven segregation of Orthodox and non-Orthodox parallelled the contemporary practice of the Ottoman Empire and various Asian states. Those regimes also treated Western European Christians as a distinct, often segregated, caste. In response, the Europeans often functioned as a single community, as we saw in the previous chapter.[56]

But despite such measures, his personal piety, society's long-standing distrust of Others, and the church's isolationist stand, the tsar was increasingly of two minds regarding Western culture. Aleksei was lured by the temptations from abroad. Already before 1670 his children learned foreign languages. Probably influenced by his young second wife Natal'ia Naryshkina, he had Western paintings hung in his palaces, plays staged, fireworks lit, and enjoyed Western-style music.[57] In 1676, Struys witnessed the tsar's delight in a concert performed by Dutch musicians taken along by van Klenk's embassy to please the Russian monarch.[58]

The New Commercial Code thus never aimed at excluding foreigners from trading with Russia.[59] Ordin tried thereby mainly to increase revenue, while the code served the added purpose of propitiating Russian merchants' claims for greater protection against foreign encroachment. Likewise, the tsar and Ordin knew that the extensive use of Western European craftsmen in various industries remained a vital necessity for Muscovy. Westerners were employed in training and commanding the army and developed branches of mining and manufacturing that only very gradually promised to replace European-imported goods with Russian-won resources and Russian-made goods.[60]

Incapable of banishing foreigners from shipping most of Russia's exports and imports, the tsarist government did try to offset foreign interests against each other.[61] Armenian-Persian[62] merchants, less of a threat on a trade route where Russian merchants were tolerated, were given advantageous terms to conduct trade through Astrakhan between the shah's realm and Muscovy.[63] On 31 May 1667, a trading agreement was concluded with Armenian merchants who represented the shah of Persia (Suleiman [Sefi] II, r. 1666–94), in the expectation of increased transport, especially of silk, across the Caspian Sea to Moscow and beyond.[64] The aim was to divert most of Persian silk exports to Astrakhan across the Caspian Sea (and some over land along its western shore via Tersk), away from the caravan route to the Levant and the overseas route via Bandar-e-'Abbas.[65] Since most silk in Iran was produced in provinces hugging the shoreline of the Caspian Sea (such as Gilhan), rerouting the silk trade to the advantage of the tsar's and shah's subjects at the expense of Ottoman Turks and Western Europeans seemed quite feasible.[66] Floor and Clawson underline the potential of this trade in noting that around 1630 silk had been the 'second-most-important European import from Asia' and that Aleppo, the port by which most Iranian silk found its way to the West, became as a result one of the largest cities of the Turkish empire.[67]

In the seventeenth century, the Iranian shahs repeatedly tried to reroute the silk trade with Europe to spite their Turkish foes, without any sustained success. Exports through the southern port of Bandar-e-'Abbas (or previously Hormuz) did not meet the Iranian monarch's expectations. In exchange for commodities the VOC shipped to Bandar-e-'Abbas and Isfahan, the Dutch trading company, Iran's dominant seaborne trading partner, preferred to receive specie rather than silk.[68] The VOC often used Iranian bullion to purchase goods elsewhere in Asia, exporting Chinese silk to Europe instead of the Iranian product. The Muscovite-Iranian agreement of May 1667 thus aimed to cut out those unpopular middlemen by redirecting the Iranian silk northward.

The Russian-Persian treaty made it incumbent upon the tsar to protect the silk import from *razboi* (brigands), who were raiding at sea and on the Volga.[69] This strengthened Aleksei's determination to have seafaring vessels built that could both transport silk and guard this potentially lucrative trade.[70] This dual economic and military purpose of the *Oryol* was traditional for sailships.[71] Aleksei's interest in having Western Europeans build a truly seaworthy Russian fleet followed his father's lead, and was far from uniquely Muscovite. In Asia, many contemporary local potentates engaged in such experiments (including Mughal emperor Aurangzeb), while in the 1670s in Europe the king of Denmark enlisted thousands of Amsterdam-based sailors and officers and numerous Dutch ships to fight the Swedes.[72]

A mere three weeks after Muscovy had struck its trade agreement with Persia, the tsar officially launched the building of a fleet by issuing an *ukaz* to ready a ship's wharf in Dedinovo.[73] This village, situated

110 kilometres south-east of Moscow on the Oka river near Kolomna, had a tradition of building rivercraft.[74] This seemed to make it most suitable as the location to construct vessels for travel on the Caspian Sea, to which they were to descend along the Oka and Volga rivers upon completion. No expenses were saved in building the first vessels of Aleksei's Caspian fleet: ultimately, the construction of the ship *Oryol*, a yacht, and several rowing boats cost, everything included, 9,021 rubles, which converts into 45,105 guilders.[75] This was a formidable sum, almost one per cent of the total tax revenue of Muscovy per annum, and about one per cent of her total army allotment of 1680.[76] This high price-tag shows how the tsar's government attached, at least for a fleeting moment, great significance to its foreign trade by way of Astrakhan.

Supervision over the project in Moscow was placed in the hands of Boyar Ordin-Nashchokin, the secretaries of the tsar's council (*dumnye d'iaki*]) Gerasim Dokhturov and Luk'ian Timofeevich Golosov (d. 1682–3) and the secretary (*d'iak*) Efim Rodionovich Iur'ev.[77] On 2 February 1667, Ordin-Nashchokin was promoted from the second highest court rank of *okol'nichii* to the highest of *boyar*.[78] This promotion was Tsar Aleksei's reward for Ordin-Nashchokin's successful truce negotiations with the Polish representatives at Andrusovo, for which he was given an additional monetary reward of 500 rubles. Ordin-Nashchokin's rise had been meteoric: originally a mere provincial *dvorianin* (gentleman), in 1658 he had been promoted, again for diplomatic accomplishments, to the rank of gentleman of the tsar's council (*dumnyi dvorianin*), and soon after to the higher rank of *okol'nichii*.[79] Although Ordin's ascendancy at the court was brief (1667–70), V.O. Kliuchevskii deemed him to have been the paramount statesman of seventeenth-century Muscovy.[80]

While Ordin was the chief manager of the *Oryol* project in Moscow, Dokhturov and Golosov – just promoted themselves to the highest commoners' rank of *dumnyi d'iak* in March 1667 and highly remunerated by the tsar's government – enjoyed the autocrat's trust only slightly less than the boyar.[81] Only three other Muscovite commoners were of equal rank to Dokhturov and Golosov in the dozens of central government departments, underlining further the initial high priority given to the project (Dokhturov received a higher *oklad* ['pay'] and was more senior than Golosov). Golosov may have been chosen by the tsar to provide a cautionary voice, for as a youth he had protested having to study Latin since it 'corrupted his soul': This scepticism regarding Western ways, not uncommon in the 1650s and 1660s Muscovy, may have lingered in him.[82] Such distrust of foreign novelties clashed with Ordin's reformist mindset, and may have played a role in the eventual collapse of the *Oryol* project. Toward the end of his government career, an embittered Ordin-Nashchokin accused Dokhturov, Golosov, and Iur'ev of incompetence and corruption.[83]

During the first stage of the naval project, from spring 1667 to autumn 1668, however, Ordin and his three assistants steered it in a unified and reasonably determined fashion. Formally, they oversaw the project as the heads of the *Prikaz* (Chancellery) of the Novgorod *Cheti* (*Chetverti*) (territory).[84] This office traditionally administered the Novgorod lands, but by the 1660s oversaw a welter of border and trading towns and regions (for instance, both Novgorod the Great, on the border with Livonia, and Nizhnii Novgorod along the Volga, as well as the littoral of the White Sea). Ordin and the three secretaries also led the *Posol'skii prikaz*, one of the most important central government departments.

In this latter capacity they conducted diplomatic negotiations with the English envoy John Hebdon in the fall of 1667 about a restoration of trading privileges for the English Muscovy Company.[85] In their rejection of Hebdon's request, the Muscovite negotiators expressed the hope that English losses would be compensated by the shipping of a growing volume of precious goods departing Russia through Arkhangel'sk. This increase was anticipated from rerouting the Persian silk trade to the allegedly safer journey across the Caspian through Muscovy to Arkhangel'sk, at the expense of the traditional overland caravan route through Ottoman territory to Mediterranean ports. Ordin's suggestion to Hebdon shows how in the autumn of 1667 the shipbuilding endeavour was paramount in his and his assistants' minds. By November 1668, when Butler and his sailors reached Moscow, the *Oryol* project was transferred from the Novgorod Chancellery to the *Posol'skii prikaz*; once launched, the *Oryol*'s activities were to fall within the purview of Muscovite diplomatic relations with Iran and thereby become the responsibility of the foreign office.[86]

Razin's rebellion prevented regular conduct of trade between Iran and Muscovy for several years after 1667; a new treaty was signed in 1673.[87] Its adjustments reflect a dimmer Muscovite view of the benefits of this trade.[88] Different from his youngest son Peter, Aleksei knew when to cut his losses and to abandon wasteful projects. The silk trade across Russia nevertheless increased in volume, especially after 1685, when it was redirected from Arkhangel'sk to the Swedish-held ports along the Baltic littoral.[89] Silk transports were, however, operated and controlled by the Iranian-Armenian merchants. The tsar's government benefited from custom fees paid by the New Julfa traders, but Russian merchants were virtually excluded from it.

6
In the Tsar's Service

In July 1658, Jan Janszoon Struys wed twenty-four-year-old Trijntje Pietersdochter in Amsterdam; whereas *Reysen* does not mention Trijntje's name, the occasion of the wedding is referred to in the book.[1] The Struys of the text can thereby be equated with the Dutch sailor Jan Struys, about whose life between 1657 and 1668 *Reysen* is otherwise silent. Some records in Amsterdam's municipal archives give us a glimpse of Struys's life during this decade, although most of the details remain beyond our grasp. His movements become much clearer once archival records and *Reysen*'s narrative reinforce each other, beginning in September 1668.

Before her death in her early thirties in December 1666, Trijntje gave birth to at least three children.[2] They were baptized in April 1659 (Jan, a boy), December 1663 (a first girl named Teuntje), and October 1666 (a second girl Teuntje).[3] Epidemics (including the *pest*) raged in Amsterdam in those days, and a quarter to one half of all women died from complications of childbirth, so neither her own nor her children's death was unusual (probably both daughters died in infancy).[4] As advanced as the Republic was, its population did not escape the high mortality common in Early Modern Europe, which was even more pronounced in cities.[5]

After Trijntje's death, two boys of parents named Jan Jansen and Trijntje Pietersdochter were baptized in a Lutheran church in Amsterdam in June 1667 (Hendrick) and January 1669 (Pieter).[6] Given the odd timing of the baptism and the commonality of Jan and Trijntje's first names and patronymics, however, Pieter and Hendrick may not have been children of our Jan Struys and the deceased Trijntje Pieters, but of a couple sharing the same (rather common) names. From his first wedding to his departure for Muscovy ten years later, Jan lived in the St Jorisstraat, a short street that still today connects the Singel with the Reguliersdwarsstraat, then located in the southernmost part of the city.[7]

It was usual for Dutch sailors to marry, belying the 'Early-Modern marriage pattern', which saw elsewhere in Europe people without significant assets remain single.[8] While he may not have owned much property, when he married in 1658 Struys's prospects were likely good compared to those of his peers. He may have saved something from his Venetian service, and there was then plenty of work for sailmakers in the city. The Amsterdam Admiralty's massive programme of building warships picked up steam in the latter half of the 1650s.[9] In expectation of future hostilities with England, and engaged in conflicts with Sweden (1658–60) in Europe and Portugal in Asia (1656–61), the Dutch were by 1658 frantically expanding their navy with specialized ships, most of which were constructed on the Amsterdam wharves. *Reysen* seems therefore accurate in reporting that the decade after 1657 was an era when sailmaking ashore was lucrative enough for Struys to settle down.[10] Struys's whereabouts in the Anglo-Dutch War of 1664–67 are unclear, but whether ashore or on board, he was primarily employed making sail.

Upon the conclusion of the second English war, a decline in shipbuilding followed, threatening Struys's livelihood apparently as well. Possibly because he could not take care of any surviving children alone, Struys already planned to remarry by the autumn of 1667 with the widow Marretje Jansdochter, his landlady.[11] Since March 1661, Marretje had owned the house in which Struys had lived since early 1658.[12] A scribble on the wedding licence of 1668 notes how Struys's goods (which included the advance he had received from Butler) were 'brought into the marriage', perhaps as a sort of collateral.[13] Marretje received thus joint title to Jan's possessions, and served as his executor while he was away, or if he died in Muscovy, for which destination Struys immediately departed after his wedding.[14]

Struys's loss of his first wife and children shows life's utterly precarious quality in the Early Modern Age. Even in the prosperous Netherlands the constant threat of indigence besieged those not born to the manor. Whereas Struys had married Trijntje partially for romantic reasons (neither spouse seems to have owned significant property in 1658), he married Marretje primarily for practical and economic reasons. He needed a guardian for his children, and, if he could hold up his end of the marital bargain and return with riches from Russia, a house awaited him that he partially owned, and thus a considerable level of material comfort. But reaping this reward appears to have been made conditional upon a considerable financial contribution on his part. The initial postponement and odd timing of his second wedding (and the addendum to the licence) seem to indicate that Marretje consented to marry Struys only if he would bring his fair share of assets into the marriage, most of which were still to be earned in Russia.

***

Reysen resumes the tale of Struys's movements in September 1668, from which time evidence about Jan Struys's wanderings can also be regularly encountered in several other sources. Recruited with more than a dozen others for service on the tsar's Caspian squadron, Struys and his companions travelled from Amsterdam to Riga by ship.[15] Encountering rough weather, the vessel traversed the sea route in a fairly lengthy thirty days. During the trip some sails ripped on the rough seas. *Reysen*'s Struys offered to repair the damage, underlining his growing skill at capitalizing on an opportunity by negotiating a handsome fee from the ship's captain. Struys had to turn to the Livonian authorities to force the skipper to pay up once they disembarked at Riga.[16]

In Riga, the party of which Struys was a member had to wait for several days before receiving transit passes from the Swedish governor, Count Philip (von) Krusenstiern(a) (*Croonenstern*).[17] The waiting time was not unexpected, for Swedes and Russians were suspicious of each other. Dutch travellers to Moscow often evaded Swedish-held territory on their way to Russia and sailed to Arkhangel'sk (as Struys would do when accompanying van Klenk in 1675), but this northern access route and port were only free of ice from late May to late September. Such a northern route would have forced the *Oryol* crew to wait for the sping of 1669 to depart Amsterdam.

The first overland stage from Riga through Livonia to the Muscovite border in October 1668 left a profoundly negative impression on Struys, if *Reysen*'s words are to be believed.[18] The Dutch party encountered people who are described as among the poorest, coarsest, and most superstitious pagans Struys had ever met. In mid-autumn, Jan Struys and his thirteen[19] fellows arrived at the Swedish-Livonian border with Muscovy.[20] Equipped with a great amount of shipwright's tools and materials transported on carts, they entered tsarist territory. The sailors carried copious sailmaking materials, including canvas-cloth, about which the Russians knew little; the Dutch, by Samuel Pepys's reckoning, made the best sailcloth, mainly using hemp and flax.[21]

The men's captain-designate was David Janszoon Butler (*David Boetler'* in the Russian record), a veteran sailor of the Seven Seas, and a nephew of a powerful Dutch merchant on Russia, David Ruts (and thus a relative of van Sweeden as well).[22] In Amsterdam in February 1667, Jan van Sweeden had contracted Butler to hire mariners who were to follow the shipwrights building the *Oryol* and its auxiliary vessels once they neared completion.[23] Butler was to command the ships himself. The twelve men accompanying Butler and Struys (*Ian' Iansen*) were identified in Russian records as *Ian' Alberts', Vilim' Vilims', Petr Bartels', Korneli Kornelis, Vilem' Popkes, Karsten Brant', Elis' Peterson, Meindert Meinderts', Danil Kornelis, Petr Arendtsen, Iakob Trappen,* and *Kornelis Brak,* names that may be found in their Dutch spelling in *Reysen*.[24] While the fourteen men would never accomplish the task for which they had been recruited, to sail a ship for the tsar on the Caspian Sea, two of these most ordinarily named Dutchmen went on to achieve extraordinary fame. In addition to Struys, the assistant constable Karsten Brandt proved instrumental in the creation of Peter the Great's navy, for which he is still remembered in the annals of Russian history.[25]

Those Dutchmen entering Muscovy near the town of Pskov were all natives of the province of Holland.[26] Seamen hailing from neighbouring villages and towns customarily maintained an informal network, joining forces in finding the best jobs and wages.[27] For men from Amsterdam and its surrounding region, Muscovy was not necessarily a faraway land of half-mythical proportions.[28] Enticed by the stories of those who had visited Muscovy, a growing number of artisans travelled from Amsterdam and ports in Holland to Russia in the course of the seventeenth century. Butler's men had probably alerted each other to the lucrative opportunity of enlisting in the tsar's service, promising a remuneration far outstripping pay on Dutch vessels. Apart from the operation of a social network and the captain's preference for fellow Dutchmen, the composition of the crew was also guided by pragmatic considerations on Butler's part, because he needed a versatile crew with advanced seafaring skills.

In *Reysen*, Struys recalls approvingly that, after crossing into Muscovy, the landscape ameliorated, for he and his fellows usually travelled through meadows and ploughed land rather than swamps and wild forest, a sign of the Russians' higher stage of development compared to that of the Livonians. Such neatly cultivated land appealed to the Dutch as it reminded them of their native country, where, in true Aristotelian fashion, human wit had fulfilled its potential in a landscape neatly organized for agricultural pursuits.[29]

From the border near Pskov, Butler's men were chaperoned by a Russian guide (*pristav*) who led them Moscow.[30] They stopped at towns and at postal stations, where the horses that pulled the carts or sledges (it soon began to freeze) were refreshed, and the passengers spent the night. Although meeting with the cold temperatures and threatened by highwaymen and wild animals lurking in the woods, the Dutch crew was unfazed by the conditions. Except for a brush-up with one of their tobacco-starved[31] Russian guides and an encounter with a couple of bothersome brigands, the Russian leg of the trip proceeded comparatively smoothly.[32] While *Reysen*'s Struys liked the countryside, the book disparages the look of the towns along the route from Pechora and Pskov to Moscow.[33] The text is especially critical of the ubiquitous use of wood for construction, which made Russian buildings weaker than the *alderslechtste* ('very worst') town in the Netherlands.[34]

Struys and his apprentice, Els (Ellis) Pietersz, raised Russian eyebrows by making brief skating tours.[35] In Muscovy, such Dutch tricks had earlier intrigued Tsar Mikhail, who had been fond of the acrobatics and juggling acts of visiting Dutch artists.[36] Novelties and curiosities enchanted both Peter the Great's father and grandfather, and Holland had a reputation for such entertaining things across Europe.[37] Although Struys believed that the Russians were amazed by his skating, it is quite possible that they were perturbed by the foreigners' audacity in venturing out on thin ice.[38] On a second outing near the village of Klin (60 kilometres from Moscow), Pietersz and Struys both fell through, and Struys had to save his young aide, who could not swim.[39] In the next few days, around the first of December 1668, Butler's crew reached Moscow.

***

The fourteen men of Struys's party had been preceded a year earlier by a group of six Dutchmen, who had been assigned to lead a number of Russian workmen in building the first vessels for the Caspian fleet. These Dutch artisans had been personally recruited by Jan van Sweeden. Van Sweeden's experience in hiring Dutch craftsmen to work in Muscovy facilitated his recruitment of shipwrights and sailors for the shipbuilding project.[40] Van Sweeden had encountered difficulties in hiring Dutch specialists. The United Provinces were at war with England until 31 July 1667, and were reluctant to part with skilled able-bodied sailors and shipwrights.[41] Some of those hired, too, disappeared with their signing bonus, for which the tsar asked the Estates-General restitution in early 1668.[42]

Nonetheless, weighed down by their tools, a small party of Dutchmen (also attracted by high wages) had reached Moscow a year before the group of Struys and Butler.[43] They had travelled to Moscow with other Dutch artisans whom van Sweeden had hired to work in his new mills.[44]

The skipper Lambert Helt, the head carpenter (and sometime first mate) Willem Klopper, the carpenter Dirk Pietersz, the sailor Jacob Trappen (who seems to have travelled to Muscovy rather with this group than with Struys's), and the diamond cutter Anthony Munster, who was also a skilled shipwright and sailor, soon descended upon Dedinovo.[45]

At Dedinovo, the Dutch shipwrights oversaw a largely Russian construction crew building a ship and a yacht, as well as some rowing boats. General command over the enterprise was given to a Dutch army officer who had served the tsar for two decades, Cornelis van Bockhoven, a member of a family of mercenary officers.[46] The van Bockhovens (father and two sons) had been recruited with twenty other soldiers by Il'ia Danilovich Miloslavskii, during the latter's visit as the tsar's envoy to the Netherlands in 1646 and 1647.[47] Many Dutch veterans, idling during the waning days of the Eighty Years' War with Spain, were then recruited by foreign employers.[48] Others served Venice in its war with Turkey.

The van Bockhovens arrived in Moscow in 1648. The following year Cornelis (*Kornilko*) petitioned the tsar to allow him to travel to Arkhangel'sk to collect his wife and children there, showing the extent of his commitment to his new master.[49] In the early years of his service, Cornelis van Bockhoven briefly served under his father's command in the tsar's bodyguard.[50] After the domestic unrest of the first years of Aleksei's reign dissipated, the Dutch officers were reassigned to organize and train modern cavalry units for the tsar (against stiff opposition by some Russians who refused to serve under foreign command).[51] Isaac van Bockhoven, the *pater familias*, also tried to improve the Russian army's use of artillery.[52]

In 1667 and 1668 at Dedinovo, Colonel (*polkovnik*) Cornelis van Bockhoven, who earned a monthly wage of forty rubles, was assisted as supervisor by other foreigners (*inozemtsy*), almost all Dutch natives like him judging by their names: His deputy Jacob Stark, Captain Hendrik van Helmond, lieutenant (*poruchik*) and surgeon (*lekar'*) Nicolaas Schak, skipper (*karabel'nyi dozorshchik*) and lieutenant Rengelt Vemin, and the interpreters Jacob Schram (Tolk) and Christoffel Koster.[53] Whereas the officers were paid high wages, the interpreters, surprisingly given the difficulty of their task, earned no more than three rubles per month each, far less than even the regular sailors of Butler's team had been promised. To handle administrative matters, the government assigned to the project the gentleman (*dvorianin*) Iakov Leont'evich Poluekhtov and the undersecretary (*pod'iachii*) Stepan Petrov.[54]

Work began in the late summer of 1667 with collecting construction materials and workers in the proximity of Dedinovo. The iron forges in Tula owned by Rotterdam-born Peter Marselis were ordered to deliver materials to the wharf.[55] But neither Marselis, nor local Russian lords and officials, nor the peasantry proved enthusiastic about lending a hand to the shipbuilding project.[56] It was difficult to hire sufficient carpenters and smiths, and to keep them working in Dedinovo for meagre wages.[57] In the fall of 1667 and following winter, Poluekhtov sent a stream of complaints to his Moscow superiors about the lack of collaboration by the local population and nobility; in response, the tsar signed a slew of decrees ordering his subjects to help the project.[58] But in following Russian traditions of *kormlenie* ('feeding') and harsh discipline, Poluekhtov did not aid his own cause: he beat the artisans working at the project and pocketed the money earmarked to buy food for the workers.[59]

Despite these problems with men and materials, the vessels were slowly built in the year prior to the departure of the Butler party from Amsterdam. One surmises significant progress in the completion of the ships when, in March 1668, van Bockhoven, Helt, and Klopper urged the tsar to acquire key supplies and experienced sailors.[60] A sailmaker had been recruited in Amsterdam in 1667 to accompany the Helt team, but he had absconded with his advance; local Russian sailmakers were not up to the task.[61] Therefore, van Bockhoven urged Poluekhtov in the early months of 1668 to underline the importance of the presence of a master sailmaker at the site, for which van Bockhoven himself also petitioned the tsar in March.[62] The solution of this problem had to await the arrival of the experienced sailmaker Jan Struys in the fall, who brought with him most of the required sailcloth from Holland as well. By late May 1668, the *Oryol* was almost finished, as were a yacht and lifeboats.[63] Still, applying the finishing touches (such as the completion of the figurehead and other ornaments) took much of the summer of 1668, and, obviously, no sails were yet available.[64] The tsar's hope that the flotilla might descend the Oka and Volga to Astrakhan in 1668 proved vain. Still, the petulant Poluekhtov and van Bockhoven had well-nigh completed their task by the early autumn.[65]

***

More than a decade after the *Oryol*'s completion, the tsar's governor-general (*voevoda*)[66] in Astrakhan, K.O. Shcherbatov, dispatched a report to the central government in Moscow regarding a mysterious abandoned ship and yacht on the shore of one of the Volga estuary's streams:

> The ship is dilapidated, the bottom and sides have rotted, it is incapable of travelling; and on it are stored three wooden masts ... ,

> decrepit ... and the end of the masts are fastened in the ship with iron rings; on its nose, at the cutwater, latticed railings, under the railings [a figurehead of] a carved lion, painted, on the bow a cabin, and on the stern two cabins, a higher and a lower one; around the cabins, at the bow and stern, railings. The ship has two decks, a higher and a lower ... On the ship is also an anchor ... At the sides of the ship are 22 gunports without shutters and among them one has a shutter ... And along this ship stands a wooden half-ship (*polukarabel'e*), on it are two closed cabins ... [67]

This description of these vessels' appearance concurs with the few hints (such as the number of gunports) about the vessels in *Reysen*. Shcherbatov had evidently happened upon the remnants of the *Oryol* and her yacht, which is born out by other historical evidence. Various features described suggest the ship was a *fluyt* (flyboat), to which type the *Oryol* likely belonged.

Although during the 1650s and 1660s the Dutch followed the British lead in building sailing ships with the exclusive purpose of warfare, the *Oryol* belonged to an earlier tradition. She was to play the traditional dual role of the sailship: a man-of-war to safeguard the trade on Persia and a merchant ship to transport cargo across the Caspian.[68] The *Oryol* was about half the size of the East Indiamen built on the Amsterdam wharves in those days; her dimensions and design rather resemble a small *fluyt*.[69] These were flat-bottomed ships that could traverse shallow waters. Even if her dimensions indicate a ship slightly shorter than most *fluyts*, the *Oryol* was most likely a version of this famous mass-produced Dutch ship, since she had the same number of masts and a similarly small crew.[70] Sporting three masts and a bowsprit, the *Oryol* measured about twenty-five metres long and six metres wide, with a cargo-hold of about three metres depth.[71] A mere twelve sailors could sail a *fluyt*: Butler's crew was slightly larger, but counted two gunners, one surgeon, and one interpreter.[72]

Like a *fluyt*, the *Oryol* could serve both as warship and as merchantman. While she carried twenty-two cast-iron cannons of two-to-five-pound calibre, her fire power was puny compared to the ships of the line built in Holland in the 1660s.[73] On those Dutch warships, the calibre per cannon was more than four times as heavy, and the number of guns was three to four times as high. Such battleships, however, would never have been able to journey down the Volga, or would have easily have become stuck in the shoals of the northern Caspian Sea (see also Chapters 8 and 9). Nor were such sizable ships necessary: a well-armed *fluyt* could easily fend off any of the primitive vessels traversing the Caspian. About the yacht we only know that she carried six one-pound cannon, and that she apparently had two cabins, as Shcherbatov noted.[74]

Although the ships were virtually ready and their crew had reached Moscow in late 1668, the Russian winter stalled the project's progress. The rivers had frozen over, which meant that no departure of the flotilla was possible before April of 1669. But whereas the climate only temporarily froze the progress of the *Oryol* project, the ultimate sea-launch of the tsar's Caspian fleet was probably fatally jeopardized by the sudden death of Jan van Sweeden, just before the Butler crew entered Moscow.[75] Van Sweeden's death, combined with the absence of Ordin-Nashchokin from Moscow in the autumn of 1668 and much of 1669, was to shut down the *Oryol* project at Astrakhan. In May 1668 Ordin had been dispatched to Kurland, eventually travelling to Warsaw, where for the remainder of 1668 and much of 1669 he was occupied with the election of the Polish king Michal and its fall-out.[76] Bereft of its two boosters, the *Oryol* was never to reach the Caspian Sea.

7
Reysen's Muscovy and Struys's Muscovy

Beginning with Struys's departure from Amsterdam in 1668, with which its third act opens, *Reysen*'s protagonist's adventures acquire a greater verisimilitude (and are often borne out by other sources) than its previous parts. Struys's oral recollections of recent events were obviously shaped through the editing process that human memory exerts, suppressing moral ambiguities and ordering them into a pattern.[1] Walter Benjamin has argued that traditionally in Europe 'seamen were ... masters of story-telling', with a knack for 'practical interests'.[2] This predilection surfaces in *Reysen* on occasion indeed, betraying in such moments the presence in the text of the historical Jan Struys (the accurate listing of his or his fellows' wages are instances of this). Many of the anecdotes rendered in *Reysen*, such as the skating forays or the tobacco quarrel on the way from Riga to Moscow, seem infused by Jan Struys's vivid personal recollections. But the book is not a diary and Struys's narrative leaves yawning gaps. His group's pause at Moscow during the winter of 1668–69 gave *Reysen*'s ghostwriter an interlude to present a chorography of Muscovy. It was patterned after the standard view of the tsar's empire rendered in various previous texts published in Western and Central Europe, especially in the work by Adam Olearius.[3]

On both occasions that Struys travelled to Muscovy, his motive was unmistakably monetary gain. While the younger Struys exhibited a spirit of adventure and bravura that was part of the sailors' world, *Reysen*'s suggestion that *Wanderlust* inspired this middle-aged father and husband to head for Russia makes little sense.[4] Money prompted the historical well-nigh forty-year-old Struys to enlist with the 'plenipotentiaries of the Grand-Duke of Moscovia'.[5] Probably because the *Oryol* project acutely needed a sailmaker, Butler promised Struys one-and-a-half times the wage of the sailmaker who had absconded with his advance in 1667.[6]

Shipbuilding had meanwhile taken a downturn after the second Anglo-Dutch War: Struys could hardly ignore the 'unusual opportunity' of four years guaranteed work for such high pay.[7] His illiteracy negated otherwise his twenty years of seaborne and shore experience. The higher ranks on board were reserved to those who were literate.[8] His employment in Muscovy on the *Oryol* was therefore a one-time opportunity: It offered the sailmaker a remuneration otherwise out of reach to him, amounting to that of a ship's officer, and three times the wage he usually received at sea.

Butler's bill preserved in Russian archives shows that, like the others, Struys was initially hired for a four-year term of service.[9] He received in Amsterdam a four-month advance totalling 240 guilders and 50 guilders of bonus.[10] For not entirely evident reasons, Struys's starting wage of 60 guilders a month was pared down to 50 guilders in the course of 1669. Butler paid meanwhile for all of the costs the crew incurred journeying from Amsterdam to Moscow, including the fees for the passes issued by the Swedish governor in Riga.[11] *Reysen* confirms Struys's earnings as fifty-seven guilders per month, fifteen of which would be kept by his captain to pay for his subordinate's food and drink.[12] By Russian standards this salary was extremely high, far above the wage of foreign metal workers in Tula's mines and forgeries.[13] In 1677, Russian infantry and cavalry *commanders* in the tsar's army were paid annual wages of 100 to 200 rubles, which translates to monthly wages of 40 to 80 guilders, remuneration in the range of Struys's income![14] Meanwhile, rank-and-file musketeers (*strel'tsy*) only earned about four rubles in money *per year*, although they received compensation in kind such as grain rations.[15]

Russian records call Struys a *master*, confirming the high prestige bestowed upon him as expressed by his wage, and indicating that the Muscovites considered him equivalent to a Western guild master.[16] His pay was now about four times the annual wage of the average Dutch worker, and about fifty per cent more than that of the average fully skilled craftsman in Holland.[17] Around 1670, the annual rent for a basement dwelling was 30 guilders in Amsterdam, and an entire house could be rented for 90.[18] Struys's second wife bought her house in 1661 for 2,000 guilders.[19] If he would serve out his four-year term and get paid, Struys stood to receive well-nigh that sum, even after the subtraction of monthly expenses for food and drink!

Clearly a promise of riches lured him to join the tsar's service. Historians of both of European sailors and of the seventeenth-century foreigners' contingent in Moscow argue that most of these men left their country down on their luck, escaping for example grinding poverty, judicial prosecution, or creditors.[20] But, given Struys's recent wedding to a rather well-to-do widow, he hardly fits this desperate image. The wedding of Marritje Jansdochter and Jan Struys shows a certain confidence that the groom would return home with his earnings at the end of his contract.

This faith was informed by their perception that he joined the service of a ruler of a country not unfamiliar to Amsterdam sailors. The printed image about Muscovy that was recycled for generations in Early Modern Europe would let us to believe that Russia constantly remained a faraway and exotic place. But the annual fleet leaving and returning to Amsterdam made Russia a known quantity not just to the leading trading houses, but also to the skippers and sailors who sailed the route in the seventeenth century.[21] When ashore in Holland, the crews bantered lore about Muscovy and its inhabitants, even if specific knowledge of the country's interior may have been sketchy compared to information about the searoute and the port of Arkhangel'sk, where annually some thirty-five Dutch vessels arrived.[22] Among those regularly travelling to Russia, it was certainly common knowledge that a kind of support network in Moscow of Dutch expatriates existed. For, as we saw in Chapter 4, Dutch artisans and mercenaries regularly departed for Moscow to serve the tsar.[23] Indeed, serving the Russian monarch on the Caspian Sea seemed less risky than embarking on the longer and more hazardous trips to West Africa and the East and West Indies.[24]

***

Reysen claims that Struys in his zeal to adapt to Muscovy had already begun to learn Russian when he reached Novgorod, mere days after crossing the border.[25] The reader is told that he acquired the language so fast that he could communicate well enough to get around without a translator in Moscow. This may have only been boasting. By August 1670, when David Butler had been in Muscovy for almost two years, he had not picked up a word of Russian, according to the Russian-speaking Faber.[26] Foreigners needed time to master Russian.[27] For many, learning the language was impeded because of their residential segregation in *sloboda*s. Most Russians kept their distance: apart from their religiously driven reluctance to mingle with foreigners, they found that Western Europeans looked and sounded strange.[28]

But precisely because he was illiterate, Struys's aural and oral skills may have been more developed than those of literate people such as Butler. His earlier travelling had attuned his ear to foreign tongues.[29]

Sailors' lives depended on a good ability to communicate in a variety of languages, as they often sailed with a diverse crew.[30] It appears as if Struys developed a decent understanding of the Mediterranean *lingua franca* when he served the Venetians and rowed for the Ottomans. Furthermore, in Moscow, unlike most Western Europeans, Struys and his comrades were housed among the Russian population in the city, at first in the house of a Russian-Orthodox couple, and then in van Sweeden's abode, where the latter's widow Maria and children continued to live.[31] This may have enhanced the crew's command of Russian, and *Reysen*'s suggestion that Struys was soon able to communicate in Russian may have had a kernel of truth in it.

Apart from the crew's astronomical wages, the men's importance in the eyes of the tsarist government can be further gauged from this permission to live in the city proper, for after 1652 Westerners were *usually* forced to live in *nemetskaia sloboda*, as we saw in Chapter 5. Nevertheless, in 1667 a number of Dutch natives owned houses in Moscow proper, near the '*foul pond*' (*poganyi prud*).[32] Van Sweeden's Moscow residence was probably one of them, a stone house built by his father-in-law David Ruts.[33] In 1670 the van Sweeden household (now headed by Maria) still used a warehouse in central Moscow's *gostinii dvor* (the traders' arcade), alongside the Russian *gosti* (great merchants), without having to pay extra fees.[34]

Such privileges were only granted by way of exception. Rather than *nemtsy* (outright Western foreigners), the tsarist government seems to have considered the men (van Sweeden's protégés, after all) *Moskovskie inozemtsy* ('Moscow foreigners'), a sort of legal equivalents of the guest-workers of continental Western Europe in the 1960s.[35] The Butler crew was involved in a project affecting Muscovite 'national security,' and van Sweeden had been a kind of honourary Russian magnate, a trustworthy long-standing servant of the tsar. Therefore, the prohibition that forced most Western and Northern Europeans to live in the *sloboda* did not apply to him, or by extension to Butler's men.

David Butler shared some of his cousin van Sweeden's enterprising spirit. Soon after Butler's arrival in early December 1668 (and his cousin's sudden death), he gave the tsar's foreign office blueprints for the construction of two galley-style ships to operate on the Caspian Sea.[36] One suspects that some of the original ideas for the galleys may have been van Sweeden's, who once had lived in Astrakhan. Butler's plans were further elaborated by a translator of the chancellery, Andrei Andreevich Vinius (1641–1717), the son of Andries Denijsz Vinius.[37] A.A. Vinius, admitting that Butler's project had inspired him while translating it, argued that oars-driven ships would be more suited to cross the Caspian. In Vinius's view, the sea sustained fewer heavy storms than others (a moot point) and was shallower than the ocean, conditions which had caused the Holsteiner ship *Friedrich* to go under on her maiden voyage during the 1630s.[38] Vinius advocated the creation of an entire fleet of shallow-draught galleys. Vinius followed Butler's original plan in proposing convicts and captives as oarsmen, thus minimizing labour costs.[39] Both Butler's and Vinius's proposals were shelved, however, even if galleys were perhaps more viable than large sailships to traverse the Caspian. It is likely that once the *Oryol* project had failed, their benefit was not thought to warrant the expense of building, maintaining, and navigating them.

Furthermore, Butler submitted to the *Posol'skii Prikaz* a set of thirty-four rules regulating the tasks of seafaring captains and their officers, which was adopted as the regimen for Muscovite sailships.[40] He also gave the Foreign Office a map of the Caspian Sea, presumably based on maps available at the time in Western Europe (although he may have received permission to consult Russian maps or data in drawing it).[41] This may have been an early version of the map that eventually found its way into *Reysen* (which carries the inscription of the year 1668).[42]

***

Jan Struys spent most of his two sojourns in Muscovy, which spanned more than two and a half years, waiting rather than travelling. Neither *Reysen* nor other sources make clear what he did to kill time in his first stay in Moscow, although Struys and Pietersz, the two designated sail-makers, must have partially passed the months by making sails from the sailcloth brought from Holland.[43] Their sojourn in Moscow dragged on longer than the Dutch sailors liked, and Struys remembered in *Reysen* how the crew rued the money they wasted there.[44] Boredom crept in, for, despite their residence in the city, the crew's contacts with the local population were limited. Even less zealous Muscovite commoners kept the 'unbelievers' at arm's length because their government ordered them to do so.[45] Some events broke the monotony of the waiting. In January 1669, the sailors watched a massive combat between bears, wolves, and dogs, a pastime that attracted 'thousands' of spectators, including the tsar himself.[46] In February 1669 (less than three months after the crew's arrival in Moscow!), Struys served as stand-in father of Cornelis Saarsz Brak at the occasion of the latter's wedding to Maria Jans, a servant of the van Sweeden family.[47] Struys's role as father of the groom shows that he was at least considered a nominal 'adherent' of the Dutch Reformed Church; otherwise he might not have been given this honour.[48]

Reysen describes how Maria Jans (whose existence can be verified in documentary sources) was an an orphan of Muslim-Tatar background who had joined the Dutch Reformed Church.[49] While she spoke fluent Dutch, her decision to marry a foreigner who had just arrived in Muscovy and accompany him on a hazardous adventure still seems bold, especially as she might *not* have been pregnant.[50] Maria's agreement to marry Brak was probably informed by the lack of eligible Calvinists in Moscow. Religion would have been paramount in the choice of a spouse for someone who had served in the household of a Calvinist preacher, where she had learned to sing all Psalms without missing a word.[51]

Maria Jans had placed herself culturally beyond the pale in Muscovy by serving non-Orthodox people (foreigners were not allowed to hire Orthodox servants[52]). She was thus a *Moskovskaia inozemtsa*, as Brak was a *Moskovskii inozemets*, belonging to a caste that would never be fully accepted in Muscovy unless they converted to Orthodoxy. Beyond impulsive romantic infatuation or pregnancy, her marriage to Brak confirmed Maria Jans's membership in the Dutch community. Sailors were not normally allowed to travel with their spouses, but the *Oryol* crew would use Astrakhan as its home port on relatively short trips, and Maria would set up her household there. While the cultural gap between the spouses appears wide, their union shows that these Dutch sailors, hailing themselves from an immigration country, were accepting of someone's changing cultural identity and not averse to this kind of exogamous union.

Reysen portrays the wedding in a benevolent and patronizing manner, especially emphasizing Maria's Calvinistic sophistication.[53] In many other passages of *Reysen*, idealized civilized Hollanders of the true faith (epitomized by Struys) are contrasted with the hostile inferior beings that encircle them at home and abroad. Most of the time, *Reysen* conforms to the patrician, Christian, and hierarchical worldview of the Dutch and Western European elite, the book's addressees. Those readers may have imagined the coarse frivolity of the wedding scene, which confirmed their cultural and moral superiority over the common folk and their simple pleasures. But the newlyweds' lusty celebration of their union for three days with their friends, dancing 'Polish' and 'Russian' jigs to the fiddle of the ship's pilot Jan Albertsz, seems an unfiltered memory of Struys. The marriage's depiction allows us a glimpse of the real-life Jan Struys, an open-minded, down-to-earth and by then avuncular man, in contrast to the staged, reflective, and superior author of the chorographic sections. In real life, Struys could ill afford to take an aloof attitude of contempt for his non-Christian fellow human beings. He always benefited from adapting to local surroundings. Only through such flexible behaviour could he survive to tell the tale of his wanderings.

Reysen's Muscovy should not be equated with the Muscovy Struys observed: at least three of his fellow sailors rather liked Muscovy and its inhabitants. Brak and Trappen married Muscovite women, and in the early 1670s Karsten Brandt chose to settle in Moscow. By contrast *Reysen* often describes Russia contemptuously.

Reysen's standardized description of Muscovy, largely based on Olearius's *Beschreibung*, confirmed among the general public a particular Western-European image (and the Dutch version resembled both the continental and the British model) of Muscovy as a state populated by barbaric, tyrannical, superstitious, and lascivious natives.[54] In this sense, 'Muscovy', similar to 'Russia' or the 'Soviet Union' later, acquired the characteristics of an ossified 'discursive formation', like the Orient. Matthew Anderson's remark of half a century ago that in contemporary England 'Russia was regarded ... , in some respects at least, as an oriental country', seems telling.[55] Peter the Great's activities in some ways altered but in other ways confirmed this latter image in Western European writing on Russia.

Both the Western Orientalist reading of Asia and its long-standing essentialized idea of barbaric Russia linger. In the West, Russians have never quite shaken the nefarious reputation that they acquired in the works of Herberstein, Fletcher, Olearius, and their epigones such as *Reysen*'s chorography of Muscovy. Nor have various Asian cultures ever been able to entirely overcome Westerners' variegated Orientalist constructions of their politics, religion, culture, and society.[56] *Reysen* does have nuances that show how discursive formations such as Russophobia or Orientalism were not entirely fixed in Struys's days. But *Reysen* and similar works strengthened Western negative objectification of non-Western others. Even if criticizing some of Struys's information, the Comte de Buffon happily mined *Reysen* as its information neatly confirmed his own theories and presuppositions about European superiority.[57] Similarly, *Reysen* at the time of its publication eminently served the purpose of justifying the Dutch maritime empire.

Reysen's chorography of Muscovy, inserted by the ghostwriter to replace most of Struys's actual humdrum activities (such as making sail), between his arrival in Moscow and his departure from Dedinovo six months later (in May 1669), covers almost thirty pages.[58] The description starts out with the location of the Muscovy's borders, followed by an account of the city of Moscow, remarks about the Russians' slavish mentality toward their sovereign, the havoc of frequent fires fueled by the ubiquitous wooden buildings, and the harsh climate. *Reysen* noted a specifically lethal Russian illness, the *Brandt* ('fire'; it is unclear which disease this is), which affected intestines and caused unbearable headaches. Turning to agriculture, the text discusses the fertility of the Russian soil and crops grown by Muscovites. Briefly, *Reysen* sketches the abundance of domestic and wild animals, and regional geographical differences.[59] Then a sort of anthropological description of the Russians and their customs is offered: food and drink, their quarrelsome nature, women, weddings,[60] and the segregation of the sexes in public.[61]

Some of *Reysen*'s wording here seems to approximate the vocabulary of a stereotypical rough-and-ready sailor. Thus, readers are told that Russians are 'strong, fat, and compact, and especially coarse of head, arms, and legs'.[62] Testifying to both cultural differences and the higher standard of living and obsessive cleanliness of Struys's native country, the author notes how '(the Muscovites) have few household goods, being a few filthy pots and dishes, wooden vessels and cups, tin brandy dishes and mead cups, which they seldom clean. They do not care to decorate the walls, like the Netherlanders, but for a couple or more of painted saints, particularly St Nicholas.'[63]

Reysen is a shade more graphic than other Western European texts in discussing the consequences of Russian imbibing, stating that some women inebriated after vodka consumption 'do not want to lag behind the men, and likewise pawn their clothes (in the *Kabak* or tavern), and, yes, leave aside all honour and shame (and) in public engage as horny beasts in all kind of lewdness, which kind of misbehaviour in the past was not considered by them to be very shameful (but) merely seen as something funny, but this is today a bit better, since the Grand Prince on the advice of the Patriarch has diminished the great number of small and clandestine (...) taverns.'[64]

In these observations about female drinking, *Reysen* shows a misogynist streak common to contemporary treatises on Muscovy. But the book mirrors specifically Dutch gender relations in its lamentation that 'it is not the sweetest life for women in Muscovy, as they are locked up hardly less than Turkish women and for the slightest suspicion are beaten, scolded and rejected by their men', behaviour apparently unacceptable in the Republic.[65] Olearius and Struys both remark on the Russian habit of removing one's cross and covering one's icons while having sex, but *Reysen* proves more censorious, or perhaps Calvinist, in its verdict: 'as if God's All-Seeing Eyes have to be spared less than the blind Images'.[66] *Reysen*'s moral condemnation of certain Muscovite customs probably reflects the 'civilization offensive' accompanying the ongoing confessionalization in the Republic and elsewhere in Western Europe during this era, a trend applauded by *Reysen*'s audience.[67]

Olearius had already remarked with repugnance on the Russian farting and belching after meals, but this disgust is largely absent in *Reysen* even if Struys recalls how garlic eaters stank.[68] An upper middle-class erudite such as Olearius, accustomed to moving around the highest circles in Western Europe, had internalized some of the refined manners of the leading courts; Struys, a sailor, may have had far fewer qualms about bodily functions, considering such things so unremarkable that he did not mention them. The elite's refined manners had not percolated to his stratum of society. Struys's silence about human bodily functions in Muscovy or elsewhere perhaps neatly aligned here with the taboos observed by the Dutch elite regarding such matters (as channelled through the ghostwriter).

Related to this omission may be *Reysen*'s lack of any mention of sodomy, which Olearius and Collins alleged to be a common Russian practice.[69] Homosexuality was taboo and a capital crime in the Republic, but sexual relations undeniably existed between men at sea, even if the sailors observed a sort of code of silence about them.[70] The silence in *Reysen* about 'sodomy' in other cultures may indicate Struys's lack of surprise at homosexual practices familiar to a veteran sailor, but it may also indicate the strengthening of the elite's taboo on homosexuality. Its mention was now beyond the pale in a book such as *Reysen*.

Russians' regular use of saunas bewildered the Western Europeans, including Struys: even the Dutch, obsessed with the cleanliness of their porches and floors, were much less worried than the Russians about personal hygiene.[71] *Reysen* recounts further fare standard in the descriptions of Muscovy such as Muscovite dress, gestures, and the rituals surrounding death.[72] An unremarkable sketch of Russian Orthodoxy and clerics follows this.[73] *Reysen* shows the usual Western European contempt for Orthodoxy, a religion seen as a primitive and superstitious form of Catholicism. The chorography of Muscovy concludes with the briefest of discussions about the monarch, his government and Russian justice.[74] This last section of *Reysen* betrays that the manuscript was still developed in 1676 as it speaks of the recently deceased '*Alexie Michalowitz Romanow, Alexius Michaels, Soon den Romeyn.*'[75]

Reysen repeats the trope of Russian slavishness, but its brief description of this trait has a different twist from Olearius:

> They are by nature slavish and as if born to serve and will seldom work through a noble or fair encouragement; but always have to be

> driven to it by blows; yes, they have so little resentment or misery about their slavery that if through death or goodwill of their lord they receive their freedom, they sell themselves again to another one; in such a manner that he who wants to be served well by them, however kind-hearted or compassionate he may be, is forced to save neither fists nor cane (.)[76]

and, in an unwitting reference to the legendary Russian spirit of freedom (*volia*[77]),

> they are, however, usually kept without enough food, and this breeds many Thieves and Murderers, so that everyone who stands to lose something should pay great heed, since despite the heavy punishments for small thefts, the Brandy- and Tobacco-lusty Slaves are thereby not kept in check.[78]

Struys emphasizes the servility of actual serfs and slaves, while Olearius had shown contempt for the slavish deference to the tsar of the elite, including even the greatest magnates of the land.[79] But *Reysen* concomitantly shows an appreciation of how Russians manipulate the system as well as they can and could be sullen and resourceful in adapting to the harsh parameters of their existence, possibly a reflection of a somewhat modern Dutch sensibility regarding the dignity and agency of the individual. Struys's tone is altogether less disdainful than that of Olearius, who takes a consistently superior vantage point and is unsparingly critical of Russian mores.[80] In the case of Olearius's *Travels* we can wholeheartedly agree with Thomas Barran's observation that, for the Westerner, 'Russians were the Other, repositories for the psychological projections of travellers' own forbidden desires and subconscious fears ... (T) ravellers' tales provide some of the raw material for the creation of an anti-world that Western Europeans assembled in order to prove the superiority of their own institutions.'[81]

Reysen's moral condemnation of Russians as inferior human beings, however, is muted compared to Olearius's dismissal. Perhaps this may be linked to the sensibilities of a Dutch sailor, less easily offended than the more usual observer of Russian ways, the well-to-do merchant, army officer, or diplomat. In addition, Struys enjoyed more liberty than them to communicate with Russians. *Reysen*'s slightly less dismissive tone about the Muscovites might derive from Struys's relative familiarity with Russians (by the time the book was published he had spent about two-and-a-half years on Muscovite soil, and visited most of the larger Russian towns) and his thoroughly Dutch lower-class sensibilities. Finally, it is conceivable that familiarity with Muscovy in 1670s Holland had also begun to complicate for the ghostwriter the traditional simplistic Western stereotype of the lascivious, ill-mannered or drunk Russians still upheld by Olearius.

***

At last, in the early spring of 1669 preparations began in earnest for the descent of the Dutch crew toward the Caspian Sea on their newly built ships along the Oka and Volga. The three-thousand-kilometre journey that commenced in May, however, would proceed at a glacial pace; Astrakhan was only reached four months later. Once more, chorographic sections in *Reysen* largely replace a description of the crew's activities during this trip.

In March 1669 Butler and his crew travelled in sleighs from Moscow in southeastern direction to Dedinovo. After inspecting the vessels at the village, they returned to Moscow, where Butler reported the ships to be seaworthy, a judgement reinforced by a merchant from Astrakhan, Ivan Savel'ev, who had accompanied them.[82] A few weeks later, after the ice on the river Oka had fully melted, Iakov Poluekhtov returned to Dedinovo to oversee the process of readying the ships; he was soon to be followed by the Dutch crew. Butler meanwhile had to fend off accusations made by the Dutch merchant Herman van der Gaeten, a business acquaintance of van Sweeden, who denounced Butler to the tsar for misrepresenting his credentials.[83] Van der Gaeten charged that Butler had never commanded a ship, knew little about sailing, inflated his expenses, and tried to swindle the tsar out of more money by pretending to have hired overseas a crew member, who had been in fact a stable-hand of the Boyar Bogdan Matveevich Khitrovo (d. 1680).[84] Van der Gaeten's motives for his denunciation are unclear, but he may have sought revenge on van Sweeden's in-law for an earlier dispute with van Sweeden, who had quarrelled with van der Gaeten and several of his partners a few years earlier.[85]

Butler admitted to tsarist investigators that he had no 'patent' for the rank of captain, but that he was nevertheless an experienced seaman who had often commanded a vessel. At first, tsarist authorities demoted Butler to *dozorshchik* (a rank below captain, more or less equivalent to first mate), and lowered the amount for his crew's wages. Butler immediately seems to have asked the tsar for a pardon and reconfirmation in the rank of captain, a request that was granted after a personal audience with the tsar in the Kremlin during Holy Week in April 1669.[86] Aleksei was perhaps more mellow than usual during this most important Russian holiday, but also had few other options than pardoning Butler, since he could not easily replace him (especially not after van Sweeden's death). Of course, Aleksei routinely encountered similar dissembling behaviour among his other subjects, and he could not afford to punish them indiscriminately for such missteps.[87]

Butler's difficulties appear to have worsened the crew's terms. Pay was lowered for Struys and some of his colleagues on the eve of their departure from Moscow.[88] On 23 April 1669 the crew (perhaps all of them) took leave of the tsar in a personal audience, a sign nevertheless of the sovereign's continued favour.[89] In an official decree, the tsar named the ship *Oryol*, or *Eagle*, probably inspired by Polotskii's poem in honour of his heir (and making the ship's carved figurehead of a lion rather incongruous).[90] Around the first of May,[91] Struys and fourteen men and two women departed Moscow for Dedinovo on a *strug* (a river barge) downriver on the Moskva, which meets the river Oka at Kolomna.[92] They arrived two days later at Dedinovo, where van Bockhoven's team met them.[93] Once on board the squadron, they fell under a ship's discipline that followed Butler's regulations as submitted to the *Posol'skii Prikaz* in December 1668.[94]

Since the water level stood high in the Oka, the fleet of ship, yacht, two smaller rowboats and *strug* (it carried supplies and equipment to prevent the ship from sitting too deeply through the water) began already on May 7 its journey toward the Oka–Volga junction at Nizhnii Novgorod.[95] Butler had nineteen men under his command. Twelve had accompanied him in the fall of 1668 from Holland:[96] Jan Albertszoon (abbreviated as Alberts or Albertsz in Dutch), the pilot and scribe; Pieter Bartelsz and Meindert Meindertsz, boatswains and rigging supervisors; Wiggert Popkes, rigging operator; Els Pietersz, sailmaker's mate; Cornelis Cornelisz (de Vries), gunner; Karsten Bran(d)t, assistant gunner responsible for powder and shot; Jan Janszoon Struys, upper sailmaker; Willem Willemsz, spindlemaker;[97] and the sailors Cornelis Brak, Daniel Cornelisz, and P(i)eter Arentsz.[98] Accompanying them were the skipper Lambert Helt, the pilot Willem Klopper, the carpenter Dirk Pietersz, the sailor Jakob Trappen, and the sailor-diamond cutter Anthony Munster who all five had arrived with van Sweeden the year previously, as well as the surgeon (*lekar'*)[99] and lieutenant Nicolaas Schak and the interpreter Jacob Tolk-Schram.[100] The crew was accompanied by a a few Russian officials and a unit of *strel'tsy*, necessary because of brigands' frequent raids on vessels travelling along the rivers.[101]

In early June 1669 the expedition halted at Nizhnii Novgorod to await the arrival of a postponed delivery of arms and ammunition. Here Iakov Poluekhtov parted company with the sailors. Despite having incurred earlier the tsar's wrath for his poor management of the project and his quarrels with van Bockhoven, he was chosen to lead the construction of other (unused) seafaring vessels in the middle of the 1670s.[102] Van Bockhoven and his deputy Stark had apparently already left for Moscow from Dedinovo. Cornelis van Bockhoven would serve the tsar for almost another decade before he fell at Chyhyryn against the besieging Turks in 1678.[103] Stark, a lieutenant-colonel by 1672, investigated the Cossack uprising in Astrakhan after the city had been recaptured by tsarist forces.[104]

In Nizhnii, the crew received six months in wages, the last payment they ever received in Russia before their escape a year later.[105] The fleet, even though still not fully equipped, left Nizhnii Novgorod on June 13.[106] Two of the crew (Schak and Bartelsz) remained behind in the Volga port to await the manufacturing of anchors as well as some rope and dragging equipment, which they were to bring with them to Astrakhan.[107] Drifting down the Volga, the expedition arrived at Astrakhan in late August 1669, almost four months after departing Dedinovo.[108] Occasionally stormy weather, a few lengthier stops at some of the fortress-cities, as well as shallow waters had led to an extremely slow journey (its average speed was less than 30 kilometres per day).[109]

Reysen closely follows Olearius's *Beschreibung* in its description of the natural and human geography along the Volga. While the occasional fortress on the river was manned by Russian-speaking *strel'tsy*, beyond Nizhnii Novgorod the inhabitants of the surrounding territory were almost exclusively non-Russians.[110] Both texts pause at Vasil'gorod (*Wasiligorod*), to describe the Cheremiss Tatars (in fact, some of them were Chuvash).[111] But a close reading of both texts leaves again little doubt that *Reysen*'s description of the Cheremiss is an uncredited paraphrasing of the learned Holsteiner's text. Olearius, indeed, offers a clear explanation of how he came to know as much about the Cheremiss-Tatar religion as he depicts (during his sojourn in Kazan he lived with a Cheremiss in the same house), whereas *Reysen* does not do the same for Struys.

Reysen embellished the alleged barbarity of the Cheremiss further than Olearius's *Travels*: whereas the Holsteiner tells of their polygamy, *Reysen* recounts a habit of fathers marrying their own daughters.[112] *Reysen* surprisingly compares this alleged Cheremiss habit to that of the Singhalese of Ceylon, whose men were said to deflower their daughters on the eve of the girls' weddings.[113] Its reference here was lifted from the travel account of the Swiss VOC servant Albrecht Herport (1641–1730), published in German in 1669, and in Dutch in 1671.[114] *Reysen*'s bizarre metaphor about alleged Singhalese incestuous practices is in fact a literal translation from Herport's German original: 'if I would plant a little tree and would not take its fruit before I would give it to someone else, that were a great evil.'[115]

Reysen's account of the use of linden tree bark by the inhabitants of Kozmodem'iansk, downriver from Nizhnii, amounts to further copying.[116] But *Reysen* offers some original touches, such as describing fruit trees along the Volga before reaching Kazan. *Reysen* suggests, after Struys and two friends explore the Kazan countryside unharmed, that this area's mainly Tatar population was neither involved in enslaving nor in robbing people, '(un)like the other Nogai, Crimean, Kalmyk, and Dagestan Tatars'.[117] In Kazan itself, though, Struys witnessed the sale of slaves by Crimean Tatars on the local market, with teenagers costing about twelve rubles each.[118] Struys's real experiences surface, too, when he recalls how Butler ordered him to bake a great amount of biscuits at Kazan; Struys had mastered baking these in his youth in Wormer, a centre of Dutch hardtack-baking.[119]

Reysen's subsequent river journey continues to borrow from Olearius, but intersperses this copying with authentic parts.[120] At Saratov, the expedition entered the region where the nomadic Kalmyks roamed, which gives the protagonist cause to depict them in the most negative of terms.[121] His palpable disgust (shared apparently by Russians[122]) precluded a thorough investigation into Kalmyk customs. Thus *Reysen* deems the Kalmyks Muslims instead of Buddhists.[123] *Reysen*'s comparison here between Kalmyks and African Hottentots shows the ghostwriter's imprint.

The flotilla passed unperturbed to Kamyshinka, a recently built fortress that aimed to halt Don Cossack forays along the Volga.[124] The fort had been erected on the spot where Cossacks hauled their boats overland between the Volga and Don rivers. The convoy arrived the next day at Tsaritsyn.[125] Via Chernyi Yar the procession finally reached Astrakhan in August, with *Reysen*'s pages covering the last stretches of the trip by mostly copying Olearius.[126]

In its third part, *Reysen* is far more reliable in its detailing of Struys's wanderings than previously, as is confirmed by comparing its rendition of developments in Astrakhan and further south with Butler's letter, Fabricius's recollections or VOC documents, as we will see in the next chapters. Nevertheless, the chronological narration continues to be interspersed with descriptions that are borrowed from other texts on Muscovy or Iran.[127] Some passages that detail Razin's rebellion at

Astrakhan may have repeated newsworthy and sensationalist anecdotes, which in the form of broadsheets, pamphlets, and magazines enjoyed widespread popularity among a Western European readership in those days (as is evident from the plethora of news about Razin in a variety of publications).[128] But in its description of this massive rebellion, *Reysen* offers a great number of authentic moments.

8
The Volga Delta and the *Oryol*'s Demise

This chapter ponders the failure of the *Oryol* project and narrates the fate of its navigators in their efforts to escape the Cossacks, for which *Reysen* appears a fairly reliable historical source. In the next two chapters, the most original part of *Reysen* will be investigated. Charting Struys's wanderings in Iran, it is rather singular in depicting the Christian encounter with the contemporary Islamic world. The significance of several crucial passages of that portrayal will therefore be explored in a rather detailed inquiry in Chapters 9 and 10.

The imaginative plan that envisioned sophisticated Muscovite sailships on the Caspian Sea was abandoned at Astrakhan in 1670. This may not have been due to the Cossack capture of the city, even if it caused the flight of the Dutch sailors toward Iran. The project's collapse was primarily caused by the tsarist government's realization that the promise of any higher returns from the trade across the Caspian Sea hardly justified the enormous outlay on ships and crews to protect this commerce. It took a generation before the effort to build a Russian navy was resumed, and the ships then built would navigate different seas than the Caspian. Whereas Peter's ostensible success in acquiring a fleet seems to negate Aleksei's Caspian failure, the question whether Russia needed a navy to prosper has meanwhile never been convincingly answered.

Neither *Reysen* nor other sources explain why the *Oryol* did not set sail for Persia in the late summer of 1669. For only toward December would frost impede the use of the waterways until the next spring.[1] Was the *Oryol* rigged too late to convoy the so-called *Gilan busa*, the tsar's barge that negotiated the Caspian Sea to Persia in October of each year?[2] Testimony by the Dutch mercenary in tsarist service, Lodewijk Faber (whose path crossed with Struys in 1669[3]), suggests geographic and other weather-related obstacles as the reason for this standstill.[4] During the season when the *Oryol* arrived at Astrakhan, the Volga estuary's water level was no longer replenished by melted snow feeding its higher reaches as it was in spring and early summer.[5] Further difficulties ensued from the river's forking beyond Astrakhan into a delta of smaller and shallower waterways streaming southward to the Caspian Sea. The ship sat apparently so low in the water that she could not even be towed toward open sea along one of these canals.

But more fundamental problems diminished the viability of the plan for a Caspian sailing fleet. In the first place, sailships with deep draught were ill suited for the northern part of the Caspian Sea itself, since two-thirds of its surface was (and is) fewer than five metres deep, with its greatest depth only reaching twenty metres: Although the *Oryol* was built with a shallow hold, it might have run frequently aground.[6] Even if the *Oryol* sat high enough on the water, doldrums and gales would have prevented it from hoisting its sails effectively. A combination of fierce winds and shallow waters had broken the Holsteiner-built *Friedrich* into two on the Caspian Sea in the 1630s.[7] Suggesting the construction of galley ships, Butler and A.A. Vinius seem to have been aware of these problems before the *Oryol* undertook her journey down the Volga, but she had remained a pure sailship.

The first sign that the *Oryol*'s progress had stalled dates from the early autumn of 1669, when the *voevoda* of Astrakhan, Prince I.S. Prozorovskii,[8] asked the Moscow chancelleries what the plans were for the ship now moored near his city.[9] Communications with Moscow were slowed by Cossacks roaming along the Volga shores, but the central government's delay in responding to Prozorovskii's query had other reasons as well. It was not just the seasonally low water level preventing the *Oryol* from sailing out. Only after half a year, the Muscovite authorities responded to Prozorovskii's query by ordering the crew to return to the capital.[10] In other words, the government ended the project in the spring of 1670.

Why did the tsarist government abandon the the *Oryol*? The ship had been constructed at prohibitive cost, well beyond the 9,000 rubles expended on its building alone.[11] Various servants of the tsar had invested a tremendous amount of time in it, from the chancellery secretaries in Moscow to people like Iakov Poluekhtov; the tsar himself personally signed a number of decrees to force his subjects to collaborate with the project. In addition, the Dutch shipwrights and crew had been hired for extraordinarily high wages, part of which would still be paid years later.

Even if the ship did not easily negotiate the Volga mouth or the Caspian, the *Oryol* could have been adapted at Astrakhan to overcome the adverse geographic conditions. In the 1630s, the *Friedrich* had at least reached the Caspian Sea. Surely, the *Oryol* could have been hauled there after some minor alterations (perhaps by unloading its cannon, for example). Prozorovskii's subsequent orders to the crew to improvise an oars-driven ship confirm that modified vessels could have performed some of the squadron's intended tasks. Indicative of the possibility of the project's resumption, Willem Barentsz. Klopper offered the tsar to traverse the Caspian with the flotilla's yacht in March 1674.[12] This suggestion was then brushed aside. Instead, the *Oryol* and the yacht remained parked on the shores of the river Kumut', where the vessels were left to rot until 1680, when Prince Shcherbatov inspected them.[13] His report led to the final dismantling of the vessels. Their wood was to serve a more useful purpose.

After the *Oryol* and its accompanying vessels departed Dedinovo in the spring of 1668, the chancelleries' interest in the squadron declined precipitously. The government's desire to bring the project to some sort of successful conclusion suddenly dissipated. Political machinations in Moscow precipitated this change of heart. By 1670, Ordin-Nashchokin, who had been absent from Moscow for prolonged periods in the previous two years, was being eclipsed as Aleksei's favourite by Artamon Sergeevich Matveev (1625–82) and by the Boyar Iurii Alekseevich Dol'gorukii (d. 1682), neither of whom considered the *Oryol* a priority.[14] Dol'gorukii replaced Ordin-Nashchokin as supervisor of the project.[15] A field general who had earned his stripes in several campaigns, Dol'gorukii was clearly less of a visionary than his predecessor and had less affinity with maritime matters. Married to a Scotswoman, Matveev was by no means anti-Western, but perhaps his scepsis about the Caspian fleet's viability was linked to his interest in disparaging his rival Ordin's policies.[16]

Most decisive was the tsar's loss of interest in the project, forgetting about it or giving up hope of its feasibility. From the summer of 1669 onward, Aleksei was preoccupied by the lengthy and cumbersome process of selecting a new wife.[17] The ultimate selection in May 1670 of Matveev's ward Natal'ia Naryshkina as Aleksei's bride confirmed the end of Ordin-Nashchokin's zenith as the tsar's favourite. Within months, Ordin retired to a monastery.[18] Upon his departure from the government, he sent the tsar a letter which argued that his service to his monarch had been sabotaged by disloyal and corrupt government bureaucrats.[19] In this memorandum, Ordin darkly warned of sabotage of the tsar's policies by *d'iaky* and *pod'iachie,* by whom he meant his former assistants Dokhturov, Golosov, and Iur'ev.[20] Thus, during

Ordin's foreign missions in 1668 and 1669, his three deputies may have colluded with Dol'gorukii and Matveev in letting the project wither.

The naval plan's prohibitive cost, the technical and geographic complications, the costly effort suppressing the Cossack rebels (see below), the changing of the guard among the tsar's favourites, and the distractions in the tsar's personal life may have therefore all played a role in the apparent lack of further interest of the tsarist government in it.[21] A lowered expectation of the fleet's benefits combined with the prospect of future sustained high costs (such as the crew's pay of some 200 rubles per month) further contributed to ending the *Oryol* project. Economically, the decision to abandon the project was probably sensible. The silk road from Iran was never truly rerouted from the Persian Gulf and the Mediterranean to the Caspian, Astrakhan, Moscow, and beyond, despite Western or Muscovite dreams.[22] Twice a concerted effort was made from the Russian side to establish a regular searoute for the shipping of raw silk in bulk, but the ships *Friedrich* and *Oryol* never loaded any fibre.[23] On a number of other occasions during the sixteenth and seventeenth century Western Europeans urged the tsar's government to actively attempt to reroute most of the silk trade between Asia and Europe through Muscovy.[24] *Reysen* reflects this Western interest when the enslaved Struys beseeches his Caucasian master Haji Mahmoud Sala to allow him to return to the Republic.[25] To compensate his owner for the loss of his Dutch slave, Struys promises to tempt Dutch merchant conglomerates to equip ships to sail the Caspian. Struys's proposal allegedly reached the ears of local Armenian silk cultivators, who met him to explore the possibilities of trade across the Caspian with the Dutch. Because Struys soon changes hands, the plan was apparently never further pursued.

The Armenians' conversation with Struys may have been invented by *Reysen*'s ghostwriter on behalf of Amsterdam merchants desirous of rerouting the silk trade via Muscovy, or shows that Struys had gotten wind of van Klenk's proposals to the tsar in 1676. Consciously or not, *Reysen* echoes the arguments Ambassador van Klenk (of whose retinue Struys was part) used in 1676 to persuade the tsar reroute the silk trade across the Caspian Sea.[26] Van Klenk offered his Russian interlocutors (then led by Iurii Dol'gorukii and Artamon Matveev) help in finding shipwrights to build a Dutch-managed Caspian fleet under Muscovite auspices.[27] The boyars relayed van Klenk's proposal to Tsar Aleksei, whose dream of a Caspian fleet revived during the last days of his life. Aleksei permitted van Klenk to hire Dutch shipwrights for this fleet, but the tsar's sudden death probably scuttled the plan. In the political limbo following the tsar's passing, Dol'gorukii and Matveev (who himself was in political trouble in 1676) proved again disinclined to pursue a similar project.

The treaties of 1667 and 1673 confirm the impression gained from Struys's conversation in *Reysen* that Iranian merchants, too, endeavoured for a protracted period to divert silk exports northward. The shah's government around 1680 requested Swedish aid in building 'strong and fast Western' ships to sail the Caspian.[28] Yet this attempt, too, was aborted. The trade between Muscovy and Persia was not as valuable as each country's foreign trade with other regions (for Muscovy with Western Europe, and for Persia with Turkey, India, and Western Europe). This modest value made the financing of an expensive Caspian fleet a dubious proposition also for the Iranian side. Meanwhile, the trade of silk across the Caspian grew regardless of government support after 1673, before Iran descended into a period of great unrest in the late seventeenth century.[29] This might even show that a government-protected merchant marine was a superfluous expenditure, as the silk trade developed irrespective of state sponsorship.

Intermittent Muscovite government's, Armenian-Iranian, or Western European enthusiasm to capture the silk trade was never matched by Russian merchants. Leading Russian merchants were consistently opposed to an increased Western presence within Muscovy. Their involvement might have made it a viable project, but they were content to leave the bulk of silk trade across the Caspian Sea in Armenian hands after the treaties of 1667 and 1673. Russian traders preferred the certainty of their traditional trading; they lacked the dynamism of the enterprising English, Dutch, Armenian, Swedish, and even Holstein traders, diplomats, and politicians who travelled in Muscovy.[30]

Besides economic or geographic complications and political intrigue, reason of state doomed the resumption of the maritime project under tsarist auspices after 1675. Between the spring of 1670 and 1675, Muscovite government expenditure was tied up in more urgent affairs than the *Oryol*. If Matveev and Dol'gorukii emphasized the further anticipated expenses of a project which served no vital defensive purpose, the tsar's frugal instincts may have been alarmed. Beyond the cost of maintaining a court befitting an autocrat, the tsar's expenditures were almost entirely directed toward warfare. Richard Hellie suggests that by 1671 the Muscovite government may have been on the verge of bankruptcy because of the cost of suppressing the Razin revolt (and of recovering from the damage).[31] After tsarist troops had barely mopped up the last vestiges of the Razin rebellion (Astrakhan was only recaptured in late

1671), Turkey invaded Poland, an attack triggered by restless Cossacks trespassing into Ottoman territory.[32] The tsar honoured his alliance with Poland by declaring war on the Crimean khan and the sultan in the fall of 1672.[33] The cost of warfare preempted the pursuit of other projects (or even supporting Klopper's initiative), especially when, in Tsar Aleksei's last years, he faced a potential war with Sweden as well.[34]

***

The *Oryol*'s crew spent many of the ten months (August 1669–June 1670) in Astrakhan idling in apparent comfort.[35] A lengthy portrayal of this inertia would endanger the exciting quality of this part of *Reysen*, but the book's silence about the crew's activities on and around the ship (and about their eventual effort to build a sort of galley) likely derived as well from a need for keeping trade plans secret.[36] Whereas this would seem to be an ideal moment in the text to describe the vessels that had been brought down to Astrakhan, no information whatsoever about them is furnished. This silence might be linked to the book's concomitant lack of any concrete information about the actual shipbuilding project at Dedinovo. In 1676, the Dutch ambassador van Klenk was trying to convince the Muscovites to restart the project of a Dutch-built-and-operated Caspian fleet, and he may have requested that *Reysen* not dwell on previous failures.[37] Because Balthasar Coyett's book about van Klenk's embassy is also silent about the ambassador's proposals to build a Caspian fleet with Dutch aid, a desire to protect strategic and business secrets appears to have played its part.[38] A wealthy and powerful Amsterdam regent such as van Klenk could easily convince van Meurs and van Someren to observe such secrecy, and may have persuaded them with a subvention toward the book's production costs.

During this lull at Astrakhan, the crew encountered a few compatriots, one of whom was Jan Termundt (d. 1704), a surgeon who awaited an opportunity to travel northward.[39] Termundt had accompanied the English envoy Thomas Brown to Iran to collect ingredients for medicine, but was forced to wait in the spring of 1670 at Astrakhan because of the Cossack rebellion cutting off the Volga route to Moscow. Termundt was a native of the Republic's province of Friesland, where he had been known as Jan van Kollum (Kollum is a village in Friesland).[40] Another Dutch servant of the tsar, the mercenary junior officer Lodewijk Faber, also met the sailors at Astrakhan.[41] Both were to play a prominent role in saving Struys from permanent servitude.

Instead of describing the men's life in Astrakhan between September and March, *Reysen* provides a chorography of the city, the people (Crimean and Nogay Tatars) and nature of the surrounding steppe, and a historical, almost journalistic, *excursus* about the Cossacks, especially Razin's followers from the Don.[42] Seventeenth-century Astrakhan was Russia's 'Window on the East', counterpart to Archangel'sk role as Muscovy's seaborne 'Window on the West'.[43] Despite Astrakhan's strategic location, the port's volume of goods was smaller than that shipped through some of the other five entry-and-exit ports (on land and at sea) of Russian foreign commerce.[44] Although the transit point for all Muscovite-Asian trade (excepting that with the Ottoman Empire), Astrakhan's income from fees and tariffs reached only half of Arkhangel'sk's revenue by 1672.[45] While it is not clear how large the southern port had been a generation earlier, the city had nonetheless more than 11,000 houses within its walls by 1707, which indicates a population of several tens of thousands.[46] This was a large town for contemporary Russian or European standards.

Astrakhan was also one of Muscovy's most cosmopolitan cities, located at the 'division' of Europe and Asia.[47] According to *Reysen*, Slavs mingled with Tatar (Crimean, Nogay), Kalmyks, Persian (mostly Armenian), and Bukharan merchants, and traders from the Indian subcontinent; its most conspicuous trade commodity was Persian-produced silk.[48] Astrakhan's homegrown shipbuilding industry dated from the early seventeenth century, but it made simplistic vessels, mostly poorly navigable *busy* ('*Boesen*') that could not sail against the wind.[49]

The sailmaker's personal recollections surface when *Reysen* remarks on the crew's delight in the town's cheap meats and fish.[50] Also betraying Struys's voice in the book are impressions of Razin's peaceful visit to Astrakhan in the summer of 1669 and the description of preparations for the town's defence against the Cossacks, as well as the subsequent siege and escape from the city.[51] More dubious is the provenance of *Reysen*'s firm rejection of the myth that the notorious *Baranets*, a regional plant resembling the form of a sheep, was fauna rather than flora, a long-standing matter of speculation among scientists in Western Europe.[52]

***

The crew's leisurely existence in Astrakhan ended in the early spring of 1670, when the central government in Moscow ordered the crew to return to the capital. But *voevoda* Prozorovskii delayed their departure by requiring them first to shore up the city's defence against the expected Cossack siege. In March and April 1670 the crew prepared (*'voorsien'*) the ship for a prolonged period of inactivity and built an oar-driven vessel (*'een groote Sloep'*) that could carry six to eight cannon and was to be rowed by fifty-two oarsmen.[53] As a result of this delay, Butler's men would never embark on their return trip northward.

In stark opposition to the court's euphoric mood of early 1667 (sketched in Chapter 5), the Russian-Ukrainian nineteenth-century historian N.I. Kostomarov argued that beneath a tranquil surface the Russian social fabric was unravelling toward 1670.[54] Two decades of social unrest and war, the definitive imposition of serfdom in 1649, a heavy tax burden, and the widespread abuse of law and corruption of the legal system spelled doom for Tsar Aleksei's foray into costly experiments such as the creation of a Muscovite navy. Measured by modern standards, most seventeenth-century monarchs exercised an ephemeral control over their subjects, especially in large empires, because of the primitive state of transport and communications. Thus, despite serfdom and the government's general effort to bind people to their place of residence (mainly for taxation purposes), many people were on the move in the time Struys traversed Muscovy. Joining nomads, soldiers, and itinerant traders were many people who were fleeing, from serfdom, army service, judicial investigations (which invariably involved torture), and taxation.[55]

Some folks became brigands (of whom Struys already met some on his way from Livonia to Moscow), while others reached frontier regions such as Siberia or the steppe of what is now Ukraine and southern Russia. There they established or joined autonomous communities of Cossacks. This borderland (*Okraina*) stretched from the Dnipro River in the west all the way to the Ural Mountains and Iaik (Ural) River in the east, to the south of the Muscovite fortification lines and to the north of the Caucasus. While this territory fell under the tsar's nominal control, it was mainly populated by free-spirited nomads such as the Kalmyks, Bashkirs, Nogays, and Tatars as well as Cossacks. All of them largely ignored tsarist authority, and their allegiance was almost purely based on the reception of regular payments in money and kind made by the Muscovites.[56] A few fortresses and fortified towns were dotted across this border region, but the steppe, largely flat grassland dissected by waterways, was a crossroads traversed by bands of mounted warriors subject to no one but the chief and elders of their local community.[57] Perhaps Kostomarov exaggerated the general restlessness of the Muscovite populace, for central Russia appears to have been fairly calm toward

1670, but in these south-eastern borderlands a specific set of circumstances led to a massive rebellion against tsarist rule.

Since the fifteenth century, Cossacks had formed autonomous communities in the borderlands of the Polish Commonwealth and Muscovy.[58] They survived by raiding and trading, through the sale of plundered goods (including captured Muslims), and by hiring themselves out to tsar or king. In wintertime, as Razin's men did in 1669–70, they usually returned to their home base (*stanitsa, sich*), where they were reunited with their wives and children who had tended to their land and cattle in spring and summer. To the tsar's slight advantage, he wielded some more influence over the Cossacks than his Polish rival, as they shared his Orthodox faith. By the 1660s, Cossacks of the Don had served the tsar as auxiliaries in war for almost a century. Most of the officially registered Don Cossacks (including Stepan Timofeevich Razin[59]), who were by treaty guaranteed a privileged status and regular payment by the tsar, were wont to obey the Muscovite ruler.[60] But Cossack *stanitsy* swelled in the 1660s with new arrivals who were not registered on the tsar's rolls and thus were bereft of such remuneration.

Aleksei Mikhailovich tried to bring areas south of the Muscovite fortification lines progressively under firmer control, but such efforts had limited success. As soon as a Muscovite army retreated from an area where it had tried to browbeat restless subjects into obedience, lawlessness and anarchy resumed. Only a patient and intricate strategy of bribery, flattery, armed force, and negotiations gradually pacified the region of eastern Ukraine and southern Russia, a process which was not completed until the 1690s (and even then major rebellions erupted here in the eighteenth century).[61] Cossack communities retained a great measure of autonomy.

Beyond its economic purpose, the plan for an armed Caspian fleet may have been part of the long-term strategy to bring these borderlands under control, for sustained unrest was not just a feature on land but also at sea. Piracy was practised by (non-registered) Cossacks, other non-Orthodox subjects of the tsar, or peoples who paid tribute to the Crimean khan, Iranian shah, or Ottoman sultan. Perhaps the *Oryol* project also aimed to counter to a surge of plundering activity by Don Cossacks. Their access to the Black Sea littoral via Azov, where they had been roaming around in search of booty for decades, was definitively cut off in the early 1660s by the Crimean Tatar khan, a vassal of the Turkish sultan.[62] Newcomers who claimed Cossack status were thereby deprived of their customary source of income.[63] Because Moscow did not pay such novices, they found a tenuous livelihood by plunder. The numerous upstarts among the Don Cossacks turned eastward.[64] Engaging in annual campaigns of freebooting in the Volga delta and across the Caspian Sea, especially these 'irregular' Don Cossacks defied a tsar desirous of greater order and tranquillity to improve economic and political ties with the shah. Perhaps Razin's Don Cossacks at first merely hoped to be bought off by one of these monarchs (and their movements in both Iran in 1668 and in Muscovy in 1669 appear to hint at such a strategy).[65]

The first news of restlessness among the Don Cossacks under Razin's lead reached Moscow in June 1667.[66] Razin's gang then captured the settlement of Iaitskii Gorodok at the mouth of the Iaik.[67] They evacuated the town before tsarist forces arrived, and in the following year crossed the Caspian Sea to plunder towns that fell under the rule of the Persian shah, along the sea's southern shoreline. The extensive Cossack devastation of the shah's territories threatened to undermine the fairly good relations between Muscovy and Iran (cemented by the 1667 treaty). The Cossacks crossed the sea again in northern direction in 1669, apparently unable to persuade the shah to enlist them as auxiliary troops (a proposal perhaps based on the knowledge that shahs enlisted converted Georgian auxiliaries or *ghulams*[68]) and incapable of managing on their own against the combined effort of Persian army and navy.[69] The tsar, assuring the shah that he had not been involved in the Cossack actions, promised the Iranian emperor to restrain the Don Cossacks from resuming their raids of Persian lands. But Aleksei did not actually impose any sanctions on Razin's men for their misdeeds against the Iranians.[70]

Over the centuries, the tsar had bought off many an armed force in the borderlands, and the tsar and his representatives turned to this strategy to placate Razin's Cossacks, when the latter returned to the Volga delta from northern Iran in the summer of 1669. Trapped by a tsarist force near Astrakhan, the Cossacks preferred to negotiate rather than fight (after a token exchange of fire).[71] They parlayed to emissaries sent by the government troops that great poverty had descended upon the Don region, because the inhabitants had been forced to abandon their usual forays on the Black Sea. Thus they justified their looting of merchant convoys along the Volga and villages and towns along the Caspian Seashore.[72] The tsar and his governor in Astrakhan, Prozorovskii, were initially persuaded by these arguments or unwilling to shed blood when a peaceful solution seemed possible.[73] They allowed Razin to return to the Don unharmed in the late summer of 1669. Before Razin's men left for home on the Don in early September 1669, they paused at Astrakhan, entering the town to trade loot they had acquired in Iran. Jan Struys bought some items from the Cossacks.[74] *Reysen* relates how during this interlude Razin threw his captive Persian concubine in the Volga as a kind of lugubrious pagan sacrifice to the river, but Struys's story (if the sailmaker was the source for this tale) was not original and he could not have witnessed the episode himself.[75]

The government's irresolute response to Razin's challenge at Astrakhan in 1669 was given in an atmosphere of palpable enthusiasm for his cause among the town dwellers of Astrakhan and Iaitskii Gorodok (even among some of the *strel'tsy*). Once in the course of the winter he gained additional support from several Kalmyk clans, various hungry refugees, and other Cossacks along the Don, Razin decided to roll the dice and attempt to overthrow the tsar (or at least his allegedly vile entourage of boyar advisors).[76] While this was a bold aim, the example of the decisive role Cossacks had performed in settling some of the succession stuggle of the Time of Troubles may have inspired him.

In April 1670, Razin moved up along the Don river.[77] He roused his followers by promising to liberate Russia's commoners from the tsar's corrupt and treacherous advisors.[78] Razin, crossing the small stretch of land from the Don to the Volga nearby, set his sights first on Tsaritsyn. His forces occupied this city without a fight because mutinous *strel'tsy* switched sides with the backing of much of its population.[79] The local governor and his few loyal supporters were killed.[80] A relief force of *strel'tsy* sent from Moscow was surprised and crushed in the city's vicinity.[81] In the following weeks, the nearby town of Kamyshin (a fortress only founded a few years earlier to stop Cossack raiding) fell into the Razinites' hands as well.[82] In Astrakhan, rumours spread regarding events upstream: clear information on Razin's movements was lacking in a city that relied on the Volga as its main communication route with Moscow.[83] In May, the Astrakhan defenders dispatched an army (including the mercenary officers Nicolaas Schak and Paul Rudolf Beem) commanded by prince Semyon Ivanovich L'vov[84] to face the Cossacks.[85]

By then, Razin's forces may have numbered 10,000, with L'vov commanding no more than 3,000 men.[86] The armies met at Chernyi Yar (which had fallen to the rebels), but before battle was given, many of L'vov's *strel'tsy* mutinied and declared for Razin.[87] Less than a hundred people, mainly officers, remained loyal to L'vov, who surrendered. Of the Muscovite commanders, only L'vov's life was spared because he had commanded the troops who left Razin at peace after his return from Persia in the previous year.[88] The young Dutch gunner Faber survived as well (although his stepfather Beem was butchered), rescued by his former orderly from Cossack beatings; he was now dragooned to operate Cossack artillery.[89] Another soldier loyal to the tsar escaped the mass execution, reaching Astrakhan in early June with the grim tidings that Razin had once more prevailed.[90]

Iaitskii Gorodok, too, once more fell to the Cossacks, who thereby encircled Astrakhan but for its Volga shore.[91] Isolated, Astrakhan's rulers tried to ready a defence against the expected Cossack siege. Ivan Prozorovskii ordered the Dutch seamen to mount the *Oryol*'s and yacht's twenty-eight cannons on the city walls, which was accomplished by early June 1670.[92] But the population and the main defensive force made up out of *strel'tsy* (as at Tsaritsyn) were sympathetic to the cause of the *hetman* [Cossack chief], partially because of his generosity when he had bivouacked at Astrakhan the previous year.[93] Ever more assertively, Razin posed as the people's champion, promising to destroy oppressors whose regime was called worse than Turkish and Tatar slavery.[94]

The Dutch crew and captain, detailed to protect one of the city's gates, wavered upon the Cossack approach to the city. The surgeon Termundt, who had lived in Muscovy for more than a decade and had seen some of the Cossack damage in Iran, persuaded the crew that flight toward Persia was a better option than to stand and fight.[95] As foreigners, the Dutchmen were likely to be massacred by the Cossacks, who were inclined to see *nemtsy* as stalwart alien allies of the boyar oppressors. The sorry fate of the foreign officers of L'vov's army seemed to confirm this murderous Cossack xenophobia.[96] The utter breakdown of the law of the land in the revolt may have intensified the Dutch crew's terror.[97]

Sheer panic is palpable from both Struys's and Butler's stories about the crew's movements during these days. Just before Razin's army pulled the noose tight around Astrakhan, the Dutchmen, leaving behind seven of their team (Butler, first mate Lambert Helt, carpenter Dirk Pietersz, the sailor Trappen and his wife, Jan Termundt, and assistant gunner *Christiaan* (Karsten) Brandt), slipped out of the town and embarked on a sloop (*sloep*) to attempt an escape across the Caspian.[98] Butler, together with his small team of bravehearts, chose fighting over fleeing only at the last moment, a decision they were not able to convey to the Struys party before it absconded.[99] Helt was to die at the city's defence during the next days; it seems that Dirk Pietersz and Jakob Trappen also lost their lives in the ensuing fracas.[100]

The Dutch group's divided loyalty to the *voevoda* seems based on a variety of personal considerations. Butler may have felt more obliged than his men to honour his commitment to the tsar's service, and Helt and Dirk Pietersz had been in Russia since 1667 and may have developed a stronger loyalty to the tsar's cause than the Dutch sailors who had arrived in the fall of 1668. Trappen's choice to stay in Astrakhan seems to have been purely based on his reluctance to expose his wife to a journey conceivably even more dangerous than the prospect of Cossack brutality.[101] Brandt and Termundt later chose to reside (again) in Muscovy. They were likely more predisposed than their comrades to defend the monarch they were to serve until their death some decades later. In contrast, those who left with Struys were fed up with the adventure, and anxious to find their way home (or better employment), even if it was unclear where they might encounter a safe haven.

Reysen clearly draws from Struys's personal memories about his flight.[102] The writings ascribed to Struys, Butler, and Faber (Fabricius) have remained a key group of sources for our understanding of the Razin rebellion, for, unfortunately, many relevant Russian records were lost through a fire in the Moscow Kremlin in 1701.[103] Surviving Russian and Dutch archival records nevertheless confirm the story of the flight of the fifteen of Struys's party.[104] In addition, a chronicle composed by the Orthodox *pope* Zolotarev mentions how 'soldiers (*saldaty*) left by sea, in boats by night, and others ran on foot across the steppe, and many died of hunger, and still others were taken hostage by the Tatars and by the Kalmyks'.[105]

Butler's and Fabricius's accounts agree in suggesting that, after the arrival on 22 June 1670 of Razin's men at the city walls, the riotous atmosphere among the city's *strel'tsy* garrison worsened because no one had received wages for a while.[106] The Nogai Tatars, who were allied with the defenders, fled, abandoning their positions outside the walls. In the early morning of June 25 Cossacks scaled the walls using ladders.[107] The Cossacks surprised the defenders by climbing up the walls in an unexpected place where they were received by sympathetic *strel'tsy*.[108] Ivan Semenovich Prozorovskii and his brother Mikhail hastened with a detachment of cavalry to the spot where the Cossacks appeared on the walls, but found themselves surrounded by traitors to the tsar's cause. In the ensuing fighting, Mikhail Prozorovskii was killed; his wounded brother was carried by loyal servants into a nearby church in search of sanctuary.[109] But the Cossacks stormed the church, slaughtering all those who resisted and capturing others who had sought refuge there.[110] Razin's violation of this ancient right contributed to his eventual depiction as a pagan (*Reysen*, too, calls him a 'father of many godless children' and reinforced his paganism by the likely fictive episode of his human sacrifice to the god of the Volga).[111]

It appears that the only serious resistance to the Cossacks was offered by the various foreigners (who included Persians and *Cherkesy* from the northern Caucasus) who had been trapped inside the city.[112] But they were heavily outnumbered, and the Europeans among them were virtually all annihilated. Three of the Dutch party fell, while it is unclear what happened to Trappen's wife. Butler and Termundt escaped the city at the last moment, while Brandt, after initially losing touch with them, was saved by his 'Persian' looks (and Faber, who survived the siege as well, served in the Cossack army, of course).[113]

The city's houses and warehouses were plundered, and the loot divided up in Cossack manner whereby shares went to all who had joined Razin's side, including Faber.[114] The city was placed under Cossack rule, consisting of a hetman assisted by *esauly* (councilors).[115] Razin stayed in Astrakhan for three weeks. A genuine class war raged, in which the have-nots mercilessly murdered the haves, along with their supposed henchmen, such as the Western foreigners. All written records of the state's administration and of private merchants were burned. In mid-July, accompanied by most of his army, the hetman himself departed on river rafts northward on the Volga, leaving the city in the hands of his partner Vas'ka Us, assisted by Fed'ka Shelyudak and Ivan Terskii.[116] Probably paranoid about plotters, this triumvirate ordered a second massacre of the local population two weeks after Razin's departure.[117]

It seemed for a moment that the Muscovite state was to be engulfed by a civil war that might end the rule of the Romanovs.[118] But at Simbirsk in October 1670 Razin suffered his first military setback since his return to Muscovite soil in 1669.[119] Fleeing to his homeland on the Don, Razin's armies were subsequently shattered by regular tsarist forces; many captured rebels were mercilessly hanged and impaled.[120] The hetman was handed to tsarist custody by the registered Don Cossack host.[121] Never supporting Razin, the established Cossacks wanted to dissociate themselves from the upstart. In April 1671 Razin and his brother were dispatched in chains from the Cossack settlement of Cherkassk on the Don to Moscow.[122] Stepan Razin was executed in a gruesome manner on 6 June 1671 on Red Square.[123]

In Astrakhan, meanwhile, rebels held out until the end of 1671.[124] After a three-month siege by an army led by Ivan Bogdanovich Miloslavskii (d. 1681)[125] and auxiliary troops led by the Cherkess prince Kaspulat Mutsalovich, the dispirited and hungry rebels surrendered on November 26. Apparently following instructions from the tsar, Miloslavskii forgave the defenders, including their last commander, although their loot was confiscated.[126] But such mildness may have been a temporary ruse. The tsar dispatched Prince Iakov Nikitich Odoevskii (d. 1697) in the summer of 1672 to seek retribution from the rebels in earnest: he unleashed a reign of terror in the city.[127]

In August 1670, the first news was published in the United Provinces about Astrakhan's fall in June to Stepan Razin's Cossacks.[128] In July of the next year, just after Razin's execution, the *Hollandtsche Mercurius* claimed that the crew which had manned the tsarist ship the *Adelaer* (Eagle) had been killed during their attempt to escape across the Caspian Sea. But the monthly periodical exaggerated the the loss of Dutch lives in the tragedy. By 1676, the ingotmaker Willem Willemsz and the assistant pilot Jan Albertsz and the sailors Arentsz and Munster had followed Helt, Dirk Pietersz, and Trappen to the grave. In May 1676, however, boatswain Pieter Bartelsz, sailor Daniel Cornelisz, and constable Cornelis de Vries were all alive, as was boatswain Meijndert Meijndertsz.[129] On May 12 of that year, Meijndertsz, together with the wives of his three fellow sailors, Albertsz's widow, and the guardians of Willemsz's orphaned children, declared before an Amsterdam notary to have given Jan Struys the rights of attorney to retrieve the arrears the tsar's government owed them.[130]

Reysen shows how the six had escaped Astrakhan in the company of the carpenter and pilot Willem Barentsz Klopper, the sailor Pieter Arentsz van Schevelingen, the interpreter Jacob Schram, the diamond-cutter Anthony Munster, the sailmaker's assistant Els Pietersz, the sailors Wiggert Poppes and Cornelis Sarisz Brak, the latter's wife Maria Jans Brak and their baby.[131] And *Reysen* and other sources attest to the escape of four of the seven Dutchmen who at first had stayed behind in Astrakhan: Captain David Butler, assistant-constable Karsten Brandt, the surgeon Jan Termundt and the gunner Lodewijk Faber.[132]

Butler and Termundt made an early unsuccessful attempt to flee across the Caspian Sea.[133] They slipped out of the city in late June 1670 as it was being ransacked by Razin's men, but ran into *strel'tsy* who had chosen the *hetman*'s side. Hauled back to Astrakhan, Faber, fluent in Polish and Russian, and now with shaved head, grown-out beard, and in Cossack dress, interceded with the Cossack leaders to spare Butler's life in exchange for the promise of a hefty ransom. Termundt survived because his surgeon's skills were needed by the Cossacks. Two months after their foiled escape Termundt, for whom Faber stood guarantee, and Butler (in disguise as Termundt's servant) were allowed to leave the city on the pretext of Termundt's need to buy medicine in Persia.[134] They joined a fleet of Indian merchants setting sail for Persia, and ignored their promise to wait on Faber (and Brandt, apparently) outside the city before travelling further south. After a trying journey during which most of their goods were stolen, Termundt and Butler reached Terskii gorod on the Muscovite border more than a month after escaping Astrakhan.[135]

Presently Faber, accompanied by Brandt, escaped Astrakhan with the help of a friendly Muslim merchant, a converted Lithuanian. Before they reached the sea they were captured by 'Tatars', under which generic name a variety of different groups fell in this part of the world. The people who enslaved Faber and Brandt were probably Turkic-speaking Kumyks, for they brought him to Tarku (Tarki; near Makhachkala today), the main residence of the Kumyk *shamkhal* (about whom more in the next chapter) on the western Caspian coast.[136] To his great fortune, Faber was soon ransomed there, either by an Indian merchant with whom he had become friends when still in Astrakhan, or by Termundt and Butler who ran into him here again.[137]

On 25 October 1670, with the apparent permission of the *shamkhal*, all four joined a caravan destined for Derbent, 160 kilometres further south along the coast, and a border town under the shah's direct rule.[138] On the road to Derbent, Butler recognized his clothes on a Persian, realizing that his chest had been looted from his crew, who had thus likely been made captive.[139] At Derbent, the foursome met Cornelis de Vries and Pieter Arentsz, who related their trials to Butler.[140] Butler pleaded with the governor of Derbent to intercede on behalf of his crew, but the governor's messages failed to persuade the *shamkhal* to release those still in captivity in his realm. The Dutchmen soon received supplies and funds from the local governor to travel on to Shemakha, a further 200 kilometres south. Butler, de Vries, Termundt, Brandt, Faber, and Arentsz reached Shemakha in early November 1670, finding the sailmaker Jan Struys there, in the retinue of the Polish envoy.[141]

9
A Dutch Slave in Asia

Reysen's description of Struys's perilous progress from Astrakhan to Bandar-e-'Abbas covers more than 150 pages. The section's detail reflects Struys's much more vivid and precise recounting of things that befell him only a few years earlier. With it, too, the publishers responded to readers' fascination with genuinely exotic territories. Additionally, Struys's wanderings in Iranian territory appealed as they represented the pinnacle of sufferings that would amazingly leave him unscathed. His patient endurance resembled that of his compatriots who had collectively overcome similar profound hardships in the years prior to the book's release.

The Iranian setting presented a convenient background against which *Reysen* profiled several key elements of the ideal Dutch self-image and morality in even sharper tones than against the preceding Muscovite background. These included the righteousness of the religious convictions of the Dutch Reformed Church tempered by tolerance of other creeds, including Islam; moral rectitude and compassion for suffering fellow human beings, deriving from strongly held Christian beliefs; a sense of cultural superiority toward cruel barbarians of varying hue; a patriotic belief in the Republic's political and economic importance in Asia thanks to the operations of the VOC; the lower ranks' respectful deference (rather than servility) to those placed higher in the pecking order; a misogyny that saw women as both physically and mentally weak and with a penchant to lead men astray from the righteous path to salvation; and a belief in basic human (male) dignity and proper judicial process.

Van Meurs and van Someren were keen to present an elaborate chorographic description of Safavid Iran, a part of the world barely known in the Republic and yet a regular trading partner.[1] After the wildly popular Dutch translation of Olearius of the early 1650s, Olfert Dapper's book about Asia of 1672 had been the sole new Dutch-language account describing Iran, but this massive and richly illustrated tome was prohibitively expensive.[2] The publisher of Dapper's book had been van Meurs, who had presented his pricey volume to a depressed market in 1672. Dapper's *Asia*, too, was less appealing as a work entirely based on secondary sources.

Contrary to Struys, meanwhile, Olearius had spent little time in Dagestan, the Erevan region, or Shirvan province, and had not travelled the route from Isfahan via Shiraz to Bandar-e-'Abbas, which omissions the Holsteiner's *Beschreibung* reflected. For Dutch and other Western European readers, *Reysen*'s description of these parts of the world were therefore exotic, exciting, and novel. From the contents of previous publications on Asia, van Meurs and van Someren surmised that Western European readers anticipated a certain standard set of horrors and delights within descriptions of this continent, even while its precise 'Oriental' character (or even its borders) was still crystallizing.[3] Certain topoi of this Orient, such as the legendary beauty of Circassian women, thus duly entered *Reysen*.

In its preceding pages on Muscovy, *Reysen* had whetted the reader's appetite by describing exotic Russian curiosities. It had already paid considerable quasi-Orientalist attention to Russians and some of the non-Slavic peoples ruled by the tsar. *Reysen* occasionally reveals in its description of Iran and its inhabitants the increasing sense of Western cultural superiority that accompanied growing European ascendance in Asia. A reversal of the 'asymmetrical' relation between Europeans and Asians was clearly in progress, ultimately leading to Westerners' contemptuous attitude of 'Orientalism'.[4] As Anthony Pagden suggests, this viewpoint was in its core grounded in the Europeans' belief that their continent had 'to play a special role in the history of creation, derived from the conviction that only those who dwelt in the kind of law-governed free urban communities of which "Europe" was constituted would ever be likely to possess the capacity to harness nature to their purposes. The others [...] remained forever in unenlightened herds.'[5] But in *Reysen* older readings of the House of Islam (for instance, as doomed infidels or the sons of Ismail) still competed with the imperialist attitude that would consider Muslim culture 'ignorant but complex' by the nineteenth century.[6]

The Dutch VOC had already been brazen enough to declare war on the shah in the 1640s.[7] Although after this conflict the relationship between the Iranian monarch and the VOC remained cordial, the speedy decline of Safavid power resumed after 1700, prompting Peter the Great to invade and occupy large stretches of the country in the 1720s.[8] Europeans thus began to probe the Iranian armour earlier than that of the Turks or Mughals, but Iran's economic importance to Westerners was modest and no major efforts at subjugation were launched before 1700. Western merchants and diplomats usually approached Safavid Iranians as if denizens of an economic and military Great Power on a par with Mughal India and the Ottoman Empire.[9] The Dutch and English (and Russian) merchants preferred negotiating with the Iranians in the guise of respectful supplicants.[10]

In their turn, the Iranians had little to no interest in the Westerners, except as merchants or, perhaps, commodities. Jean (John) Chardin (1643–1713), who travelled like Struys in Iran in the 1660s and 1670s, discerned one key reason for the decline in power of the Safavid empire relative to the European maritime empires: 'But you would scarce believe how little Curiosity the Eastern People have in such Remarks and Observations [on the origins of tobacco and sugar]. There's scarce a Person among their Learned Men, who keeps a Register of the Discoveries that are made in the Arts and Sciences.'[11] In contrast, the tsars, less able to isolate themselves from new knowledge, began to '[r]egister' Western scientific and technological advances at precisely this time. By embracing such innovations, Peter the Great and his successors kept Russia from being reduced to a colonial dependency of the West.[12]

The prevailing parity of East and West before 1700 is reflected by Struys's enslavement. At a time when Dutch traders dominated the burgeoning transatlantic slave trade, Muslim masters lorded it over Christian slaves around the Mediterranean, Black and Caspian Seas. The geopolitical and economic centre of gravity was shifting to Europe, but this is more apparent to us than it was to contemporaries: in the borderlands of the Caucasus, this European advance was as yet imperceptible.

Brief in historical time, Jan Struys's spell as a slave (June–October 1670) in the Caucasus is a lengthy one in *Reysen*.[13] *Reysen* thus displays with emphasis Struys's steadfastness in the face of physical abuse and other forms of persuasion. The horrors confronting him are heavily detailed, confirming the region's alien barbarity expected by the readers. Although often based in fact, *Reysen*'s story of Struys's slavery and eventual release presents a version of a 'captivity narrative', a subgenre within travel writing then popular in Protestant Western Europe.[14] But while his suffering is emphasized for dramatic effect and moral instruction, and while its description follows a certain formulaic pattern, the general veracity of Struys's lot as rendered in *Reysen* is confirmed by other accounts and archival documents.

***

Struys and his party escaped a possibly brutal end at the hands of the Cossacks in Astrakhan in June 1670, but their suffering had only just begun. After leaving the besieged city at night, the Dutch escapees laboriously made their way toward the open sea, some 70 kilometres south of the city.[15] The Volga's delta is a complex network of streams, and the Dutch, attempting to remain unnoticed, were forced to travel under the cover of darkness, which hampered their sloop's navigation. Barely evading slave-hunting Crimean Tatars, they were lucky to meet friendly Tatars who helped them reach the Caspian Sea. Floating on an overloaded leaky vessel with a primitive sail and a few oars, they hugged the coast until they crossed the Bight of Kizliar (*'Golf van Kieselarke'*) by way of the open sea.[16] Making only slow progress, their meagre supplies (mainly bread) quickly dwindled. Thunderstorms with heavy rainfall and high winds had the crew furiously bailing water from their craft. Sighting land again (the western Caspian shore near the river Terek's mouth), the Dutch encountered a boat manned by friendly Dagestani Tatars, who shared their food with them. The two vessels eventually anchored near Tersk, the Russian garrison town that guarded the border between the tsar's territories and those nominally under Muslim rule.[17]

Soldiers from Tersk boarded the Dutch boat for inspection.[18] Their behaviour and countenance made the crew doubt the inspectors' loyalty to Tsar Aleksei. Ignoring an invitation to meet the 'local governor,' the sailors hauled anchor and sailed south-westward from Tersk. But they thus became trapped in a rather narrow waterway (on the maps of the Caspian Sea in both Dutch and English versions the water is called the *'verdwaelde inboght'* [lost inlet]).[19] Outmanoeuvring a Russian ship sent by the Terki rebel governor, the *Oryol* crew managed to head back to open sea. They spent the night ashore on allegedly 'Circassian' land.[20] Here lived the southernmost allies of the tsar. *Reysen* tells how Struys was swept off his feet by the 'local' women, whose friendliness was tantamount to lasciviousness (underlined by an engraving of a bare-breasted woman in *Reysen*), but this description borrows plainly from Olearius's observations.[21]

The Terek river did not rigidly demarcate a religious border between Christianity and Islam, for sizable Muslim communities lived under the tsar's rule north of the river, and innumerable Christians lived under Ottoman and Persian rule to its south and west. The North Caucasus was 'a vast middle ground', in which no political or religious side dominated.[22] Dagestan's population had thus only recently become immersed in Islam, and local beliefs still had some unorthodox traits that set them apart from mainstream Sunni teachings.[23]

Even though seventeenth-century borders between Europe and Asia were fluid in practice, contemporary Western geographers considered Muscovy west of the Ural mountain range more European than Asian, whereas territory from the Caucasus mountains southward was Asian rather than European.[24] The Terek formally separated Christian-ruled territory (to its north) and Muslim rule (to its south). This geographic and political dividing line was also culturally marked in the Western view. By the second half of the seventeenth century, Western Europeans no longer imagined the tsar's realm west of the Urals as the 'Orient, ... a place of romance, exotic beings, haunting memories and landscapes, remarkable experiences'.[25] After 1650, periodicals such as the *Hollandtsche Mercurius* regularly reported about Muscovy, sign of its entrance into the European cultural landscape. The tsar's realm was increasingly defined as European (nonetheless, like the Mediterranean, Muscovy was clearly more exotic to the Dutch than states in closer proximity to the Republic).[26] But in leaving Muscovy the sailors crossed into what seventeenth-century Christian Europeans considered Asia, a continent ruled by various non-Christians that was geographically discrete.

Still bearing south, Struys's party managed to elude a final pursuit by Cossacks loyal to Razin.[27] Along the shore of Dagestan[28] (in the lower parts of the eastern Caucasus) lived the Turkic-speaking Kumyks and their subjects (usually given the more generic name of Dagestani Tatars in *Reysen*), ruled by their *shamkhal* who resided in Tarki, the town where Faber would be ransomed.[29] The *shamkhal* was an Islamic ruler with his own religious Leader of the Faithful (*sheik-ül-Islam*) to assist him. The *shamkhal* had stubbornly maintained independence from all three neighbouring empires surrounding his lands. The Kumyks spoke a Turkic language and adhered to Sunni Islam, which culturally linked them to the Ottoman Turks (and yielded them the generic name of Tatars), but because of its remoteness from Istanbul (and the terms of the Treaty of Kasr-i-Shirin of 1639), it was the shah and tsar who vied for the *shamkhal*'s political allegiance.[30] Concomitantly, while local rulers such as the *shamkhal* nominally exercised power over their Dagestani subjects, many of the communities over which they presided enjoyed far-reaching autonomy, because the conditions of the terrain prevented effective central control.[31]

By 20 June 1670, Struys's sloop was navigating along Dagestani-Tatar land.[32] The Dutchmen were apprehensive. The Kumyks had a reputation of being 'great people's thieves', as *Reysen* calls them.[33] Kumyk men predominantly earned their livelihood by capturing people and selling them as slaves or receiving ransom for their release, while Kumyk women herded their clan's cattle. The Dutch vessel had passed the coast near the settlement of '*Boynak*' (Buynak, the residence of the *shamkhal*'s designated successor), when a storm stranded it 35 kilometres north of Derbent.[34] As the first outpost of Safavid Iran under direct control by the shah in Isfahan (rather than a tributary realm), Derbent might have meant safety, for Persia and the Republic had official agreements that mandated respectful treatment of each other's subjects.[35] Although such guidelines could be somewhat whimsically enforced, the Iranians, who were on reasonably good terms with the local VOC officials, were inclined to abide by them.

Only briefly did Struys's crew evade *Tatar* (here meaning Kumyk, Dargin, or Lezgin, all Dagestani ethnic groups) slave hunters.[36] When some crew members proposed to leave Brak and his family behind as they hindered the party's search for safety, *Reysen* has Struys scold his fellows for their un-Christian behaviour toward the Braks. Persuaded by their veteran companion, who extolled traditional bonds of amity toward one's fellows as well as Christian and Dutch solidarity, the entire group decamped together.[37] Soon after, mounted men, subjects of the *shamkhal*, surrounded them. The Dutch party surrendered without resistance. Captured in territory formally ruled by the overlord of the Dargins (the *Osmin* [*utsmii*]), their Kumyk captors drove them northward to the land ruled by the *shamkhal*.[38] Once the Kumyks reached their homeland, they robbed the Dutch crew and stripped and raped the unfortunate Maria Brak, after which they temporarily disappeared.

Reysen's mention of nudity and rape underlined the Dagestani's inhumane barbarity that now confronted the sailors, since both were largely taboo in seventeenth-century Western European writing.[39] But *Reysen*'s demonization of the Dagestani is later complicated when the book notes that the captors' brutality was driven by a thirst for revenge against people they held to be Cossacks.[40] In 1668 and 1669, Razin's men had wreaked havoc on Dagestan's coastal settlements, for which the *shamkhal*'s men now exacted retribution.[41]

The same *Tatar* assailants returned to rob the marooned Dutch of their few remaining belongings.[42] After the mayhem of this double raid, the crew was divided into three parties. Struys, his protégé Els Pietersz, and Jacob Schram (*Tolk*, the interpreter) lost contact with the rest. Klopper, de Vries, Willemsz, Meindertsz, Arentsz, Brak, Maria Brak and their child were taken to the Kumyk settlement of Boynak (Klopper and Meindertsz were soon separated from the others, and Arentsz and de Vries escaped).[43] Later Struys learned how the *shamkhal* sold Brak (as he probably did with his comrades) to serve another master for a three-year term, while the potentate added Maria to his harem after raping her, too. Brak eventually fled, but his ultimate fate is uncertain. Albertsz, Bartelsz, Munster, Poppes, and Cornelisz were dragged to another settlement. Two of them succumbed in the next months, but it seems that Bartelsz and Cornelisz were alive in 1676.[44] Poppes's fate is unknown.

VOC records provide insight into the stranded crew's further fortunes. The Company offered its aid in securing the sailors' release in the spring of 1671, after Butler had reached Isfahan. VOC officials experienced great difficulty in ransoming some of the *Oryol* crew in the years following. By 1673, the Persian VOC branch was still trying to ransom five Dutch 'nationals', among whom were a woman and her child, held captive '*in Chammagij*' (perhaps Shemakha, but it might indicate 'by the *shamkhal*').[45] This, despite gaining the shah's offical support for their cause by way of a *firman*. It shows how the shah did not exercise much genuine control over the various Caucasian communities and their overlords.[46] In their actions toward the Dutch captives, neither *utsmii* nor *shamkhal* seemed mindful of the cordial relations between the Republic and Iran. But most of the Dutchmen nevertheless found their way back to freedom.

In the summer of 1670, however, for the trio of Jan Struys, Els Pietersz, and Jacob Schram-Tolk prospects were grim. Stripped to their underwear, they underwent further tribulations at the hands of other Dagestanis.[47] These were Caucasian-speaking Dargins, who were subjects of the *utsmii* residing at Kaidag (Kaytak/Kaitagi).[48] After apprehending the three Dutchmen, the Dargins first dragged them across the ground tied to their horses. They then bound Struys to a tree and fired blunt arrows at him, a scene depicted in the engraving of *Reysen*'s title page.[49] The marauders justified their cruelty by communicating to the Dutchmen that they were exacting retribution for all the murder and plunder the Cossacks had committed in previous years, but they may have been primarily trying to force Struys to betray friends suspected of hiding nearby.[50]

Did the Dargins interrogate Struys in Russian, Turkish, or Tatar? Beyond the fairly narrow plain along the shores of the Caspian, Dagestan is a land of deep valleys and high mountains, allowing its inhabitants to maintain a great variety of cultures and languages. Most speak and spoke so-called Caucasian tongues, such as Dargin, Lak, Lezgin, or Avar, but they communicated with neighbouring non-Caucasian speakers usually in Turkic Tatar (the language of the Kumyks), the *lingua franca* in these borderlands.[51] Perhaps some basic Russian was used at first (with Schram translating), since the Dutch trio was ignorant of the Tatar tongue. But Struys may have only fully understood afterwards (when he encountered Russian speakers at the *utsmii*'s court) that his torture had been provoked in part by vengeful sentiments. Ending Struys's torture without learning his comrades' whereabouts, the Dargins tied Struys and Pietersz together and brought them before the *utsmii*.[52] This chief refused to believe the captives' avowed assurances that they were Dutch. Russian slaves were called in who swore that the three were indeed '*Duytschen*' rather than Cossacks.[53] Their lives were therefore spared, but the men became slaves in shackles.

The *utsmii* then sent Struys to his son in Erevan.[54] Struys parted ways with his protégé Els Pietersz, who at the *utsmii*'s palace began to suffer from delusions, and with Schram (who disappears without a trace from *Reysen*'s pages). Struys was tied up, placed on a mule, and brought to the court of his new master *Mahumeth Sultan*.[55] Struys arrived at Erevan during the early days of the summer of 1670. *Reysen* estimates that the settlement was as large as the Dutch town of Alkmaar (about 10,000 people). Erevan was populated by a minority of poor Christian Armenians and a majority of Muslims, but it also harboured a Roman Catholic monastery. Its main economic activity appeared to be the slave trade.

The section of *Reysen* that followed became one of its most notorious parts, for it has Struys ascend Mount Ararat.[56] Christian tradition held that Erevan was located at the foot of Biblical Mount Ararat, which *Reysen* suggests (erroneously) to be the highest mountain of the Caucasus. The episode begins with two Carmelite friars somehow mistaking Struys for a surgeon, and renting his services from Prince Mahumeth.[57] The prince promised Struys his freedom if he accomplished the task for which the monks had hired him: mending the broken leg of a hermit who lived on Ararat. 'Hoping for God's grace and a bit of luck', Struys accepted the task and, accompanied by the Carmelites and a peasant guide, ascended the mountain (its summit is at 5,165 metres), with a mule carrying firewood and provisions.[58] Above the clouds, Struys arrived at the hermit's cave, situated in an apparent temperate microclimate free from rain and wind. The wounded recluse told Struys of living at this spot for a quarter century. Struys set the broken bone, wrapped the limb in a binding (or *Truss*) and proceeded to rub four times daily the damaged area with a concoction made of boiled egg yolks.[59] Struys's regimen included having the hermit lay prone for two weeks, which more than anything may have helped a healing process that was well advanced by the time Struys left. He prescribed the hermit to wear the bindings for another year and to continue rubbing the ointment into the skin around the breach.

Out of gratitude, the hermit gave Struys several precious presents: a brown-red piece of hardwood from the Ark (said to be still lodged in the mountain above), a piece of rock the hermit claimed to have carved from beneath it, and a wooden cross (supposedly made from the ship's interior wood) on a silver chain. The recluse, a native Roman whose name was *Domingo Alexander*, explained to Struys that if he ever went to Rome he would receive enough money for his relics to live comfortably for the rest of his days. The hermit also supplied the Dutchman with a statement written in Latin affirming the artefacts' authenticity, a document duly displayed in both Latin and Dutch in *Reysen*.[60] *Reysen*'s Struys tells his readers that he managed to salvage both cross and wood (and the testimony, apparently), but in 1673, at St Helena, the English stole his rock after they entered his ship.

Voltaire, always interested in exposing superstition, argued in his *Dictionnaire philosophique* that *Reysen*'s account could not be true, for Struys claimed to have climbed five miles a day.[61] At that rate, the *philosophe* claimed, the mountain would be thirty-five miles high, allowing one almost to reach the moon from its summit. Yet Voltaire was a careless reader, for *Reysen* merely notes how Struys travelled five miles a day in ascending the mountain. In addition, Voltaire protested more convincingly that Struys's account of the mellow climate on Ararat stood in sharp contrast to that of the French traveller Tournefort, who had been sent by Louis XIV to explore the mountain and had depicted a generally forbidding natural environment.

Still, peculiar microclimates have been identified in many mountain ranges. And not just for that reason is Voltaire's dismissal of *Reysen*'s passage on Ararat too categorical. Wood purportedly from the Ark has been diffused across the globe, even if according to archeological carbon-dating its origin can never be Noah's Ark.[62] Pieces of wood have been collected since time immemorial on a mountain identified dubiously as Mount Ararat, the Buyuk Aghri Daghi, rising to the south of Erevan.[63] Struys may thus have very well visited a Roman hermit on the Buyuk Aghri Daghi and even helped to heal him. The Carmelites' request to Struys to treat the hermit may have been informed by the good reputation Western Europeans surgeons enjoyed in seventeenth-century Asia.[64] The sailor Struys had witnessed the work of ship's surgeons in setting broken bones, and he was a quick learner. Professional surgeons, who normally doubled as barbers in those days, learned their skills by apprenticeship, not by study. 'Bonesetters' practised without formal schooling.[65] Additionally, it bears some relevance that during the 1690s Jan Termundt instructed Peter the Great in basic surgical techniques, bloodletting, the dressing of wounds, dentistry, and anatomical section.[66] Had Termundt or Schak, the surgeon who fell at Chernyi Yar in May 1670, taught Struys some similar rudimentary medical skills?

Upon his return in Erevan, *Reysen* has Struys again placed in leg irons to prevent an escape.[67] Physically suffering from the chains, he bemoaned to his master the injustice of not releasing him after he had accomplished the Carmelites' task as he had been promised. Impervious to such lamentations, Prince Mahumeth baited Struys with two women as future wives in exchange for his conversion to Islam. Despite the women's best efforts (they stroked his body to arouse him), Struys relates how his loathing of both Islam and divorce made him reject these overtures. He told the women he could never be unfaithful to his wife and two children in his native land.

While *Reysen*'s narrative here was shaped by the conventions of the captivity narrative, its text reflects Western European attitudes toward sexuality.[68] *Reysen* follows the tale of the temptresses with an account of Mahumeth's lucrative ownership of public baths in Derbent, Shemakha, and Isfahan.[69] The narrative notes how such establishments were heavily frequented by Muslims, since they were obliged to wash themselves after sexual intercourse. *Reysen* further expands on the sexual theme in explaining how Muslims were barred from sex with pregnant women, which led them to father many children with different wives.[70] Through the mirror of Iranian sexual *mores*, *Reysen* provides thereby insight into Dutch prudery about sexuality and misconceptions about bodily hygiene.

In *Reysen*, Struys also declined Mahumeth's offer to make him '*hopman*,' that is a commander of his master's troops. In a dialogue supposedly reported verbatim, the sailmaker stated that he would rather turn a knife in his own heart than forsake his religion for Islam.[71] In this conversation between Mahumeth and Struys, *Reysen* casts the Dutchman as a stalwart religious patriot. This dismissive response, however, was an ideal often belied by Christians in practice. Individual conversions by Christians to Islam were a common phenomenon throughout the first millennium of Islam's history.[72] Muslim slaves enjoyed a far higher status than

Christian slaves, and conversion offered a better prospect of manumission.[73] Many Dutchmen became 'renegades' in the seventeenth century, as we saw in Chapter 2.[74]

It can be suspected that Struys, an obvious opportunist, might have eventually converted to avoid permanent slavery.[75] A few months later, when Struys traversed Iran from north to south, he had already learned the local habitus well enough to pass for a Muslim.[76] While the book at times portrays him as an unwavering Dutch Protestant, the historical Jan Struys seems to have shared the lenient, tolerant, and often indeterminate religious outlook of many of his compatriots.[77] *Reysen* records several times how he enjoyed good relations with various Catholic monks. At various moments, *Reysen*'s Struys even surpasses this ecumenical outlook by praising Muslims such as Haji Biram Ali (see below).[78] More than anything, the brevity of his separation from fellow Christians explains why Struys retained the two key markers of his identity, his Christianity and his Dutch culture and allegiance.

Although not always consistently (here both the book's lack of rigorous editing surfaces as well as the shifting perception of the Muslim world), *Reysen* draws a sharp division between seventeenth-century Islam and Christianity. Mahumeth and Struys's debate aimed primarily at providing the reader with a proper moral, religious, and patriotic lesson: genuine Dutchmen were not to forsake Christianity and fatherland for worldly delights. Matar suggests that 'in the popular imagination ... apostasy pointed towards a fearsome historical inevitability: as Christianity had replaced Judaism, so would Islam replace Christianity'.[79] Any Muslim efforts at converting Christians thus needed to be rejected, even at the pain of death. *Reysen*'s Struys demonstrates exemplary 'Calvinist' self-discipline toward Mahumeth's temptations.[80] Whether he was ever baited by temptresses is highly doubtful, for Muslim men would not quickly force female believers to abase themselves to convert a Christian.[81] Suspicions about a deliberate invention here are reinforced by the knowledge that in the Republic similar stories of steadfast behaviour in the face of Muslim temptation were current about the sea heroes de Ruyter and Maarten Tromp.[82]

By the insertion of this temptation episode, *Reysen* offered a compelling patriotic message to its Dutch audience. When the book was published, the Republic was still in shock about the war which had threatened its very survival in 1672. In this moment of crisis, allegations of treason had been rampant, and a mob had lynched the brothers de Witt as the scapegoats responsible for the Republic's torment.[83] Struys's defiance of slavery, female wiles, and Muslim temptation symbolically represented Dutch defiance of France, England, Münster, and Cologne: in the wilderness of the Caucasus, Struys stands firm before the threat of spiritual and physical death, after which his own ingenuity, resolve, and his friends' aid bail him out. *Reysen*'s story ends in the fall of 1673, when Struys comes home to a Holland which itself had survived the worst of dangers of its demise at the hands of Catholic France and its allies. The sail-maker's personal fate and that of his country, both on the upswing at the time, neatly aligned here.

***

Reysen relates how Struys threatened suicide anew if he remained chained, which made Mahumeth fear bringing a curse upon his house and, more practically, for a loss of property.[84] He thus sold the Dutchman to a Persian (probably meaning a Farsi speaker), Haji Mahmoud Sala, for 25 *Abbas*. *Reysen* notes that the 25 *Abbas* (a mere seventeen and a half guilders in Struys's reckoning) for which Struys changed hands was a small price for a slave compared to rates prevailing in Turkey and on the Barbary coast.[85] Haji Mahmoud Sala traded in dyestuffs from his homebase at Derbent along the Caspian shore. Master and slave travelled the more than 500 kilometres from Erevan to Derbent in July and August 1670. Sala spoke some Russian, and thus Struys's communication with his master improved significantly. He worked first as a stevedore, carrying goods between Derbent and his master's barge, moored in a sheltered bay nearby. Haji Mahmoud Sala promised to take Struys to Isfahan, where he knew many European merchants among whom surely one would buy Struys his liberty. But instead Sala would sell Struys in Derbent.

At this point, *Reysen* turns to the geography of the region. The book offered an engraved map of the Caspian Sea that was supposed to be bound here between the pages of printed text.[86] As the sole map in the book, this engraving is unusual. Perhaps it anticipated a sort of proprietary or imperial desire, inspired by the various attempts (such as van Klenk's in 1676) to establish a Dutch-organized or Dutch-controlled trade across the Caspian Sea.[87] Although not recognized as such in its time, *Reysen*'s map of the Caspian was the most accurate depiction of that sea available in Europe before 1700.[88] It is superior to those published in the works by Olearius and Witsen, even if it still misrepresents the sea's shape and clearly guessed at the shape of its eastern shores.

While the map's caption notes the basis of the engraving was a sketch by Struys of 1668, this is an unlikely scenario, for he only arrived at

Astrakhan in 1669 (and in Derbent in 1670!). The written identification of certain places such as *Meijndertsz's eiland* (an island thus named after the boatswain Meindertsz) makes it further unlikely that the illiterate Struys was the draughtsman. Other people, such as Butler, who had brought maps with him to Moscow, or the *Oryol*'s pilot and scribe Jan Albertsz, could have sketched this map, but the identity of the map's creator remains uncertain.[89] To pretend that a sketch by the sailmaker Struys provided the draught for the engraver was a slick publisher's touch, for the map authenticated Struys's travels. Jacob van Meurs, meanwhile, who had been an engraver of prints and maps before establishing himself as a publisher, may have engraved *Reysen*'s map from the original drawing.[90]

Reysen's text could not explain a conundrum that had troubled previous geographers: the destination of water from the many rivers that drained into the Caspian Sea. The text suggested that it disappeared in the Gulf of Gilan into two whirlpools draining into subterranean streams.[91] More plausibly, in keeping with A.A. Vinius's and Butler's plans, *Reysen* argued that the Caspian Sea was best suited for flat-bottomed ships carrying 40 to 50 *lasts* (60,000 to 75,000 kilograms) without much of a keel, because the sea was very shallow in many places, especially in its northern reaches.

***

Struys's fate as a slave illustrates the changing global constellation in the late seventeenth century and the limitations of Dutch capitalist 'hegemony'.[92] The Dutch were keenly engaged in the transatlantic slave trade, in which they may have been European leaders in the second half of the seventeenth century.[93] In Asia, in the various outposts of the VOC (including Persia), the Dutch often owned their servants, too.[94] Yet, this still was an age when the Dutch ran considerable risk of becoming commodities themselves, usually as slaves of Muslim masters.[95] We can agree therefore with Aune, who warns that '[t]he dominance of European economic and military power and the stability of binaries such as East/West and Christian/Muslim that characterize [later] cross-cultural encounters cannot be assumed relevant in the early modern period'.[96]

Western European sailors and villagers from coastal regions regularly ended up as galley slaves on Muslims ships in the Mediterranean.[97] This was so common that Western European governments maintained permanent representatives at the courts of the most notorious Barbary

Coast chiefs in North Africa and in Istanbul to seek the manumission of their compatriots.[98] Across Christian Europe, secular and religious authorities collected money for the ransoms of those who had become Muslim slaves.[99] Holland's villages regularly raised money for the release of seafaring neighbours and others from villages and towns further away.[100] Some of the most famous Dutch admirals of the age commanded punitive expeditions to destroy the North-African potentates' ships.[101] And enslavement by Muslim raiders threatened Europeans across the continent: Khodarkovsky points out that after sub-Saharan Africa, contemporary 'Eastern Europe was [...] the second largest supplier of slaves in the world'.[102] In Iran, one could find Polish, Russian, German, English, and Dutch slaves as well as enslaved Georgians and Armenians.[103]

Reysen's depiction of the Derbent slave market where Struys was sold in the summer of 1670 is infused with a moral outrage and reflects the experience of a former captive:

> It is very lamentable to see how they undress, move, and touch the miserable [male and female] slaves, which is rather like we would do with an oxen or horse in the fatherland; they do not make much work out of beating a male or female slave to death like a dog and if it were not for their fear of losing the money they have spent, [the slaves] would often not live long; because they are very quick-tempered and angry people and oddly jealous, thinking that even if a slave only laughs in the presence of a woman that there is something more to it; but as far as my patron went, I had no reason to be concerned about this. [...] The Persians are usually [...] much more friendly toward the slaves than the Turks, as I experienced on the galley [and] as we daily hear from the manner by which slaves in Barbary are hurt [;] even if they made it difficult for some, this was primarily intended to try to have them abandon Christianity rather than innate cruelty. [...] The thirtieth [of July 1670] my patron brought me to the market [...] and traded me to one Haji Biram Ali. This was a very rich merchant in gemstones and other precious goods.[104]

This might have made uncomfortable reading for those invested in the Dutch West India Company, for its employees treated Africans often worse than Islamic masters their slaves.[105] But given the company's indifference to the well-being of the crews on the slave-hauling ships, stockholders and board members probably soothed their guilty consciences by grouping Dutch sailors under a rubric that was not much different from African slaves, as a faceless, barely sentient horde of beings, only fit to work for their masters.

Reysen's anguished commentary of Derbent's slave market and other renditions of Struys's indignation at his enslavement represent to a considerable degree the viewpoint of the underprivileged. At a later moment, in observing the proceedings on the slave market in Shemakha in 1671, *Reysen* notes how 'people who are brought there to market are looked into the mouth like horses and undressed into the nude, [and] felt everywhere, especially the joints, to judge whether or not they are strong'.[106] The sailmaker was evidently appalled by this violation of the human dignity, even if, as in this case, none of the various Christian or pagan slaves were Western Europeans. Although channelled by the ghostwriter, the sailor's 'proletarian' solidarity and fury at the humiliation of fellow human beings rises to the surface in these lines.[107]

Reysen recounts how not long after Struys was sold to Haji Biram Ali, a fabulously rich master with nine wives, he saved his new owner from drowning in a rip current in the sea, and resuscitated him on the beach.[108] Out of gratitude, Haji Biram Ali promised to take Struys to Isfahan, where he would release him to the care of the VOC. Struys meanwhile quickly became a confidant of Haji Biram Ali's first wife Altijn. She had been born in Poland of mixed Polish-Dutch ancestry, and had been captured in a Crimean-Tartar raid in the mid-1650s, after which she ended up as Haji Biram Ali's spouse. Altijn warned Struys that rather than releasing him in the Iranian capital, her husband probably would force him to accompany him to Mecca on the *haj*; as a Christian, he would likely be killed during this pilgrimage. She proposed to Struys to abscond with her on one of her husband's vessels across the Caspian Sea to Muscovy. If Struys's wife was dead after they returned to Amsterdam via Muscovy, they would be free to marry; if not, Altijn would survive on the support from her uncle and friends of her late father who resided in the Dutch capital.

Sensibly, *Reysen*'s Struys objected that the Cossacks controlled the northern part of the Caspian Sea. He ultimately rejected a plan which originated in jealousy, for Haji Biram Ali was spending the nights with two barely pubescent Georgian female slaves of recent acquisition. Altijn did not want to live 'like a twenty-six-year old widow'.[109] In rendering another instance of female duplicity, this story falls within the general misogynist orientation of the book, although here in an unusual form of a (secretly) Christian wife betraying a Muslim spouse.[110]

Reysen relates how, precisely at the time of his discussions with Altijn, Struys heard about the arrival in Derbent of the *Oryol*'s constable Cornelis de Vries and the sailor Pieter Arentsz, who had escaped their captivity at Boynak.[111] De Vries and Arentsz had received the protection of the Iranian governor (*sultan*) of Derbent, and sought to enlist the *sultan*'s aid in seeking the liberation of their friends.[112] Accompanied by his master, Struys was allowed to pay his friends a visit. De Vries and Arentsz had news about Els Pietersz, who still languished in captivity in Boynak.[113] Struys convinced Haji Biram Ali that Els was a son of his, begging his master to aid Els Pietersz. Haji Biram Ali sent a Russian renegade to Boynak to try to ransom Els Pietersz, but the young man had been moved elsewhere.

Awaiting the promised trip to Isfahan, according to *Reysen*, Struys worked for Haji Biram Ali as a cattle herd on the plain and lumberjack in the mountains.[114] In a surprise raid, hundreds of brigands fell on the lumbercrew; Struys managed to emerge unscathed. In another incident, robbers stole Struys's cows; fortunately his life was saved at the last moment by some of Derbent's residents who accidentally happened upon the scene. Thereafter, Haji Biram Ali assigned Struys to domestic chores, pitying him as a magnet for bad luck. The Dutchman began to enjoy a very comfortable life, as good as could be for a Christian in 'Turkish' slavery.

Reysen is likely faithful to Struys's reminiscences in suggesting that he (as well as his enslaved fellows) was allowed to move freely around the town and its surroundings.[115] This latitude with most 'infidel' slaves was usual in Muslim cultures and accords with centuries-old Islamic customs and legal permissiveness toward the 'People of the Book'.[116] In Safavid Persia and Ottoman Turkey, Jews and Christians were free to practise their beliefs because they were based on Holy Scripture.[117] Although occupying an inferior legal and political standing compared with Muslims, these *dhimmi* or *ahl ad-dimma* were permitted to develop largely self-ruling communities coexisting parallel to the dominant Muslim community across the House of Islam. Slaves were of course subject to many more restrictions than those at liberty, but were not often placed under strict supervision. Barriers presented by language, dress, and conditions of terrain made escape a difficult option, even if several of the *Oryol* sailors succeeded in fleeing.[118]

In September 1670, Haji Biram Ali set out with Struys for Shemakha.[119] *Reysen* narrates how Struys's master had assembled a huge caravan that mainly carried supplies to rebuild the houses he had lost in Shemakha through a fierce earthquake in 1667. The trip was risky, since highwaymen abounded in the province of Shirvan through which the caravan travelled.[120] While *Reysen*'s emphasis on the uncivilized unsafety of barbarian surroundings was informed by a Classical trope, frequent reference to the dangerous travel conditions in Persia also reflected reality. In the first half of the seventeenth century, the roads of the Safavid empire were famous for their safety, but after 1650 brigandage increased with the erosion of the shah's government's power, as Struys had already experienced.[121]

In an odd transfigurative moment in *Reysen*, Struys is said to spot the place at Niasabad where the Holstein Embassy (of which Olearius was secretary) was stranded in the 1630s.[122] At the nearby *Barmach* mountain he saw petroleum bubbling to the surface, which was collected in a fairly simplistic manner.[123] After the caravan reached Shemakha, Struys worked as a carpenter on his master's houses.[124] His luck had definitively turned, for Shemakha was a settlement through which passed most north-south land traffic in the eastern Caucasus.

Reysen describes how in late October 1670 Struys met here two Franciscan friars with whom he discussed his faith.[125] This was not Struys's first encounter with monks in *Reysen*, and once again his Catholic interlocutors are presented as reasonable and friendly human beings. This positive portrayal may indicate Catholic involvement in *Reysen*'s preparation, of which the St Sebastian-like frontispiece may be another sign.[126] But Struys's dialogue with the monks can also be understood as an effort to contrast a tolerant (or even ecumenical) Dutch worldview (as personified in Struys's friendly interaction with Catholics, Orthodox, and Muslims) with the exclusive religious fanaticism of the Dutch foe, Louis XIV, or even Dutch Calvinist zealots. In this same period, the French king increased the harassment of Huguenots, which was to culminate in the Revocation of the Edict of Nantes in 1685.

In Struys's conversation with the Franciscans in *Reysen*, he outlined the main traits of his religion to the Franciscans in his passable Italian (acquired in earlier journeys). The friars accepted his creed as 'good', having 'nothing against it' (unlike a certain French monarch, therefore!).[127] The episode also echoes Luther's stand of 1517, when he had encountered Franciscans as his first opponents. Rather than Luther's condemnation, Struys received Franciscan approbation, hinting at the ecumenical and tolerant Amsterdam milieu in which the book was conceived.

The Franciscans took pity upon their fellow Christian and began to devise a plan to liberate him from his Muslim master.[128] They persuaded a Polish envoy present in Shemakha, the Georgian-born Bogdan

Gurdziecki,[129] to advance the ransom needed to gain Struys's freedom. *Reysen* agrees with other sources in its portrayal of Gurdziecki as a cruel and self-serving opportunist.[130] By the time he left Iranian territory in 1672, this corrupt boor was thoroughly detested by both Poles and Iranians.[131] Gurdziecki's embassy had vainly tried to conclude a Polish-Persian alliance against Turkey, to which Muscovy might be invited.[132] In early 1670, Gurdziecki began the trek home from Isfahan, but Razin's revolt barred the Georgian Pole from departing for Russia, and at Shemakha he awaited developments further north.

Struys conveyed the Franciscans' efforts to an upset Haji Biram Ali, who reiterated his promise to release Struys without compensation to his Dutch compatriots at Isfahan.[133] Recalling Altijn's counsel, Struys distrusted his master's words and replied that he was too impatient to wait any longer to return among the Christians. Haji Biram Ali warned Struys that the Polish envoy had misbehaved so badly in Iran that the shah was planning to lodge a complaint about his behaviour with the Polish king. The consequences of this diplomatic row would be unpredictable and might keep him much longer in Shemakha. Struys ignored this caution, partially having convinced himself that returning home via Muscovy might be much faster than via Isfahan and, then, Batavia (and potentially more lucrative). Haji Biram Ali sold Struys to Gurdziecki for 150 *Abbas* (about 125 guilders[134]), a price that had increased sixfold since he was first sold to Prince Mahumeth. Struys appears to have sworn an oath of loyalty to the Pole that bound him to Gurdziecki as long as this ransom was not repaid.[135] Although now in Christian custody, he did not yet enjoy unencumbered freedom.

10
Liberation

Extraordinarily high wages had attracted Jan Struys to work in Muscovy, but, rather than riches, he had found himself in the summer of 1670 a precious commodity as slave to Muslim masters in a place far from home and hearth. His endurance and pluckiness, however, and a good dose of luck and human compassion came to his rescue when all seemed lost. As the previous chapter showed, man was not always wolf to man even in this age in which have-nots everywhere were routinely brutalized by those in power. Soon after his transfer to the Polish envoy's service, men far above Struys in the Dutch social hierarchy bothered to seek his full release. Struys regained his freedom (his servitude to Gurdziecki lasted altogether an entire year) thanks to the far-reaching influence and money of the Dutch East India Company. A Christian and Dutch sense of obligation to one's fellows in need, and a patronage system in which the stronger had the duty to protect the weaker, underscored this effort. But the VOC's service was not free of charge: he was obliged to pay the Company back for the money it had paid enabling his release. Reimbursing the Company for its debts, he came home after five years of travels, in October 1673, at which point *Reysen* comes to a close.

In the fall of 1670 in Shemakha, according to *Reysen*, Struys soon discovered from the sidelines how Gurdziecki maintained a debauched court.[1] Drunkenness and gluttony were enjoyed by the Georgians of his retinue, who barred native Poles (and Struys) from their daily carousing. The ambassador's treatment of the Poles became so abusive that they began to hatch plans to murder Gurdziecki, only to wound him in the end. Gurdziecki ordered the execution of the plot's ringleader, the native Pole '*Paniegros*', and the confinement of all other Poles in the retinue.[2] The execution order was countermanded by the Shemakha *khan* (the

Iranian governor of Shirvan province, residing in Shemakha), but the treatment of the Poles, of whom Struys was considered part, continued to be appalling. Struys only avoided starvation thanks to his former master Haji Biram Ali, who generously supplied him with food.[3] When Struys began to hatch plans to escape Gurdziecki's service by fleeing to Smyrna, the arrival of Butler's group brought solace.[4]

On the request of Termundt, Brandt was lodged with Gurdziecki, while Butler, de Vries, Termundt, Faber, and Arentsz fended for themselves (perhaps offering medical services).[5] Struys's treatment now improved, possibly since his compatriots had promised Gurdziecki repayment for the costs he had incurred in ransoming Struys and lodging Brandt. But perhaps there was another reason: The shah was applying pressure on the *persona non grata* Gurdziecki to leave his realm expeditiously.[6] Gurdziecki feared having to account for his misbehaviour before his Polish superiors and preferred self-imposed exile for the time being. The envoy suddenly needed his Dutch servant's help: he inquired with Struys about the option of sailing to Holland on VOC ships (presumably by way of Bandar-e-'Abbas).

Meanwhile, both Gurdziecki and the Shemakha *khan* refused to aid Butler in ransoming the rest of the crew.[7] Not unreasonably, *Reysen* suggests collusion between the Shemakha *khan*, the *sultan* of Derbent and the *utsmii* and *shamkhal*, who were all unwilling to confiscate the precious human commodities captured by their subjects.[8] Even if these chiefs did not cover for each other, the local Caucasian communities were wont to ignore orders from above.[9] Given such resistance and the rough conditions of the terrain in the Caucasus and Iran proper, any chances of success in liberating the crew may seem prohibitive. Iran was an immensely large country given the era's simplistic means of transport and communication. From Shemakha to Isfahan the distance by road was about 1,500 kilometres, and from Isfahan to Bandar-e-'Abbas another 800 kilometres.[10] The shah's huge empire, however, had a population of a mere seven million. Iran was traversed by only a few highways. Those of the shah's subjects who congregated in the various towns along these roads met a great variety of itinerant people: it proved therefore possible to discover the whereabouts of the marooned Dutch. In addition, the shah's government preferred to maintain the good terms it enjoyed with the VOC.

And it was of course crucial that the crew's captors traded in human commodities for a living. Rather than using their slaves for themselves, the Dagestani wanted to exchange their prizes for a fair sum, and sought out prospective buyers to that end. Some of the difficulty encountered by

Butler, Termundt, or the VOC in ransoming the Dutch slaves may rather have resulted from a deliberate Dagestani strategy to drive up the price of their wares than from an effort to keep the sailors in permanent captivity. The attempts to obtain the release of sailors from their various captors, while cumbersome, thus eventually bore fruit in several cases, even if some of the mariners were broken by the experience. Probably before their captors realized the value of their prey, several of the Dutch slaves escaped because of their captivity's lax conditions, finding with relative ease the southward way out of the lands of the *utsmii* and *shamkhal*.

After Butler in Shemakha failed to liberate the rest of his crew, he travelled together with de Vries to Isfahan in November 1670.[11] VOC representatives there might persuade the shah to personally mediate in the affair. Termundt stayed behind, standing as guarantor for the repayment of the money Gurdziecki had expended on the Dutch sailors he had ransomed, which Butler was to collect from the VOC in Isfahan as well.[12] Faber now joined Gurdziecki's party (it is unclear under which conditions), while Arentsz remained seconded to Termundt.[13] The Russian-speaking Faber (perhaps accompanied by Brandt) left the others in early February 1671, eager to rejoin the tsarist forces who were on the offensive against the Cossacks, but he could not at first reach Muscovite territory and returned to Shemakha.[14]

Not long after Butler's departure from Shemakha, the first mate Willem Barentsz Klopper resurfaced there.[15] Gurdziecki, who appears to have paid his ransom on the spot, added Klopper to his retinue (and Klopper was eventually to follow Gurdziecki back to Astrakhan, see Chapter 8). In March 1671, Termundt, Klopper, Arentsz, and Struys received a note from Anthony Munster who beseeched its recipients to ransom him. But his comrades lacked funds and Gurdziecki was not willing to send money before he had received the man. Two months later, VOC officials observed Munster's arrival in Isfahan, after he had apparently bypassed Shemakha; still enslaved, he had gone insane, and he died within days of reaching the capital.[16] Around the time of Munster's plea, news about Meindert Meindertsz also reached Shemakha. Meindertsz worked as an indentured smith in Derbent and was eventually bought by a Persian from Baku, who sold him on to Jan Termundt (and he would reside in Amsterdam in May 1676).[17]

The Dutch contingent waited in vain for the return of Butler. When by the early spring of 1671 nothing had been heard from Butler, Struys began to despair about ever being able to leave Shemakha. *Reysen* has him send sent letters to the Dutch consul in Izmir (Smyrna), and to his old acquaintance, the merchant Giacomo Molives, who was Livorno's chargé d'affaires in this Ottomon port.[18] Struys implored them to forward his letters to Holland. In early June Termundt, accompanied by Arentsz and an unnamed Polish Jew who had escaped 'Tatar' slavery, decided to travel to Isfahan.[19] Apparently, Gurdziecki concluded that allowing the surgeon to depart offered the best chance for a return on his investment in the Dutch captives, since Butler's reappearance seemed unlikely after seven months.

After a difficult four-month-long journey, Butler and De Vries only arrived in Isfahan in March 1671.[20] There Butler immediately seems to have written the letter that was later published in *Reysen*.[21] VOC officials in the Iranian capital now sprang into action, arranging for Cornelis de Vries to depart with a caravan for Bandar-e-'Abbas, and seeking the shah's mediation.[22] Butler, persuaded that the residing VOC employees would exert every effort to obtain his subordinates' release, joined de Vries somewhat later in Bandar-e-'Abbas. In November 1671, they sailed for Batavia.

Struys's waiting period in Shemakha provided the ghostwriter with another logical interlude to entertain *Reysen*'s readers with chorographic detail about Iran, laced with some titillating anecdotes.[23] The civilization of Iran was an alien novelty to Europeans, and no template for its description existed akin to the detailed and organized model of the contemporary descriptions of Muscovy. Beyond a consistent subtext celebrating the Dutch worldview, *Reysen*'s chorographic discussions of Iran's nature and culture seem randomly selected and appear imbalanced or incomplete, confirming the absence of a hegemonic model for this section (whereas the Herberstein-Olearius model supplied one for Muscovy[24]). *Reysen* presents striking but superficial details about politics, religion, and culture without placing them within a linked and balanced framework. *Reysen* sequentially outlines Iranian practices such as worship of sacred places, religious holidays, the presence of eunuchs, and the investiture of provincial governors. But the book does not coherently explain the country's religions (especially Shi'ism) or its political and military organization. For his presentation of Iran, *Reysen*'s ghostwriter clearly lacked models as rich, organized, and successful as existed for Muscovy.

Contemporary readers would likely overlook such shortcomings and be captivated instead by a series of fascinating anecdotes. One particularly spectacular incident, a murder that supposedly occurred during Struys's stay in Shemakha, deserves our attention, too, for it provides telling insight into the mindset that informed *Reysen*'s composition.[25] Rendered on an engraving in the book and incorporated into *Reysen*'s frontispiece, this gruesome murder was advertised as a highlight of the narrative. It epitomized Muslim barbarity and reiterated the text's distinctly misogynist message. *Reysen* tells the reader how during Struys's residence at Shemakha a former Polish slave tried to escape her Persian husband by seeking sanctuary with Gurdziecki. Under pressure of the Shemakha *khan*, the Pole returned her to her husband, who proceeded to nail her to a sort of cross, and then skinned her alive. After this, he dragged her body to the fields near the town to be eaten by birds of prey, and nailed her skin to one of the walls in his house as a warning to his dozen other wives. *Reysen*'s Struys attributes the horrific episode to the merciless jealousy of the Muslim Persians, suggesting that similar insanely jealous men made women wear *burka*s or *chador*s.[26] The text adds that the shah and the grandees not only had their harems guarded by castrated eunuchs, but even ordered part of every eunuch's penis amputated to ensure that no sexual intercourse between castrates and wives took place.

Even if this murder happened when Struys resided in Shemakha, he could only have witnessed part of it.[27] This flaying story may be an outrageous mariner's yarn, or an anecdote that Struys merely overheard. Whatever the case, *Reysen's* publishers believed the brutal tale would appeal to the lurid obsessions of European readers. The gory murder's prominence in its frontispiece undoubtedly resulted from marketing strategies.[28]

Beyond demonizing Muslims, this episode reflects pervasive *European* misogyny.[29] Previously in *Reysen*, Orientalist temptresses had tried to lure Struys to Islam.[30] A Muslim convert woman, Altijn, had tempted Struys with a plan that might have led him to his doom. Struys's incorruptibility in the face of such temptations supported the Dutch national myth of determined and heroic Dutch*men* who could not be broken by their foes. By contrast (quasi-)Dutch women as exemplified by Altijn and Maria Brak were bereft of agency and destined to be subjugated. In the flaying scene, *Reysen* shows how Struys's loyalty to Christianity and his country had been the proper (and manly) choice. Those who veered from the straight and narrow might await the terrible fate of the Polish female renegade, who deserved to be penalized for betraying her faith and her country (as well as her husband!). The episode underlines the conviction that even the most outwardly alluring female outside concealed an abhorrent inside: to use Laura Brown's words, *Reysen* sees the woman's body as the instrument of the 'essential corruption', of, indeed, Original Sin.[31] Thus *Reysen* presents here a moralizing tale in which the puritanism and male chauvinism of Dutch culture complement each other.

The remarkable popularity of flaying tales further explains the story's lure. In the Netherlands, the story of the Achaemenid *Persian* king Cambyses flaying the corrupt judge Sisamnes was often represented in paintings.[32] Christian tradition held that the flaying of the martyr St Bartholomew had occurred in the very same region of Classical Media. Olearius reported the custom of flaying people in Iran, while Struys's contemporaries Thomas and Edward Browne recount its practice in the Turko-Iranian wars of the early seventeenth century.[33] Flaying was meanwhile a punishment still occasionally applied in contemporary Europe, albeit not in the Netherlands.[34] In the midst of their rebellion against the Polish crown, Ukrainian Cossacks had skinned the Jesuit priest Andrzej Bobola in 1657, and, in contemporary Muscovy, flaying of criminals seems to have sporadically occurred.[35] Perhaps, then, *Reysen*'s gruesome story was at least based on an actual incident. In the seventeenth-century Netherlands, however, judicial torture of suspects and punishment of convicts by bodily mutilation made way for less violent, and less public, methods.[36] In the prominence of its flaying tale, *Reysen* held up a starkly barbaric East to 'civilized' (Western) Europe. More emphatically than any other part in the book, *Reysen*'s Iranian section provided a window on practices seldom encountered in the Republic and alien and reprehensible to Dutch eyes, thus underlining the allegedly higher civilization of the Netherlanders.

Reysen followed the tale of the flayed wife and the emasculated eunuchs with a description of further sufferings inflicted upon subaltern groups in Iran. The book presents Struys's visit to a great slave market held at Shemakha in the summer of 1671, in which some 500 people were traded by captors who included Muscovite subjects.[37] The slave trade in the borderlands of the Caucasus and the steppe to its north was indeed not just one-way traffic, as near the Terek river Russians often bought captured Dagestani from the 'heathen' Circassians.[38] But in *Reysen*'s narrator's view the misdeeds of Muslim, Slavic, or pagan Caucasian slave traders paled before the Christian Georgians, who had no qualms about selling their own children into slavery. Although Georgian depravity was deliberately emphasized for effect, it was in fact a long-standing custom for Georgians to surrender some of their children as *mamluks* to the Muslim world, a version of the better known *devshirme* in the Balkans.[39] *Reysen* consistently portrays the Georgians negatively, showing no effort to fathom the reasons for this practice.[40] Georgians functioned in *Reysen* as ideal scapegoats, since their depiction as morally corrupt and treacherous Christians offended neither Protestant nor Catholic readers.

***

When Faber returned to Shemakha after his first fruitless attempt of 1671 to travel back to Muscovy, the young gunner became Termundt's proxy in rescuing Struys from bondage. From Isfahan, Termundt somehow provided Faber with funds received from VOC officials.[41] Faber proceeded to lend Struys money to buy an Arabian horse, which Gurdziecki considered sufficient compensation to dismiss Struys from his bond.[42] Termundt continued to seek the release of Struys's still enslaved comrades.[43]

In the company of Faber and Brandt, Struys set out for Isfahan around 1 November 1671, according to *Reysen*.[44] By this time, the first news relayed by their Persian agents about the diverse fortunes of the *Oryol*'s crew reached the VOC's governors in Batavia. A letter of 19 December 1671 by the VOC's Council of the East Indies to its governing board in the Dutch Republic merits a lengthier quotation, as it verifies the movements of the sailors rendered in *Reysen*:

> Several Dutchmen, numbering about twenty people, of diverse talents, with as their commander one David Butler of Amsterdam with the rank of captain, having entered the service of the Muscovite Grand Prince to sail on a ship equipped there for that purpose, [descended] along the well-known river Volga to the town Astrakhan and, after arrival there, [were] to follow the orders of the Muscovite governor of that town. But the further aim [of this] is being kept secret by the Muscovites. The aforementioned captain believed to have been assigned to navigate across the Caspian Sea to the silklands of Persia and on behalf of the Grand Prince to begin trade there. But this voyage did not proceed any further than the town Astrakhan, because rebellious Cossacks, who had become very powerful in its surroundings, surprised and violently took the same town, murdering all who were found inside of it. [... T]hey took prisoner the aforementioned Dutchmen and sent them as slaves to Derbent, a place somewhat subject to the Persian king. Thither the aforementioned captain and also a constable, named Cornelis Cornelisz. of Amsterdam, escaped and after many troubles and suffering much poverty came to ours [the VOC officials] in Isfahan. The others, as far as we are advised, were transported to another place, named *Shamkhal* [*sic*] and have further been sold into slavery ... in this region. Ours had obtained from the king in Isfahan a *firman* for the release of those lamentable people, but with little hope that it would lead to something fruitful, because [earlier efforts had failed].[45]

The VOC's Persian agents continued to inform the Company's council in Batavia about the marooned crew's fate. Six weeks after the above letter, in January 1672, the Batavia Council recorded how Jan Termundt had purchased the freedom of the sailmaker '*Jan Struijs of Durgerdam*'.[46] These documents thus confirm the outline of the wanderings of the *Oryol* crew in Muscovy and Iran as rendered in *Reysen*.

Reysen recounts how in November 1671 Struys had embarked on the 2300-kilometre trek across Iran that would bring him to its port on the Persian Gulf, Bandar-e-'Abbas. Although now a free man, he was still exposed to great risk as an infidel foreigner who travelled with valuable goods throughout his five-month voyage. The caravan that Faber, Brandt, and Struys joined in November 1671 had been organized by Haji Biram Ali, who had invited the Dutchmen to accompany him to Isfahan.[47] By Struys's reckoning, the caravan destined for Isfahan numbered 2,000 people and a thousand camels and horses. It wound its way slowly through barren highlands, usually spending the night at the free *caravanserais* along route.[48] On the suggestion of a Russian travel companion, Struys disguised himself as an Iranian and pretended to be a mute.[49] Highwaymen especially saw foreigners as valuable prey in Iran.

According to *Reysen*, Struys's disguise meanwhile allowed him to visit holy places from which Christians were barred. Haji Biram Ali took Struys on a visit to the sacred grave of the Shi'ite saint and ancestor of the Safavids, Said Jibrail.[50] In Ardabil, like Olearius before (a fact that makes *Reysen*'s tale rather disputable), Struys saw the graves of Shah Safi (Shaykh Safi-al-Din, *c*.1252–1334), the real founding father of the Iranian ruling dynasty, and of his successors.[51] He also enjoyed the healing baths at Ardabil. The sojourn in Ardabil spanned about two weeks. Such large expeditions as Haji Biram Ali's customarily engaged in some trading during these lengthy stopovers, which allowed other travellers to join the caravan; it is moot, however, whether Struys was able to see all the sights as reported in *Reysen*.[52]

Once it resumed its slow progress, the caravan was shadowed by brigands who tried to pick off stragglers. Furthermore, many of the roads wound treacherously along narrow paths, high passes, and deep ravines. In mid-December, the caravan reached Sultanye, halfway between Isfahan and Shemakha.[53] Recalling the scene with proper moral indignation, Struys supposedly witnessed a week later in Qazvin how Iranians practised prostitution, but this passage of *Reysen* was wholly copied from Olearius's work.[54]

Around 1 February 1672, the caravan reached Isfahan: Struys, Faber, and Brandt immediately hurried to the *Hollandtsch Huys*, the headquarters of the VOC in the Persian capital.[55] They were warmly welcomed by the 'upper-merchant' Frederik Bent, the 'under-merchant'

Reynier Kasenbroot and their assistant Huybert Balde, who invited them to lodge in the *Huys*.[56] Although the VOC's share of the Persian trade to Holland (and their virtual monopoly on Asian silk import) had been threatened by the *Oryol* project, the Company's agents did not blame the sailors for this badly failed attempt at competition.[57] The Company's dogged attempts to seek the crew's liberation clearly demonstrate a sense of responsibilty for compatriots in a time of need.[58] Struys now found opportunity to tell his stories to those residing on the premises of the VOC office. Possibly here someone suggested for the first time that he have his tales written down and published.[59]

Reysen presents a lengthy albeit uneven description of the impressive Persian capital and its suburbs, which largely summarizes Olearius's even more detailed description.[60] The book also contains an engraving of Isfahan, which grossly distorts the dimensions of the capital and its stunning Maydan, the large central market square.[61] The picture enlarges out of all proportion the residences of the English East India Company and the VOC, simplistically embellishing the importance of the European presence in Iran.[62] According to *Reysen*, Isfahan was so large that it took an estimated sixteen hours or at least a full day to walk around its perimeter. Its actual circumference was about twenty-four imperial miles (and it may have had some 500,000 inhabitants in those days), which *Reysen* increased two-and-a-half times.[63] The engraving shows, however, a far smaller town which might be rounded on foot in no more than an hour.

In February 1672 Struys reentered the service of the VOC for a monthly wage of sixteen guilders, which would eventually allow him to reimburse it for all of the 285 guilders the Company paid Termundt to achieve his liberation.[64] Struys worked for the VOC only until October 1673, when the remainder of the sum he still owed was waived as compensation for his tribulations after his capture by the English at St. Helena.[65]

In his capacity of Company agent, Struys travelled from Isfahan to Batavia at the first decent opportunity, after about a month in the capital.[66] Struys joined a caravan headed for the port Bandar-e-'Abbas,[67] which carried a great amount of VOC merchandise (silk, velvet, carpets, and tapestry were the main staples exported by the VOC).[68] He departed Isfahan much too late to join the VOC-fleet that each year left the Iranian harbour for Batavia in the late winter.[69] Thus he was destined to stay for months in the scorching heat of the Persian Gulf port, awaiting the next departure of a Batavia-bound flotilla in the early summer.

In the late winter of 1672, Struys's caravan slowly crossed the snow-covered Iranian highlands.[70] To fortify the travellers, Bent had provided generous supplies of which he appointed Struys overseer. But this made

Struys vulnerable, and he was set upon by some of his travel companions. They robbed him and beat him up when he tried to regain his wares; soon after, brigands attacked the caravan and knocked Struys unconscious, but he somehow managed to stay with the troop on both occasions. Somewhere in March 1672 Struys visited the ruins of Persepolis (some 450 kilometres south of Isfahan). His impressions were crudely reflected in an engraving.[71] Wholly unexpected from an unsophisticated sailor, Classical Greek authorities are invoked in *Reysen* here to explain that a drunk Alexander the Great, egged on by the whore Thais, had precipitated the destruction of the former Persian Empire's capital's splendour.[72] These erudite references reveal once more a ghost-writer well-versed in the Classics.

In Shiraz, Struys lodged in a Carmelite monastery.[73] *Reysen* implies that the Carmelites were extremely hospitable to Struys because of his aid to the hermit on Ararat, but the Shiraz monastery was a usual resting place for Christian travellers.[74] Since Struys's previous caravan companions refused to venture further with him, he had to organize another team of camel drivers, guides, and pack animals, in which he succeeded with Carmelite help. Two of its resident monks, the Neapolitan Fellisello and the Pole Iadislav (Vladislav), would accompany him to Bandar-e-'Abbas.[75] The French surgeon 'Hakim Robin', whose real name Struys never learned, promised to dispatch a letter to Struys's wife via the route to the Levant.[76]

Reysen tells how robbers attacked Struys's expedition when resting at night in a *caravanserai*, some ten days after leaving Shiraz.[77] Five of Struys's group were cut to pieces in their sleep, but the rest managed to defend themselves and corner the intruders, who surrendered under a hail of bullets. The next morning the prisoners were suspended by ropes upside down from trees. Struys's team cut off their hands, noses, ears, and penises, placing the latter in the mouth of the brigands.[78]

Relating the marauders' barbaric punishment, *Reysen* adopts a philosophical tone:

> Their penalty albeit not [issued] by court or bench was of the same scale as their crime: because these scoundrels, knowing that they were wanted and never would be pardoned, did not refrain from insulting travellers with the mentioned cut-off body parts without any respect or compassion.[79]

Christian forgiveness or compassion was not extended to criminals in Early Modern Europe.[80] In *Oronooko*, the contemporary novel by Aphra

Behn about a noble slave in pre-Dutch Surinam, the same abject cruelty (including castration) is applied to the allegedly rebellious African slave Oronooko-Caesar.[81] Even if in seventeenth-century Holland incarceration sometimes replaced bodily mutilation in punishing criminals and castration was altogether uncommon, courts still sentenced convicts to physical penalties. Struys's unease as expressed in the quotation is rather about the absence of judicial process than about the severity of this brutal retribution, and may hint at an editorial hand: the statement underscores the general Dutch respect for due process, even if in Holland mobs occasionally roughed up culprits caught in the act before the authorities apprehended them.[82]

Editorial prudence also informs the tableau of a subsequent scene in *Reysen*. It depicts Struys and his Christian travel companions celebrating Brother Fellisello's birthday, enjoying a copious meal accompanied by Shiraz wine, and hoisting their glasses to the King of Spain (a Dutch ally in the 1670s) and the Prince of Orange.[83] The scene was the publishers' and ghostwriter's subtle bow to the Orangist mood in the Republic at the time of *Reysen*'s publication, for at the time when this scene takes place, early in 1672, William III had not yet assumed the post of Holland's stadtholder, and a toast to Johan de Witt might have been more likely.

Before Struys and his company reached 'Abbas's port, another skirmish with highwaymen saw people killed on both sides, but ended with Struys's group beating the assailants back.[84] *Reysen* may seem to exaggerate the frequency of robberies and attacks to highlight the dangers of the route, but Safavid Iran had entered an economic decline by 1670. Hardship became acute when the country was visited by adverse weather, epidemics and the like; by the end of the decade people died in considerable numbers of famine.[85] Desperation gradually replaced the stability and tranquillity of the days of Shah 'Abbas the Great (1571–1629), whose successors' lethargic decadence, not unlike that affecting their Ottoman neighbours at the time, did nothing to overcome the economic difficulties. People's destitution may have increased the frequency with which caravans were ambushed.

Somewhere in the early spring of 1672 Struys reached Bandar-e-'Abbas, where he was received by the VOC's Persian Director, François de Haze.[86] De Haze allowed him to stay in the Dutch lodge until ships departed for Batavia. Struys, here too, was asked to tell everyone the story about his travails. *Reysen*'s description of Bandar-e-'Abbas is possibly more authentic than most of the other urban geographical depictions in his book. The text refers directly to *Reysen*'s engraving of the port.[87] Its proportions are not as exaggerated, and, intriguingly, engraver Coenraet Decker portrays a Dutch yacht identified in the caption by the name (*Nuysenburgh*) it is given in the text.[88]

Epidemic disease, one of the scourges of the Early Modern Age now befell Struys: in July 1672, he was taken ill while awaiting the VOC ships to load and depart. *Reysen* estimates that the unbearable heat in summertime cost the lives of about half of the foreigners who visited the Iranian port.[89] In his enfeebled state, *Reysen*'s Struys found a way to leave the scorching city as quickly as possible, believing that only a change of air could save him, voicing the common belief in Europe that disease spread through the air.[90] Director de Haze allowed him to embark early on the *Nuysenburgh*, which was preparing to sail for Batavia. The pilot, who was a native of the village of Akersloot near Wormer, let Struys sleep in the ship's cabin and supplied him with all the remedies he could find to heal him.[91] At the critical moment of his illness, the ship's surgeon let Struys bleed. The fever began to gradually wane. The ship's carpenter,[92] suffering from a similar disease, was not so lucky and died.

The ship set sail at the end of July 1672, reaching a scorching Muscat on August 1.[93] From there, sailing along the Malabar coast of India and Ceylon and through the Sunda Straits, the *Nuysenburgh* reached Batavia within a month.[94] Struys met Governor-General Johan Maetsuycker (1606–78) and other members of the VOC's Council of the Indies, thanked them for their aid in ransoming him, and promised to pay the Company back the remainder of the sum owing, which *Reysen* claimed he did in 1673.[95] Struys was reappointed as sailmaker on the VOC ship the *Hollandtsche Tuyn*, for a monthly wage of eighteen guilders (the mandated wage of 'upper' sailmakers in the VOC at the time).[96] He soon petitioned the Councillors Cornelis Speelman and Pieter van Hoorn to allow him to leave for Europe on the flyboat *Europa*.[97]

With six other ships, the *Europa* departed Batavia on 4 February 1673, reaching the Dutch Cape Colony in mid-April.[98] Struys recalled how only at the Cape he heard of the calamitous French invasion of the Dutch Republic a year earlier.[99] The governor of the Cape Colony, IJsbrand Godskens, directed the *Europa*'s captain, Pieter Koker, to set sail for St Helena to offer assistance to a Dutch contingent which had conquered this Atlantic island on the English in January 1673.[100] The other ships would follow a week later. Nearing St Helena in May 1673, the *Europa* unexpectedly encountered seven English warships, supported by some fireships and merchantmen. Utterly outnumbered, the Dutch surrendered to their foe, who, it transpired, had recaptured the island just before the engagement.[101] Not for the first time, *Reysen* has Struys robbed of virtually all his earthly possessions, which included his prized chest. He was left merely with his testimonial and relics from his stay at Ararat. At this juncture in the narrative, he remarks how greed trumped compassion as much among Christians as among non-Christians.[102]

But at least English captivity meant a return to Europe. The English flotilla captured two more Dutch ships – the *Wapen van Veere* and the *Alphen* – when they arrived near St Helena from the Cape in subsequent days.[103] The English carried the Dutch captives to Ireland, where they were released in the harbour of Baltimore in August 1673.[104] By this point in time, much of English public opinion had turned against the war with the Republic. Perhaps the lenient treatment of the Dutch prisoners-of-war was connected to this sentiment, but no clear common code of conduct among the Western Europeans toward captured (Christian) enemies existed in any event.[105] From Cork, according to *Reysen*, Struys sailed to Bristol (*Brest* or *Bresto*), and walked from there to London, pausing to marvel at the fine manner in which the city had been rebuilt since the 1666's fire. From Harwich he crossed the North Sea,[106] finally crossing the threshold of his home, and 'being reunited with his wife and children', on 7 October 1673, the moment which concludes *Reysen*.

After five years full of adventures, Struys returned to his homeland hale, and we have no reason to disbelieve *Reysen*'s somewhat happy ending. Although he arrived empty-handed in Amsterdam, he had gained a confidence and a resilience that was to become apparent from his actions in the following years, and would yield him the rewards he had so vainly pursued before 1673. He capitalized on several opportunities that enabled him to leave behind the uncertain existence on the edge of indigence usual for the seventeenth-century commoner. The first of these was his successful sale of his stories to van Meurs and van Someren, who enlisted a ghostwriter into transforming them into *Reysen*, as we will see in the next chapters.

11
Reysen's Readers

Somewhere between October 1673 and July 1675 Jan Struys's tales came to the attention of the Amsterdam publishers Jacob van Meurs and Johan van Someren, who decided to publish a version of them as a book.[1] Before 1675, both partners had published descriptive cosmographies, but neither had ventured yet in the direction of adventurous travel accounts. This chapter will investigate the kind of reading audience they tried to reach with *Reysen* and the other two books of their trilogy, van der Heiden's *Vervarelyke schipbreuk* (*Calamitous Shipwreck*) and Schouten's *Oost Indische Voyagie* (*East Indian Voyage*). Chapters 12 and 13 address the manner in which *Reysen* was actually composed under these publishers' auspices in 1675 and 1676.

The production of *Reysen* shows how age of the scholar-printer was long past in the 1670s Dutch Republic. Seventeenth-century Dutch publishers primarily considered books to be commodities that needed to sell on the market on which their economic survival depended. During a pronounced downturn in the Dutch economy and with their country at war with mighty France, van Meurs and van Someren sought to broaden their audience beyond the erudite and wealthy readers who usually bought their products.[2] But they also aimed to appeal to their usual readers, who in those uncertain times may have had less money to spare or hesitated to buy such non-essential consumer goods as books. The plethora of *Reysen*'s reprints during the first century after its publication attests to van Meurs's and van Someren's business savvy. Evidently, *Reysen* successfully tapped into the demand for adventure stories set in exotic locales that went beyond the borders of the Republic. Chapter 13 assesses why *Reysen* already sustained severe criticism after 1690, signalling a transitional stage in European intellectual history, without it losing appeal to readers for another century.[3] After 1800, however, *Reysen*'s readership dwindled to that a small community of historians of Muscovy and of Asia.

It should be noted that *Reysen* was not intended to appeal to those searching for a sort of roadmap or early Baedeker guide.[4] Neither was it intended to supply prospective traders with exhaustive data on conditions of commerce in various countries, nor to provide diplomats with details about politics. Such information was handled with circumspection. The VOC's secrecy about its information regarding its commercial operations, the copious secret diplomatic correspondence of the day, the mere handful of copies printed of both editions of Witsen's *Noord en Oost Tartarye*, or the survival of Witsen's and Kilburger's notes on Muscovy solely in manuscript suggests that a select audience of 'insiders' in the Western governments and leading Western businessmen had access to unpublished reports far more detailed and accurate on various parts of the world than *Reysen*.[5]

Instead, texts such as *Reysen* were intended to entertain and intrigue a general reading public that had no serious intent to visit these places.[6] Like Henry Fielding's Parson Adams, many among its audience found comfort in the idea that reading books 'was the only way of travelling by which any Knowledge is to be acquired', as a reading of the Biblical story of the prodigal son happily confirmed to them.[7] Apart from those fairly rare passages when the sailmaker's voice surfaces and the equally rare instances when his worldview coincides with that of his country's elite, of course, 'Struys's' *book* voices the mentality of the hegemonic group to which most of its readers belonged (or aspired to belong); channelled by *Reysen*'s ghostwriter, their presuppositions and prejudices inform its pages.[8] As the preface subtly admitted, the contents, presentation, and tone of *Reysen*'s text mostly reflect *'een beschaafder Penne, als de mijne'*, that is the ghostwriter's 'more civilized pen', which was well attuned to *Reysen*'s readership.[9]

Reysen, along with other contemporary texts of which the entertainment value surpassed the scientific accuracy, was expected to conform to a sort of template rather than be wholly 'authentic' or 'original', in our modern understanding of these terms. In order to sell, *Reysen* had to appeal to these readers' expectations and presuppositions. Of course, some novelty was welcomed, for otherwise the reader might prefer a previously published text on the same topic. In this vein, the Iranian section especially appealed to this curiosity, although it had also been a while since the last book about Muscovy had been published in the Republic, allowing for *Reysen*'s recycling of older accounts on the tsar's empire.

Given the extensive contacts of Dutch merchants, soldiers, sailors, and artisans with Russians, it is remarkable that few works on Muscovy written by Dutchmen had been published before 1676, and only a few books on the tsar's realm had been translated into Dutch. Works by Anthonis Goeteeris and J.P. Danckaert as well as translations of those by Adam Olearius and Sigismund Herberstein were at least decades old by 1676.[10] Hessel Gerritsz's *Beschryvinghe vander Samoyeden landt* (1612) had depicted mere parts of Muscovy.[11] The works by Isaac Massa (of the 1600s) and Nicolaas Witsen (of 1664–65) only existed in manuscript; otherwise merely some brief pamphlets had been published about extraordinary events in Muscovy, such as those on the first False Dmitrii (in 1606) and on Razin (in 1671).[12]

After 1650, the increasing claim for scientific rigour in the Republic challenged the credibility of several derivative compilation texts on Muscovy. A treatise republished in 1646 claiming that *most* Russians were sodomites began to lose credence in the increasingly sceptical Dutch Republic.[13] The most recent publication before *Reysen* had been the sensationalist *Het ellendigh leven der Turken, Moscoviters en Chinesen, aende Christenheyt vertoont*, a second-hand compilation of 1664 summing up the alleged worst aspects of the tyrannies of Ottoman Turkey, Muscovy, and China.[14] A political treatise by Georg Hornius (1620–70) published in Leiden in 1667 analysed the Muscovite state and its society, but as a Latin text this work did not reach a wide leadership.[15] It may have been the source for the section on Muscovy presented by Pieter (Petrus) Val(c)kenier in his compilation *'t Verwerd Europa ...* , published in 1675 (when *Reysen* was already in preparation), but neither author had visited the tsar's realm.[16] Many Dutch readers had the ability to peruse foreign accounts because they read French, Latin, German, English, or Italian, but no new detailed account on Muscovy originally written in Dutch had been published after the 1610s. The sober treatment of Muscovy in Dutch that *Reysen* supplied with its appearance of first-hand observations seemed thus overdue.

Despite fairly important trade contacts beween Iran and the Dutch East India Company, an even greater textual void existed in the case of Safavid Persia. Although several foreign accounts (such as Olearius's work and some compilations) were available in translation, no first-hand Dutch traveller's account of Iran had ever been published before *Reysen*'s publication.[17] In 1672, van Meurs published one compilation which included a description of Persia, but its author, Dr. Olfert Dapper, had never travelled to any part of the Asian continent.[18] Persia was altogether much less known in North-Western Europe than Russia and thus even more fascinating to the European reader.[19]

In other words, before *Reysen* was published, information about those parts of the world to which it dedicated most of its pages had thus been limited, second-hand, and obsolete.[20] *Reysen*'s publishers thus aimed to satisfy readers' demand for detailed, updated and new, descriptions of exotic and barely known regions.[21]

Apart from their appeal to curiosity, the partnership of van Meurs and van Someren strategically capitalized on the surging patriotic mood of the Dutch, who were defying the odds in their war against Louis XIV (1672–78). The Dutch empire's vast overseas reach in Asia, more pronounced eastward than westward by the 1670s, was charted in the three books by Schouten, van der Heiden, and Struys and bolstered Dutch morale after the narrow escape from French subjugation. The rather brisk sales of *Reysen* further highlight the appeal of the crucially important maritime trope of the Dutch seventeenth-century self-image. As a rugged and intrepid sailor, Struys performed heroics similar to those of Linschoten, Barentsz, or Bontekoe, other Dutch seafarers whose adventures in print had previously become wildly popular.[22]

Apart from its chorographies of faraway locations, *Reysen* also appealed to readers' desire for a window on the life of the lower classes.[23] In the 1670s, the Dutch elite still partook in some expressions of popular culture, which explains readers' interest in tales of emblematic mariners, even of such humble station as the sailmaker Jan Struys.[24] Publication of seafarers' writings were popular in the Republic, but their authors were usually ship's officers rather than sailors or sailmakers such as Jan Struys.[25] In the 1670s, the Dutch John Joneses (*Jan Janszoons*), burdened by work and hindered by halting literacy or illiteracy, did not often tell of what they saw and heard in manuscript or print.[26] No more than two manuscripts written by seventeenth-century Dutch mariners have been found by Rudolf Dekker.[27] *Reysen* thus seems unique as the only *published* travel account that appeared under the name of a Dutchman of such modest background. While depicting the humiliation of a man of means, status, and power would have upset the contemporary sense of hierarchy, *Reysen*'s engraved frontispiece could portray the book's protagonist as violated after his capture in Dagestan because he belonged to a subaltern social group.[28] In this sense, *Reysen*'s frontispiece and the scandalous treatment of the book's protagonist echo the contemporary patrician interest in the stylized portrayals of the boisterous lower classes seen in Dutch genre painting. And *Reysen* also showed exemplary endurance and perseverance on the part of this scion of the lower classes, without a hint of rebellion against his subordinate station.

Religious sentiment was also duly taken into account, for *Reysen* underlines the righteousness of the usually tolerant Dutch Reformed worldview of the Republic's hegemonic elite.[29] In its depiction of a sort of series of 'pilgrimages' whose 'pilgrim' always returned to the Holy Land of the Republic, *Reysen* appealed to the same Calvinist sensibilities as did John Bunyan's contemporary *Pilgrim's Progress* (1678) across the North Sea.[30] Struys the protagonist, confronted with temptations that could have led him astray, remains unwavering in his faith. But religious tolerance and quasi-scientific curiosity are equally present in *Reysen*.[31] The book therefore reflected the religious ambiguity or broad-mindedness of the Dutch patricians, while its awkward compromise between religious and empiricist strains reveal an intellectually uncertain and probing period. *Reysen* appealed to 'a long-lived desire to comprehend the divine scheme for creation and to classify and order all of God's creatures', which had led to Olearius's work's enormous popularity across Western Europe.[32] This desire to chart God's plan is reflected in *Reysen*'s patron Nicolaas Witsen's quasi-theosophical wish 'to show that all known as well as newly-discovered people and their religious systems shared some common basis in Judaism and early Christianity, and that some knowledge of the One, True God could be discovered in all important cultural systems, thus exhibiting the oneness of Creation, the essential unity and comprehensibility of the world.'[33]

Jacob van Meurs, as the most renowned publisher of geographic and travel accounts in the Republic during the 1660s and 1670s, contributed profitably to the exploration of this religiously inspired quest.[34] Van Meurs's prior publication of cosmographies on Asia, Africa, and America targeted a reading audience interested in acquiring a comprehensive, 'encyclopedic' knowledge about the world, deriving from an urge to discover Witsen's longed-for divine pattern and a more pedestrian curiosity.[35] The selection of the various ethnographic-geographic exposés contained in *Reysen* and its sister texts aimed to further this endeavour toward an exhaustive catalogue of the world's cultural and natural variety. Some of the lacunae remained in the cosmographies van Meurs had issued between 1668 and 1672, which Struys's, van der Heiden's, and Schouten's travels filled. In the case of the trilogy, the publishers concomitantly supplied a sort of 'thick description' of the specific regions and peoples in the Old World (Europe, Asia, and Africa) with whom the Dutch traded.[36]

The trilogy's form as adventurous travel accounts and the books' printing in quarto with fewer engravings (rather than the folio editions predating the 'crash of 1672') show how van Meurs and van Someren tried to entice readers who had less money to spare in the taxing 1670s.[37] Previously, van Meurs had his geographical books printed in folio format and included dozens of engravings, but such works were too expensive to appeal beyond the most affluent audiences.[38] Van Meurs may also have possessed a surplus of engravings, for which he had not recovered the costs (see the remarks on Kip's rendering of Tenedos in the next chapter), and which were now added to *Reysen*.[39] One could therefore infer that van Meurs and van Someren decided to tweak van Meurs's formula for a serialized description of the world in book form to reach better sales in the depressed economic circumstances of the war-torn 1670s.[40] Inspired by the initial success of van der Heiden's dramatic tale of a shipwrecked Dutch crew, the trilogy was also decidely less arcane than van Meurs's earlier cosmographies.

In a highly sophisticated manner, *Reysen* thus endeavoured to appeal to a complex set of readers' interests, while its publishers tried to offer it for a price within the means of its potential readership. Given the book's plentiful reprints in a variety of languages, van Meurs's and van Someren's joint venture had altogether more success than van Meurs's earlier lavish folio productions of Dapper's and Montanus's books (about which more in the next chapter), which underwent few reprints.

12
Reysen's Creation and its Creators

Roland Barthes once wrote how 'a writer can only imitate a gesture that is always anterior, never original [; his] only power is to mix writings, to counter the ones with the others, in such a way as never to rest on any one of them.'[1] Barthes' imitating writer is aptly reflected in the invention of *Reysen*. *Reysen* was far from an authentic piece, and the involvement of its alleged author Jan Struys in its *writing* was rather ephemeral to boot. The book's creation instead resulted from the combined efforts (beside those of the printers and their personnel[2]) of two publishers, two engravers, a mapmaker, the two authors of the letters included in the book, two patrons, a ghostwriter, and the seafaring sailmaker. This capitalist venture is described in this chapter, while its success in delivering a profitable product is topic of the next.

Before the age of mass advertising, the selling of tomes such as *Reysen* began at the bookshop (while smaller books were sold by itinerant pedlars, for example). At the store the potential buyer could inspect the sheets awaiting binding. Stacked in a pile, the frontispiece and title page lay on top. *Reysen*'s top leaf demonstrates the sophisticated marketing of the Republic's publishing industry and betrays particularly van Meurs's experience as an engraver.[3] The engraving showed people in turbans and savages (the men in fur were a shorthand for this) firing arrows at a bearded St Sebastian-like figure tied to a tree, who in the background was seen to be dragged over the ground by a horse. Located at the top of the picture, the book's title (*Voyagien door Moscovien, Tartaryen, Oostindien en andre deelen der Werelt*[4]) was fashioned to look as if written on flayed human skin, which concomitantly framed all sides of the picture. The flaying of a woman was hinted at by a depiction of turbaned people pointing at a nude body prone on a bench to the left of the 'martyr'.[5] To the exclusion of the rest of *Reysen*'s contents, the frontispiece depicts events that unfold in the Caucasus and Iran, the most exotic and least known setting for the readers, evidently a place of extreme (and rather intriguing) torments. Promising the prospective purchaser exciting tales of horror in faraway places, this baroque tableau was hard to ignore.[6] Clients in the store would want to find out how the snapshots of events pictured in the engraving unfolded in the text. Very aptly, Elio Biancaforte compares it to a 'good movie poster'.[7]

But the engraving also referred to the era's tropes more subtly. It followed a tradition of depicting evil 'Eastern' cruelty in Western-European print at the time. The picture resembles those published in contemporary books such as Pierre Dan's work on the Barbary coast.[8] This portrayal of Eastern (often 'Turkish') 'evil' may have evolved from the demonization of the Other in European Reformation's iconography, as can be seen when comparing the depiction of Eastern cruelty to that of Catholic or Protestant cruelty.[9] *Reysen*'s frontispiece also followed earlier models of seventeenth-century works on Persia, which in themselves were developed from earlier popular templates by artists such as Peter-Paul Rubens (1577–1640).[10] This intertwining of two stereotypical print images was thus a calculated ploy appealing to Europeans' deeply ingrained (and predominantly negative) attitudes toward Others. In addition, we saw in Chapter 10 how Persian flaying had been a long-standing popular topic in Netherlandish iconography. The frontispiece's framing of an engraved picture by hanging skin probably found its origins in a work by Lucas Cranach the Elder (1472–1553).[11] The picture thus drew from a palette of Netherlandish aesthetic traditions.

The book buyer's attention was then to turn to a title page announcing 'a precise and elaborate description of the named countries [Italy, Greece, Livonia and so on] and what pertains to their nature, very wondrous and truthful incidents that the Author sustained through Shipwrecks, plunder, slavery under the Tatars and Persians, intense famine, torture, and other difficulties [*ongemakken*],' unfolding over the course of twenty-six years.[12] Additionally, it promised two letters by other writers about the sack of Muscovite Astrakhan in 1670, and copper engravings based on the author's drawings, of which the consumer had already been given a foretaste by way of the frontispiece.

The overflowing style of those first pages matches the form of the rest of the book. *Reysen* abounds with factual details in various lettertypes and engravings, showing the lack of moderation of late baroque bookmaking. In this respect, the aesthetic sensibilities of creators and purchasers converged. Tidily ordered narratives were not as common in the seventeenth century as they are today. This was an age when genres were still crystallizing or even emerging, and especially the smooth and seamless narrative of the Victorian novel was still far in the future.[13] The writing's uneven quality in *Reysen* may have even confirmed the author's identity as an unsophisticated sailor to contemporary readers.

As was customary, the engravings were printed on separate sheets that were to be inserted between the text pages.[14] Van Meurs and van Someren's desire to turn a profit is also expressed by their choice of engravings. The selection of the particular scenes for the pictures reinforces the emphasis on Muscovy and Iran of the frontispiece and the text, and alternate between sensationalism and exotica. If we include the two letters, three-quarters of *Reysen* deals with events in those two empires, and about the same proportion of images is dedicated to them. Altogether, the book's twenty images depicted parts of Thailand (two images), Astrakhan and its sacking (three), Razin's sacrifice to the Volga, 'Tatars', the flaying scene (in a diptych), three Greek islands, Muscat, the Caspian Sea's map, and another seven images of Iran. *Reysen* lacked pictures of the Indonesian Archipelago, Africa, Italy, Livonia, Formosa, Japan, or of the Tsar's vast territory besides the Volga delta. Therefore, well-nigh half were engravings of Iran, and five of Astrakhan and its surroundings. This selection underlines how the publishers anticipated readers' interest to be in the calamities that befell Struys as part of the *Oryol* crew in those little known parts of the world. Readers' curiosity about these regions was further sated by the inclusion of the map of the unfamiliar Caspian Sea. Additional testimony about the Cossack terror at Astrakhan in the form of two letters spiced up the text further.[15] The longest and most interesting of these was David Butler's, which was accompanied by a lively engraving of the pillage of Astrakhan in 1670.

The translations of *Reysen* in French and German (see also Chapter 13) may have partially derived from the publishers' wish to use the book's costly engravings again (typesetting was a comparatively cheap operation); after a French version was printed in Amsterdam, van Meurs's widow probably sold the plates to the English publisher(s), as she would subsequently sell the engraved copperplates for the illustrations of one of Olfert Dapper's works to a German publisher in the 1680s.[16] Meanwhile, Johannes Kip's engraving of Tenedos (Tenos) in *Reysen* resurfaced in Olfert Dapper's last work, a chorography of the Mediterranean Islands.[17]

The client in the bookstore who came across *Reysen* may have been equally intrigued by its author's humble background, for few of such station wrote such bulky books. This was another instance of slick marketing by the publishers. At the age of forty, the historical Jan Struys could not sign his name on his wedding licence.[18] It is beyond belief that seven or eight years later he wrote a weighty volume of almost 400 pages about his adventures. Even if Struys had developed some rudimentary literacy to record his impressions, whatever notes he might have made can only have been fragmentary and scarce at best, given the scope of his tribulations as related in the book (and corroborated by the VOC archives).

Other Netherlandish publishers are known to have engaged scribes to prepare journals or logbooks kept by seafarers and other travellers for publication. This method was used to compose the pioneering Dutch work on travelling to remote regions, Jan Huygen van Linschoten's *Beschryvinghe* (part of his *Itinerario*).[19] In *Reysen*'s case, it is likely that Struys transmitted his narrative entirely in oral form to his scribe. This oral origin occasionally seems to surface in *Reysen*, as when Prince S.I. L'vov is named '*Elbof*'.[20] Dutch ears at Astrakhan heard the Russian name '*L'vov*' as '*Elbof*', incapable of imagining the Russian sequence of a soft letter L followed by a V. The ghostwriter lacked means and time to check this spelling and rendered the name phonetically, following Struys's pronunciation of it.

The creation of *Reysen* went therefore beyond having a 'more polished pen than mine' ordering, elaborating, and editing Struys's *written notes* (*Geschriften*), as its foreword disingenuously claimed.[21] As Struys nicely aligned with the archetypical Dutch seafaring hero, the publishers maintained the pretence that he wrote the text.[22] An anonymous ghostwriter wrote *Reysen*, to whom Struys orally conveyed the details of his adventures.[23] Almost beyond a doubt, van Meurs and van Someren's choice as ghostwriter was Dr. Olfert Dapper, whom they trusted as a skilled hand.[24] Dapper was a native of Amsterdam who had trained as a medical doctor in Utrecht and read a number of languages; in 1663 he had made his name with a description of the city of Amsterdam, one of van Meurs's first successful publications.[25] After this, Dapper wrote a series of cosmographical works which were published by van Meurs, with whom he thus enjoyed a close business relationship.[26]

A prolific and well-published author, Dapper had been forced to recede from the limelight as a result of the political developments of 1672. In the early months of that year, Dapper had dedicated a new work, a description of Asia, to Cornelis de Witt (1623–72), the older brother and confidant of Johan de Witt.[27] The latter was *Raadspensionaris*, or Grand Pensionary, of the Estates of Holland and West-Friesland from 1653 to 1672. In the absence of a stadtholder of the house of Orange, de Witt was the senior political leader of the province, and, by extension, of the Republic during that period. In 1672, the de Witts became the scapegoats who were held responsible for the Republic's embattled position (in its 'year of disaster' [*Rampjaar*]), when it faced a combined assault by the armed forces of the kings of France and England and the bishops of Cologne and Münster.[28] The lynching of the brothers in August 1672 was followed by a universal purge of de Witt followers from city councils across Holland.[29]

The dedication of Dapper's *Asia* underlined his close association with the disgraced de Witt faction. Its fall landed Dapper in a political limbo that jeopardized his writing career. Significantly, one passage in *Reysen*, strategically inserted at the end of the book, shows sympathy for the brothers de Witt, calling their death 'horrific' ('*soo schrikkelyk waren omgekomen*').[30] Its insertion shows how the dark skies over Dapper's head were beginning to lighten when he was finalizing *Reysen*'s text in 1676. In that year, his name publicly reappeared for the first time in four years, as the author of the (unchanged) second edition of his book on Africa.[31] The next year van Meurs finally published a new original work under Dapper's name. Five years between new publications was a long hiatus for a prolific author who, prior to the de Witts' murder, had turned out a lengthy tome (his books of the 1660s reached usually more than 500 folio pages!) every other year or so, and who mainly lived by his pen.[32]

Dapper's political rehabilitation in the mid-1670s was facilitated by the good offices of Nicolaas Witsen, one of *Reysen*'s patrons. Witsen's father, Cornelis (1605–69), an Amsterdam mayor in the 1660s, had patronized the young Dapper during the heyday of Johan de Witt's political leadership.[33] Luckily for Dapper, Nicolaas Witsen sided with the Orangist faction in 1672, and was actually elevated to the Amsterdam City Council as part of the anti-de Witt purge. Witsen subsequently developed a good relationship with stadtholder William III (1650–1702), the Prince of Orange, the central political figure in the Republic after the demise of de Witt.[34] Witsen's protection allowed the author's name to reappear in public when the greatest anger against the de Witts had subsided. *Reysen*'s reference to the 'undeserved support, and ... friendship' by Nicolaas Witsen in the dedication may express Dapper's rather than Struys's gratitude.[35]

Dapper hailed from a very modest background (his father was a ropemaker) and remained familiar with the seafarers' world, using testimony of sailors for his other books.[36] He was thus capable of simulating the voice of a man of the people, if filtered for a much more refined audience than the usual humble listeners to sailors' lore.[37] Dapper proceeded to weave into Struys's oral narrative chorographic sections about various exotic regions designated for such treatment by van Meurs and van Someren. Some descriptions in the first two sections of *Reysen*, such as those of Madgascar or Thailand, were selected not because Struys had much to say about them, but because they were thought to appeal especially to readers' curiosity. Because *Reysen* was published as part of a trilogy, certain territories already covered in the other two books were omitted.[38] This explains why accounts or engravings about a number of regions in South Asia, South Africa, and South-East Asia were largely left out of *Reysen* even if Struys evidently visited some of those parts on his voyages.[39]

To prepare the text of *Reysen*, Dapper relied heavily on 'academic' cosmographies and chorographies available in Dutch and other languages. As we saw in previous chapters, Olearius's work was particularly mined for *Reysen*'s description of Muscovy and Iran. Dapper did not need to seek far, for his own *Asia* included a number of passages on the eastern Caucasus and Iran based on Olearius.[40] Struys's adventures formed a lighter counterpoint to the somewhat ponderous text by the Holsteiner, whose more detailed discussion of Muscovite, Caucasian, and Iranian customs is also abridged in *Reysen*.[41] Dapper then combined information contained in Olearius's *Beschreibung* with additional texts in composing *Reysen*. Dapper's creative hand becomes manifest on a number of occasions, as when *Reysen* further embellishes the alleged barbarity of the Cheremiss than Olearius.[42]

While Olearius's description provided much material for the description of the lands through which Struys travelled before he reached Isfahan, Dapper turned to his own work (and the notes he had collected for it), Pietro della Valle (1586–1652)'s volumes, and other books for the description of the route from Isfahan to *Gombroon*.[43] For *Reysen*'s accounts of Madagascar, Siam, and Formosa, Dapper used a variety of sources in the same manner.[44] *Reysen*'s section on Thailand indicates how Dapper even seems to have used hand-written sources.[45] Throughout his career as a cosmographer, Dapper worked in private libraries and used unpublished manuscripts for books that appeared under his name, collecting a vast store of information.[46] Information from both manuscript and published sources thus was combined with Struys's tales in writing *Reysen*. Additionally, as the next chapter will show, Dapper also imposed the conventions of a number of genres upon the text.

In 1675, the modern understanding of copyright was still in its infancy, even if, ironically, van Meurs and van Someren did protect their rights to *Reysen*'s reprinting in that year.[47] The unattributed use of other people's work in the seventeenth century was seen rather differently than plagiarism is today (even if Barthes's remarks quoted at the beginning of the chapter show how authenticity for modern authors, too, may be an impossibility). Dapper's unacknowledged copying from other writers' texts was a customary practice in the age. Other Western European writers also routinely duplicated the image of the Russians (and other exotic peoples) as supplied by leading authorities such as Herberstein and Olearius.[48] Some of this habit was rooted in the anonymous tradition of monastic chroniclers, who neither claimed authorship nor accounted for their sources in their copying, while some of it was based on the idea that good writers should imitate authoritative Classical models.[49]

The Dutch burghers who could afford the money and time to read a book such as *Reysen* were educated at *gymnasia*, *athenea* and universities, where they were taught the canon of the classical Greek and Roman texts.[50] This widespread knowledge of the *Iliad* and *Odyssey* and of Herodotus's *Histories* among the elite undoubtedly made van Meurs and van Someren even keener to present Struys's stories, for they carried echoes of Odysseus's tortured return home to Ithaca. Olfert Dapper's familiarity with the classical tradition was profound: his second work had been the first integral Dutch translation of Herodotus's *Histories*, which included the Greek historian's biography of Homer.[51] In Dapper's own books and in *Reysen*, the ethnographic-geographical exposés closely follow the classical model pioneered by Herodotus. Other telling traces of Dapper's classical erudition can be detected in *Reysen*. The book has Struys visit Troy and various places associated with Alexander the Great, which had little meaning for an illiterate sailmaker, but all the more significance for his ghostwriter and *Reysen*'s audience.[52] Levelling a charge of 'bestiality' against the Cheremiss reveals an editor familiar with the geographers of Antiquity, for whom bestiality was part of the trope of the barbarian.[53]

But how could a skilled and learned writer such as Dapper have produced a book in such a patchwork form? First, *Reysen* was his first (and only) attempt at composing a work of an eclectic combination of genres.[54] But besides the hybridity of the genres and baroque overabundance, the publishers' rush to maximize their profit caused *Reysen*'s untidy appearance. Whereas the frontispiece had 1677 as year of publication, the printed page shows 1676.[55] Already the front-matter of the book demonstrates therefore the lack of polish encountered throughout the volume. The inaccurate dating of events in the text's margins is similarly a blatant flaw, although this was also the consequence of Struys's inability to date events.[56] The use of Gothic printletter for Struys's voyages in *Reysen* is oddly combined with Roman script in its marginalia and for the preliminary two letters.[57] Several of the book's engravings are hardly linked with the text at all, and several look strongly alike. Their artistic quality is uneven, adding to the book's quilted appearance. The dedication's text was left unchanged from the version prepared before Struys's return from Muscovy: Ambassador van Klenk's representation of Struys's grievance in front of the tsar was therefore solicited as a future favour, although by the time the book appeared van Klenk and Struys had returned to Holland from their Moscow mission months earlier.[58] It all suggests the publishers' great hurry in delivering the final version to the printers.[59]

In September 1675, the Estates of the Province of Holland published an *octrooi* (licence) to protect van Meurs's and van Someren's copyright of the trilogy of Struys's, Schouten's, and van der Heiden's books.[60] *Reysen* must therefore have been in preparation by July 1675, when Struys departed on his second tour to Russia, from which he only came home in October 1676.[61] The text was finalized after Struys's return, since it contains entries referring to events occurring in early 1676.[62] During Struys's absence from Holland, the ghostwriter assembled the book's chorographic descriptions, but the manuscript apparently never advanced far enough to consider its publication before Struys's reappeared in Amsterdam.

Upon Struys's arrival in the autumn of 1676, *Reysen* was hastily finished. It became evident that Struys's travel companion Balthasar Coyett was preparing the publication with a rival publisher of an account of Klenk's embassy's progress that would cover most of the topics on Muscovy that were to be included in *Reysen*.[63] The book thus had to be issued expeditiously, which explains the numerous sloppy elements in *Reysen*'s presentation. The publishing team's speed paid off. Struys's breezy adventures and *Reysen*'s various chorographies appealed much more to Western European readers than Coyett's *Historisch Verhael*, which never made it beyond one Dutch edition. Quickly authorized and pirated translations of *Reysen* were released in German (1678) and French (1681).[64] We may surmise a sale of at least 3,000 authorized copies of the book by 1681. Two years later, an English edition (issued with permission from the Amsterdam publishers) appeared in London.[65] This represented a veritable success and vindicates van Meurs and van Someren's efforts to issue *Reysen* as speedily as possible.

Even if Olfert Dapper never received public recognition for his labour on *Reysen*, the writing of the book gave him an opportunity to excel in his chosen profession of master compiler at a time when he had to avoid drawing attention to himself. The project probably provided him with a stipend that sustained him just before his political rehabilitation. Jan Struys, too, received a reward for allowing his stories to be published in the book, although the payment was likely in the form of a one-time lump sum.[66] Struys and Dapper's fame proved enduring. *Reysen* and Dapper's openly authored books are still used as a source for the historical study of some regions.[67] And the unlikely figure of Jan Struys, the intrepid Dutch traveller of humble origin, became a celebrated bestselling author.[68]

13
Genre and the Test of Time

Besides allowing the Western European reader to fill in blank spots on the map of the Old World, *Reysen* transcended contemporary travel accounts in Dutch and other languages both because it was compiled from an unusual number of sources. Its structure around a mere sailor's memories reveals a further debt to a specifically Dutch publishing tradition, although the alleged author's humble post was a unique twist. In an eclectic manner, the work also was influenced by the conventions of a great variety of genres. Some of these had found defined form in the course of a long tradition, whereas others were still developing their shape. *Reysen*'s hybrid genre resulted from both intentional and unwitting efforts to reflect evolving Western European aesthetics. It combined elements of autobiography, travel journal, chorography, quasi-Biblical parable, Dutch epic, picaresque story, sensationalist pamphleteering, early journalism, and odyssey, while the letters with which the book opens herald the arrival of a new fiction genre.[1] They reflect an already existing subgenre of travel accounts in letter form, which would develop into the epistolary novel after 1700.[2]

Conventions common to these various genres were superimposed upon the narrative skeleton of Struys's stories, connecting each of the three main sections of the book (the two additional letters included in *Reysen* are original, and fairly straightforward, eyewitness accounts that seem to have been left largely untouched by the editor). Although this potpourri of genres probably derived in part from the calculating efforts by publishers to have Dapper produce a book in which there was something for almost every reader and thus increase their sales, *Reysen*'s richness, untidiness, as well as its overflowing and liminal quality were to the taste of the baroque. *Reysen*'s iconography especially expresses the aesthetics of the late baroque.[3] Following its predilections, *Reysen*'s engravings appeal to readers' emotions, tend toward monumentalism and, theatricality, and use religious symbols (for instance, Saint Sebastian's legend).

Beyond the influence of baroque formulas, *Reysen*'s probing transcending of the conventions of the travel genre reflects the unravelling of traditional rules and the burgeoning search for new manners of creative expression that gripped Europe by 1670. Paul Hazard suggested in the 1930s that the debate about traditional paradigms indicated a 'crisis of the European mind'; more recently, Jonathan Israel has identified this growing doubt about old religious and cultural certainties in the Dutch Republic as the beginning of the Enlightenment.[4] Scientists and philosophers based in the Netherlands in the 1670s and 1680s abandoned received wisdom for new truths, but, with the exception of Spinoza, few proposed an all-encompassing novel philosophy, theology, or scientific theory.[5] The last third of the seventeenth century was more an age of radical questioning than of system-building, more of debunking scientific myths than of suggesting comprehensive explanations. Schmidt suggests that in the wake of the definitive loss of New Netherland, topics such as 'American innocence' had become exhausted by the 1670s, but nothing immediately replaced it.[6] Similarly, the old cliché that all Muscovites were homosexual, still acceptable in the first half of the seventeenth century to a readership inclined to accept myths, is no longer found in *Reysen*, a sign of an emerging Weberian disenchantment.[7] Even by its peculiar form *Reysen* shows the hallmarks of an era that was in the process of discarding old intellectual and cultural axiomas and in search of new certainties. Perhaps *Reysen*'s alternate use of Roman and Gothic typefaces mirrors this play of old and new during this transitional phase.

Reysen's hybrid genre also reflects a period when the separate genres of autobiography, fictional novels populated with rounded characters, and scientific natural and geographical descriptions were only beginning to crystallize. Only in the next century, travel accounts took on an unambiguously 'scientific' guise, as in James Cook's and Bougainville's works.[8] More than anything, *Reysen* belonged to this latter category, '[a] favourite literary genre [...] the description of a foreign country, which combined details of a person's journey with information of an encyclopedic nature'.[9] As we saw, both Jan ten Hoorn, the seventeenth-century Amsterdam publisher, and Henry Fielding articulated the profound attraction of such texts to their readers.[10]

Reysen's combination of travelogue and chorography was common to the writing about Muscovy in Western Europe at the time.[11] *Reysen* followed earlier models in this respect, and was virtually the last Dutch installment of the many chorographies of Muscovy that found their origins in a work by the learned and careful observer Baron Sigismund von Herberstein (1486–1566). In the first half of the sixteenth century, this Habsburg diplomat furnished such a convincing and definitive sketch of Muscovy that it conditioned subsequent Western European travellers to a disposition to see the same things the Baron had observed on his pioneering trip.[12] Brenner's remarks about Columbus apply here, too: the Western travellers (even those only familiar with the oral version of its clichés) saw what they wanted to see. Whatever they observed in Muscovy before 1700 confirmed the tropes laid out by the Imperial diplomat generations earlier.[13] Parts of his portrayal remained unassailable until Peter's Great Embassy of 1697–98. Herberstein's rendering of the tsar's realm ingrained itself in Western European imagination; perhaps even Peter the Great's whirlwind transformation of Russia did not quite dissolve this image.

Conventional seventeenth-century rules for writing travel accounts strongly influenced the ghostwriter's composition of *Reysen*.[14] But nowhere does the book abandon 'fact' altogether and become entirely 'fiction', it seems. The quotation marks here are chosen deliberately, for Erica Harth has noted that before 1690 at least under the French definition of the word *histoire* both factual and fictive, or possible, events were grouped.[15] Readers preferred 'facts', but were willing to accept 'fiction' in travel accounts and histories as long as it had verisimilitude (they usually had little opportunity in any case to verify the authenticity of textual information), as is evident from Madame d'Aulnoy's case (see Chapter 2). The most dubious parts of *Reysen* are encountered when sailor's lore or tall tales are presented as if they befell, or were observed by, Struys himself: the sacrifice of Razin's mistress to the Volga and the flaying of the Polish-Persian woman are among the more glaring examples of this.

Encyclopedic in scope and informed by a proto-scientific mindset, Olearius's *Beschreibung* surpassed Herberstein's treatise even if it still borrowed some of its framework and clichés from the Habsburg diplomat. The previous chapter explained how Olearius's *Beschreibung* served as the preeminent source for *Reysen*'s long sections on Muscovy and Persia.[16] Because it was so highly respected, the *Beschreibung* was so elaborately plundered: its towering authority is reflected by many pages of *Reysen*. We also saw how Dapper used other works on Muscovy, Persia, and different regions visited by Struys, and followed certain classical models. Mund points out that most chorographies were cast in the mould of works by the highly respected classical authorities such as Ptolemy, Strabo, Herodotus, and Pliny.[17] *Reysen* echoes Herodotus's writings in an era when many of the names were still in use that had been given to regions and waterways by Herodotus two millenia earlier.[18] References to Troy (and perhaps the tour around the Greek islands) are contrived insertions, a Classical veneer giving *Reysen*'s words greater authority.[19]

J. Paul Hunter describes the formulaic parameters of Early Modern travel accounts that clearly apply to *Reysen* as well, with its sensationalist print on the cover and its interchange between 'chronological' descriptions of spatial movement and 'topical' description of locations.[20] Its sole organizing principle was chronology. But in giving significant attention to events, and by its personalized passionate tone (especially during the third journey), *Reysen*'s contents transgress the travel genre's borders suggested by Hunter.[21] Additionally, *Reysen* veers distinctly from any rigid formula of Early Modern travel accounts in several other ways. Its attempts at transgressing formulaic boundaries probably drew from readers' ennui with the genre in its highly clinical incarnation. This fatigue prompted Defoe a generation later to present a fictional travel-cum-adventure novel in *Robinson Crusoe* (1719) and led Swift to parody the travel genre with his *Gulliver's Travels* (1726).[22] In its liveliness, *Reysen* has a distinct journalistic quality, a forerunner to Cornelis de Bruyn's work in the next generation.[23] This is most pronounced in the part on Astrakhan, both in Struys's rendition of the Cossack threat, and in the two letters that precede the three journeys. Journalism, though in its infancy before 1700, was soon to channel news and information to the curious and shape public opinion in a plethora of periodicals.[24]

Furthermore, *Reysen* has much in common with the picaresque genre, then highly popular in the Dutch Republic and across Europe, of which Cervantes's *Don Quijote* (1605 and 1615) is the most famous example.[25] Many Dutch adaptations and translations of Spanish and French picaresque originals were published in the 1650s and 1660s.[26] Struys does not quite fit the prototype of the *pícaro*, because he is bereft of the characteristics of a rogue. But *Reysen* ridicules other cultures in an unadorned and realistic manner, resembling the lampooning of various social groups in a picaresque tale.

A Dutch national standard for seafaring tales affects and contextualizes *Reysen*'s contents.[27] Struys's book, like the books by Schouten and van der Heiden, clearly owed something to the writings of other famous Dutch globetrotters. All these works celebrated an emblematic Dutch spirit, enterprising and indomitable. This Dutch tradition is perhaps one reason why *Reysen*'s hero is more personalized than Hunter suggested as normal for the genre.[28] *Reysen*'s publication in 1676 came in the wake of the greatest moment since 1572 of Dutch tenacious survival in the face of overwhelming odds. Struys's defiance and endurance symbolized the Dutch Republic's defiance in the face of a phalanx of foreign foes in 1672.[29] Other elements of the book may be linked to this peculiar Dutch quality, such as *Reysen*'s tale of the Venetian-Turkish conflict. It resonated with readers familiar with the long-term maritime struggle between Dutch and English. Further fortifying its readers with national pride, *Reysen* also reflects the tail end of the early 'heroic' years of the Dutch global empire's establishment in Africa and Asia during the 1630s and 1640s. But as we saw in Chapter 2, *Reysen*'s success can also be attributed to the resemblance of Struys's wanderings to traditional maritime epics predating the development of a collective Dutch identity, such as Homer's *Odyssey*, Sindbad's adventures, or the Biblical tale of Jonah; Struys's sufferings echo the Book of Job and even the people of Israel in the Old Testament, and his ultimately futile wanderings those of the prodigal son.[30]

Different from diaries which only began to be kept during the age, works in the autobiographical genre predated the seventeenth century, but lacked their modern form.[31] *Reysen* reminds of this type of writing, sharing some traits with St Augustine's *Confessions* or Bunyan's book, even if its references to Divine Providence often seem oblique.[32] Early Modern Europeans had little stomach for self-analysis outside of religious treatises. Unless it was an allegorical tale, seventeenth-century readers commonly deemed the publication of a living person's autobiography an 'arrogant' act.[33] But in the Republic accusations of hubris were suspended if the autobiographical story was a seafaring adventurous tale such as Struys's, Linschoten's, or Bontekoe's.[34]

Reysen's narrative's focus is on exotica and thrilling anecdotes, not on introspection, however. Struys the protagonist does not interpret his psychology; his emotional life lacks complexity.[35] The text is a far cry far from a self-reflective autobiography in the style of Rousseau's eighteenth-century work. In *Reysen*, Struys remains a largely flat character, upon whom much suffering is inflicted, which leaves his psyche unscarred.[36] His strength of mind is contrasted with Els Pietersz's tragic insanity resulting from the latter's enslavement. Struys shows few other urges than a primal survival instinct, a sporadic fraternal inclination, and a longing for material comfort. Noteworthy is his keen attention to monetary matters, evidence less of a typical Dutch materialist spirit than of their central importance for the historical Struys. Even if his adherence to marital fealty, appeals to human dignity, and compassion are not merely Dutch Christian-humanist moralistic admonitions inserted by Dapper, *Reysen* does not present a comprehensive and nuanced portrayal of Struys, and the scarce other evidence about him makes it difficult to sketch the historical sailmaker's psychology.

It is evident that *Reysen*'s Jan Struys does not speak as the Jan Struys of history. The ghostwriter polished the sailmaker's language into a civilized (*beschaafd*) narration of Struys's voyages. Even if the book's few dialogues and quotations do occasionally approach the colloquial language found in contemporary ribald plays of the day, this may be attributed to Dapper's familiarity with the language of sailors and other common folk than to Struys's speech.[37] It is difficult for us to fathom if readers recognized an authentic popular voice in some of the more colourfully crude expressions in *Reysen*.[38] Even if they recognized its artifice, this seems to have hardly diminished the book's great contemporary popularity. Thus *Reysen* sports the form of an autobiography, but utterly lacks the self-analysis and psychological introspection with which we associate the genre.

Reysen has a form that probingly transcends earlier travel accounts, but while it borrows elements of a variety of genres that began to find form toward 1700, it merely experiments with these writing categories, giving it also in this respect a patchwork quality. It presages the popularity of many new genres that emerged in the eighteenth century, and hesitantly heralds the end of the traditional Early Modern travel account in the style of Herberstein and Olearius.

***

Toward the end of his life, Struys's apparently peaceful retirement was upset by a literary attack which called his good name into question. Despite this criticism, the first of a series that continued long after his death, *Reysen* found a ready audience for decades afterward. To pinpoint exactly the appeal of *Reysen* across Western Europe is impossible, as it is for any book, but its success testifies to a shared European literary culture transcending the Dutch language. In fact, all three books of the trilogy sold well for a considerable period across Western Europe, indicative that van Meurs and van Someren read the market well. They had anticipated this success, for they would otherwise have refrained from seeking copyright protection.[39] Licences were usually only sought for books of which Dutch publishers expected good sales.[40] Usually, publishers declined such extra expense, for it cut into their profit margin.

Reysen found an eager audience, as the slew of translations and the many Dutch reprints indicate. Given the many Dutch, French, German, and English editions, the book's uneven form was of no apparent concern to Dutch or other Western Europeans readers during the late seventeenth and early eighteenth centuries.[41] Today, the Russian National Library in St Petersburg possesses 22 or 23 editions of the book in Western European languages: five or six in Dutch, eleven in French, four in German, and two in English.[42] In Dutch, *Reysen*'s quality was easily surpassed by the meticulous and detailed publication of Cornelis de Bruyn's work in 1714.[43] But *Reysen* continued to be bought by Dutch readers until the 1740s at least.[44] It seems the harrowing quality of Struys's adventures compared to de Bruyn's (who travelled as a sort of guest of honour in Russia), and *Reysen*'s cheaper editions were preferred over the richly illustrated and thus expensive work by de Bruyn (a trained painter).

Reysen was not printed in equal measure in all of the three main North-Western European languages.[45] Most of all, the European elite's growing use of French caused this imbalance.[46] The first translations into German and French were conducted under the auspices of van Meurs (and after his death his widow), and van Someren.[47] We saw in the previous chapter how they may have printed a total of about 3,000 copies of *Reysen* in three languages between 1676 and 1681. The official German translation published in Amsterdam in 1678 was followed by a pirated edition published in Zürich in 1679.[48] It took however over a century before two more German editions were published.[49] This limited popularity was probably due to a few factors: many a German reader and publisher knew Olearius's works and could easily discover that in chorographic terms *Reysen* had little to add to them. Struys's tale might appeal most to the reading audience of the Hanseatic city-states such as Hamburg, Bremen, or Lübeck, whose merchants traded overseas, but most literate people in those maritime regions were capable of reading Dutch before 1700 and French thereafter.

Judging by the reviews and criticism of the book, it was *Reysen*'s descriptions of exotic animals, plants, and natural phenomena that, despite their cursory nature, especially interested contemporary readers. In this respect, age-old curiosity about strange natural occurrences combined with the spread of a popular-scientific mindset that found its origins in the Baconian emphasis on experimentation and observation. This thirst for the outlandish found also expression in the great popularity of curiosity cabinets (shared most famously of course by Peter the Great), the most renowned of which were Dutch.[50] That *Reysen*'s fame should partially be attributed to the burgeoning appetite for scientific observation and the traditional fascination with 'freaks of nature' across Western and Central Europe is apparent from the initial response to it in France and England. The appearance of the French translation *Les Voyages de Jean Struys en Moscovie* in Amsterdam in 1681 led to a review in the *Journal des Sçavans*.[51] This discussion emphasized the quasi-scientific curiosities reported in the book, such as the tails of certain inhabitants of Formosa, the omnipresent gold in Siam, the Great Bell of the Kremlin, the *Baranets* zoophyte ('plant–animal') of the Volga region, the peculiarities of Mount Ararat, and the high incidence of infanticide on Madagascar.[52]

The popularity of *Reysen*'s French version was concomitantly linked to a sudden boom in published accounts of French travellers to Asia during the 1670s.[53] Their popularity, in turn, was related to the contemporary French attempts to carve out a colonial and trading empire in Asia. This endeavour began in earnest with the foundation of the *Compagnie des Indes* under the auspices of Jean-Baptiste Colbert in 1664. It received a boost with the dispatch of a French armada (the so-called *escadre de Perse* under the command of de la Haye) to India in 1670, and gained renewed strength after 1683.[54] Muscovy never seems to have been within the sights of French merchants or French economic policy (which was usually pro-Swedish, pro-Turkish, and pro-Polish in the seventeenth century), but the French were interested in the fortunes of their Dutch (and English) competitors and foes in India and Persia.[55] Texts on Persia, especially, enjoyed a spectacular vogue, and underscored the interest in Struys's account.

The *Journal des Sçavans* review was translated into English and published in the first issue of the *Weekly Memorials for the Ingenious*, appearing in London in January 1682.[56] This review alerted local publishers to the possibility of an English translation of Struys's work. Struys's *Voyages* in English enjoyed only a brief flurry of popularity during the 1680s.[57] Perhaps in the English-language world, the rich Hakluyt and Purchas tradition of travel accounts already provided abundant reading materials about travels in exotic countries. Struys's story had little to add to them. Furthermore, as Matthew Anderson points out, the British interest in Russia remained rather minor and intermittent before Peter's Great Embassy, and seems to have been equally slight with regards to Iran.[58] Finally, readers' taste was shifting to novels set in exotic locales, such as Aphra Behn's *Oroonoko*, Daniel Defoe's *Robinson Crusoe*, and Jonathan Swift's *Gulliver's Travels*.[59] English parodies of the travel account genre, such as the latter two works, suggest that the English audience was becoming bored with mariners' tales.

It was the French translation of 1681 that made Struys's name most famous, which is not merely evident from the great amount of French reprints, but also from references to it in other French-language publications. The review in the *Journal des Sçavans* helped the book's widespread diffusion among French readers (including of course those who read the language outside of the Kingdom of France), but it eventually met with a hostile response from critics. *Voyage en divers états d'Europe et d'Asie* by the French Jesuit Philippe Avril took Struys to task for allegedly fraudulent and contradictory descriptions contained in *Reysen*, specifically the geographical location of Mount Ararat and Astrakhan.[60] Avril's textual criticism followed the lead of his fellow Jesuits, the Bollandists, who engaged in sophisticated text criticism, especially scrutinizing hagiography.[61]

In 1694 a Dutch translation of Avril's treatise appeared in Utrecht. Stung by Avril's accusations as rendered in a Dutch review annoucing this translation by Pieter Rabus, the editor of the bimonthly journal *Boekzaal van Europe* (*Bookhall of Europe*), Struys himself travelled from Holsteiner Friedrichstadt to Rotterdam to repair his name and reputation.[62] In his review, Rabus had followed Avril in questioning the veracity of Struys's account, while casting his own doubt on the sailor's *sole* authorship.[63] In a meeting with Rabus, Struys underlined that he had depicted the location of the two places correctly, leading Rabus to publish a sort of retraction in *De Boekzaal van Europe*.[64] Struys appears to have remained silent about Rabus's suggestion regarding the 'shared' authorship of his book. Of course, Struys had admitted such aid already in 1676, by thanking the more 'civilized pen than [his] own' in *Reysen*'s Preface.

Thus, provoked by what he clearly thought unfair criticism, not long before his death (it cannot have been easy for him to travel at his advanced age of 65), Struys himself came to the defence of *Reysen*'s authenticity. His indignation shows, before anything else, that Struys himself at least was convinced of the book's truth. That Dapper had used other sources to embellish his descriptions escaped Struys. Otherwise his indignant refutation of the Avril criticism on points of detail would seem downright silly, since *Reysen* was rife with plagiarized passages. Struys's illiteracy made him fail to point out to Rabus that in *Reysen*'s first Dutch edition of 1676, the typesetter had never even entered the number of miles that separated Mount Ararat (Erevan) from the Caspian Sea, about which Avril claimed *Reysen* had erred; he left instead a blank space, and the Jesuit had criticized the invention of *Reysen*'s French translator![65] Of course, from our vantage point Struys's behaviour shows a lack of scruples, claiming authorship of a book that he never wrote. His perspective, however, was informed by a less narrowly defined definition of authorship than we normally use today (although celebrities' ghostwriters are not unknown), as we saw earlier in Chapters 2 and 12.

A decade after Avril's accusations, François Halma (1653–1722), a well-known Dutch publisher of the late seventeenth and early eighteenth centuries (and publisher of Witsen's *Noord en Oost Tartarye!*[66]), levelled another devastating criticism at books of *Reysen*'s kind:

> One knows from experience, that many travellers, incapable of using a pen themselves [...], after just here and there noting something down, without organization, continuity, or coherence, when they come home seek one or other scribe to compose [the story of] their travels using many [other travel accounts about] the same region; and thus again present as something new to the world what former travellers have observed and have described long ago, and elaborate that [the traveller] himself has discovered these many marvels, and strange things, as are recorded in the travels as published under their name. [...] How [well] do [these] others climb the summits of mountains, and how many further wonders do they encounter on their steep journey[!][67]

Halma's words were included in a foreword to a book based on notes taken by E. Ysbrants Ides, a Danish servant of the tsar whose ancestry was Dutch; for its publication as a book, Ide's manuscript, which had been written in a peculiar hybrid language between northern German and Dutch[68], had been translated into Dutch by Nicolaas Witsen, the regent who had been Jan Struys's patron a generation earlier.[69]

Whereas neither Witsen (as Ides's editor) nor Halma names Struys, Halma's words reveal a precipitous decline in *Reysen*'s reputation among those engaged in science and scholarship. *Reysen*'s plagiarism of ethnographic and natural descriptions had become reprehensible. A concomitant attack on *Reysen*'s engraving of Persepolis was levelled by Jean Chardin, a famous French-English traveller, who rejected the depiction as utter fantasy.[70] Chardin was not far off the mark: if it was not a version of previously existing pictures of the ruins, the engraving had been at best based on a vague oral description or primitive sketch by the sailmaker. This image (and virtually all the others included in *Reysen*) hardly justified the book's title page's claim that the copper plates were based on Struys's own drawings.

It is ironic, however, that the map of the French translation of Ides's book (the only edition supplying one), based on the meticulous scientist Witsen's maps of Siberia, shows the Caspian Sea in a far more distorted shape than the map published in *Reysen* in 1676.[71] Witsen refers in neither edition of his monumental *Noord en Oost Tartarye* to *Reysen*, although he had rather critically appreciated Struys's shipwrighting skills in the second edition of his work on shipbuilding that was published just prior (in 1690) to the first edition of *Noord en Oost Tartarye* (in 1692).[72] It seems Witsen, who possessed a copy of *Reysen*, disowned by 1692 the book he had sponsored in 1676.[73]

Witsen's silence and Avril's criticism had thus set in motion a series (that has continued until recently) of harsh judgements about Struys's book as being untrustworthy and fictitious. Cornelis de Bruyn, another of Witsen's protégés, travelled much of the same route through Muscovy and Persia as Struys had covered a generation earlier, but his account of 1714 omitted any mention of Struys's book.[74] This omission confirms how *Reysen* was increasingly seen as inaccurate and unreliable. But in the eighteenth century dismissal of the book was not universal, not even among those with scientific credentials or pretensions. For his *Histoire naturelle*, a celebrated multi-volume text from the heyday of the Enlightenment purporting to be a 'biography of the world', the Comte de Buffon mined *Reysen* as a source for especially anthropological information (upon which Buffon based some proto-racist theorizing).[75] If *Reysen* had been comprehensively discredited by then, it is unlikely that Buffon (then considered a scientific genius equal to Carl Linnaeus [1707–78] and an unwilling *philosophe*) would have deigned to consult it.

Until 1730, *Reysen* continued to be read widely. The many reprints of the work throughout the first decades of the eighteenth century underline the business acumen of van Meurs and van Someren in publishing *Reysen*: among the widening audience of eighteenth-century Western European readers, less discerning than the purists associated with the Royal Society (such as Witsen) and the Jesuits, Struys's work remained popular long after the 1700s, with a particular flurry of republication occurring during the 1710s and 1720s, when six French and two Dutch editions were produced by a variety of publishers.

It is conceivable that those French-language reprints of Struys were initially prompted by French readers' heightened interest in Russia generated by Peter the Great's visit to France in 1717. Also, in 1721 Montesquieu's widely popular *Lettres persanes* were published, in which Iran became an imagined exotic country that served as touchstone to criticize the absurdities and flaws of European society and politics.[76] It may have led to a final upsurge in interest in *Reysen*'s description of Iran among French readers. But even among this group the appetite for *Reysen* began to wane after 1730. Hence the reissue of the text became episodic, with a last wave of two French editions and one German edition around 1830. In 1788, the book was still listed as one of the best known and most widely circulating Dutch books, but by the 1820s, the interest in *Reysen* had vanished, certainly among the Dutch reading audience, and the sailmaker was by then largely forgotten in his native country.[77] Many reasons can be suggested for this decline in popularity such as the text's poor organization, the lengthy interruptions of the hero's actions, and the narrator's naiveté, all objectionable to the sensibilities of nineteenth-century readers. Whereas the genre of the travel account continued to be popular, its protagonist was expected to be more sophisticated or realistic than Struys, whose occasionally coarse treatise and boorish expressions also offended the delicate Victorian reader.[78] For such readers the style of *Reysen's* tales had become antiquated.

In the Netherlands, two efforts were made to transform *Reysen* into a children's book during the twentieth century.[79] Van Saerdam appears to have planned a three-part reissue of Struys's work, but only the first part was ever published. Perhaps this was linked to the Wall Street Crash, which happened in the same year as the first volume was published. A more obvious cause for this failure was van Saerdam's inability, like Douwes's later, to forge an exciting event-filled adventure tale out of the unwieldy material.

***

But while the book lost its appeal as a 'trade book' in Western Europe after 1800, *Reysen* was discovered in Russia as a historical source which, despite its flaws, allowed scholars to understand aspects of Muscovy's past not obtainable in other sources. The book's first public diffusion in Russian occurred in 1773, when Russia's 'first journalist', Nikolai Novikov, referred in one of his magazines to a Slavic (*po slavensku*) translation of *Reysen* printed for Tsar Peter the Great's personal use in 1719.[80] Novikov summarized this excerpt in short form for his readers. Much later Petr Iurchenko discovered that this translation was linked to Tsar Peter's interest in collecting and preserving documents testifying to Russia's past.[81] Iurchenko identified the first Russian translator of *Reysen* as Venedikt S(c)hilling, likely a Swedish prisoner of war, who had translated part of the text around 1701.[82] Unhappy with this translation from the Dutch, Peter ordered more than a decade later a new translation to be made from the French Lyon edition of 1684; this was the printed edition which Novikov presented to his readers in 1773.[83] That in the eighteenth century no Russian version existed does not mean that *Reysen* was entirely unknown, of course, because, as in the Holy Roman Empire and Holland, French was the language of the eighteenth-century Russian elite, and some of its members may have possessed copies of *Reysen*'s many editions in that language.

The academic interest in *Reysen* can be traced to Russian scholars working around the middle of the nineteenth century. Russian historians then discovered *Reysen* as a historical source for their portrayal of the reign of Tsar Aleksei, especially of Stenka Razin's rebellion and the pioneering efforts to build a Russian navy. In 1857 A.N. Popov accurately summarized the opinion of his day in suggesting that 'Struys's travels to Persia, Turkey, India and other countries are seen as mythical in European literature.'[84] Historical source criticism (*Quellenkunde*) had dismissed *Reysen* as a reliable text.[85] But Popov disagreed with this categorical rejection, becoming in the process the first modern historian to rediscover the value of *Reysen*, especially in rendering the Razin rebellion.[86] Bereft of many 'private, personal' sources for the age, the great pre-revolutionary historians Sergei Mikhailovich Solov'ev (1820–79) and Vasilii Osipovich Kliuchevskii (1841–1911) proceeded to use *Reysen* in their work, and many scholars have since followed in their footsteps.[87] For Russian researchers, 'Struys's' words enriched the dry documentary archival materials (such as the massive paperwork generated by the *prikazy*, in their own lifeless language) that were published in the course of the nineteenth century. As both Kliuchevskii and Solov'ev understood, *Reysen* should not be dismissed in its entirety as derivative and plagiarized.[88]

Popov's pioneering attempts to present parts of the book to a modern audience were followed in 1880 by the first printed Russian version of the entire Muscovite section of the book prepared by Iurchenko for the journal *Russkii arkhiv'*.[89] Both the anonymous letter and Butler's, which are included in the first Dutch edition of 1676 were also translated in the same issue.[90] Iurchenko had translated a French edition of *Reysen* published in Lyon in 1682. An integral Soviet version of *Reysen* based on the original Dutch was published in 1935 as *Tri puteshestviia Ia.Ia. Streis* (*The Three Voyages of J.J. Streis*).[91] Since the fall of the Soviet Union, *Tri puteshestviia* has been republished, showing a sustained Russian interest in *Reysen*.[92]

Following the footsteps of their nineteenth-century Russian colleagues, many recent historians have taken a close, albeit guarded, look at *Reysen*, and gained fresh understanding of the seventeenth-century world.[93] Very few seventeenth-century Europeans who occupied so low a place in the social hierarchy left a description of their lives abroad in those early days of European expansion.[94] Before *Reysen*'s publication, no Western commoner left behind an account of his experiences in seventeenth-century Russia or Persia. Even if it is often impossible to assess where Struys's recollection ends and the work of his ghostwriter begins, we are occasionally able to glean the historical sailmaker's response to the Dutch empire and lands beyond. Therefore, besides unique information on contemporary societies, Struys's actions as they appear in *Reysen* offer a glimpse of some of the lost world of disenfranchised groups in the Early Modern Age, especially when they text holds up to the evidence from other sources.[95] But *Reysen* allows us most emphatically to ponder the Dutch elite's view of the world, and, therefore, the mindset of the world's first hegemonic capitalists.

14
'Any Soil is the Fatherland for a Courageous Man'[1]

When Struys returned to the Netherlands in 1673, he was about forty-five years old, an advanced age for Dutch mariners who had not made ship's officer. Surviving another five years away from home defied the odds, since most Dutch sailors did not return to Holland from even one trip to the Indonesian Archipelago; neither did more than half of his *Oryol* comrades return to the Republic. While he had thus been fortunate, the prospect of a destitute old age loomed. Over the years, Struys had not acquired any substantial assets (even if he gave his second wife title to his 'possessions' when he remarried in 1668). Never a member of a sailmakers' guild, he could not rely on the modest provisions such organizations offered their elderly members.[2] Although his second marriage made him conditionally co-owner of his wife's Amsterdam house, his actions after his return in 1673 attest to a continued quest for financial security. They suggest that he lost his tenuous title to co-ownership of the house, perhaps since he failed to bring back any wages with him.[3]

We are not sure how Struys became involved in the writing of *Reysen* in 1675, but financial gain was his driving incentive. For probably a few hundred guilders, he told his stories to Dapper.[4] Although a clever fellow, Struys was an illiterate sailor, who would have known little about proper remuneration for such work. And even prolific writers could hardly survive off the proceeds of their writings in the Dutch Republic. The reward for consenting in 1675 to have his adventures published thus only brought temporary relief from looming indigence. Sources reveal how Struys between 1675 and 1680 embarked on a sustained quest for money.[5] But when Struys appears in the historical record for one final time in 1694, his financial circumstances had improved markedly. In the *Boekzaal van Europe*, Rabus portrays Struys as a man of some means and status. Indeed, Struys's grave was still worth a mention in a description of Schlesvig a century later. Documentary and printed evidence allows some insight into this shift of his financial fortunes.

While the *Oryol* project had been aborted, its crew did not give up on the arrears the Muscovite government owed them. The men's dogged pursuit of their unpaid wages seems to demonstrate the Dutch belief in the inviolability of labour contracts. Their determination may have been strengthened by their knowledge of Ivan Bogdanovich Miloslavskii's announcement at Astrakhan in December 1671 that 'those true subjects who had been compelled to flee the country could now return and would be compensated by the Tsar for the misery they had suffered, each according to his merits and standing.'[6] Miloslavskii proceeded to pay in early 1672 Lodewijk Faber a first installment of forty rubles (200 guilders) compensation upon the latter's return from Persia to Astrakhan.[7] Rumours about redemption payments to some of his *Oryol*'s companions had undoubtedly reached Struys in Amsterdam by 1675. When a golden opportunity to retrieve some of his lost Russian wages beckoned, he left for a second time for Moscow, delaying the completion of *Reysen* for more than fifteen months.

The chance to present his case to the tsarist authorities offered itself when in the spring of 1675 in Amsterdam the illustrious merchant Koenraad van Klenk organized an embassy destined for Moscow: Struys could now attempt to 'find a smouldering ember under the ashes', as *Reysen*'s dedication would call it.[8] Van Klenk was destined to conduct most of his negotiations in Moscow with officials overseeing the *Posol'skii Prikaz*, the very department that had supervised the *Oryol* project from the spring of 1669 until its collapse. It is unlikely that Jan Struys expected full payment for the remainder of the four-year term he had accepted in the summer of 1668 (or its renegotiated terms of the spring of 1669). But he evidently counted on receiving some remuneration.[9] If the tsar was willing to pay Struys for one year's arrears (perhaps for the period between his last payment in Nizhnii Novgorod in June 1669 and his escape from Muscovite territory in June 1670), he stood to receive 600 guilders, enough money to survive adequately in the Republic for three years.

Officially, Struys would serve as palfryman and constable on the embassy, both somewhat honorific posts, but promising him some remuneration, even if he failed in his mission to get paid by the tsar's government. Serving as gunner during the seaborne stages of the embassy's progress constituted a promotion from senior sailmaker in the ship's pecking order, too.[10] But van Klenk may have hired him for more than just sympathy for the sailmaker's cause, genuine need for

Struys's services as ship's gunner, or his skill grooming horses. Struys was familiar with Moscow, and had at least a halting command of spoken Russian. This skill would allow him to operate as interpreter-informant for the Dutch envoy.

The ambassadorial journey by sea to Arkhangel'sk, and from there over land to Moscow, was altogether uneventful compared to Struys's earlier adventures. The embassy's two ships departed Holland on 31 July 1675 (Julian Calendar); after disembarking in Arkhangel'sk on 6 September 1675, van Klenk's party reached Moscow on 21 January 1676.[11] The Russians applied deliberate stalling tactics to slow the embassy's progress covering the approximately 1,500 kilometres between the Russian port and capital. The Muscovites tried to stagger the parade of foreign envoys visiting them in 1675 and 1676 to prevent the Western ambassadors from colluding with each other.[12]

Once in Moscow, van Klenk's embassy essentially failed to accomplish its main two goals, the conclusion of an effective military alliance against France and Sweden and the lowering of tariffs charged to Dutch traders under the terms of the Muscovite New Commercial Code of 1667.[13] Despite this, Dutch-Russian relations were exceptionally cordial during the visit, helped along by van Klenk's massive amount of gifts for the Muscovites. Russian desire to please van Klenk was not just the result of the ambassador's generosity. In the past, van Klenk had rendered invaluable services to the tsar in facilitating the flow of arms to Russia during the Thirteen Years' War. On minor points, therefore, the Russians proved accommodating. Most striking among these was the Muscovite promise to address the Republic's Estates-General hence with a more prestigious title than previously.[14] Between January and May 1676 van Klenk enjoyed four times the honour to 'see the tsar's bright eyes', the phrase used for a personal audience with the tsar.[15] Van Klenk's advocacy of rerouting part of the silk trade across Muscovy (really a private project) was also lent a willing ear.[16]

Van Klenk, too, received satisfaction in the matter of the unpaid wages owed to his constable and palfryman. In the 'chancellery Russian' (*prikaznyi iazyk*) usual for government records, it is noted how during one negotiating session between van Klenk and the Boyars Mikhail Dol'gorukii (a son of Iurii) and Artamon Matveev, *d'iak* Emel'ian Ukraintsev and several of his colleagues,

> the envoy spoke [how] in former years by the order of the great tsar his tsarist majesty of blessed memory [the recently deceased Aleksei]

> was sent a foreigner [*inozemets'*] mariner [*karabel'shchik'*] to *Astrakhan'*, and during the brigandage [*vorovstvo*] of Stepan Razin [he] left across the sea and went across many lands to the Holland country [*galanskaia zemlia*], and from the Holland country travelled with him the envoy[;] and that the mariner asked for mercy and petitions [literally, beat his head on the ground, *b'el chelom*] the great tsar his tsarist majesty [for] the wage [*zhalovan'e*] for his brothers-mariners who were with him in Astrakhan' and his wage, [in other words] asked that the great tsar his tsarist majesty orders the payment of the arrears to him the mariner and that the great tsar [paid for his brothers], and [the envoy] gave a petition to [the tsar's] close [*blizhnye*] boyars. And [his] close boyars received the petition from the envoy, and they said that they would report it to the great tsar, to his tsarist majesty.[17]

Ambassador van Klenk thus raised Struys's and his comrades' case before his high-ranking Russian negotiation partners.[18] As with other minor points the Muscovite officials yielded to van Klenk, they appear to have paid to Struys (and those supplicants of the *Oryol* he represented) some of their arrears. This is all the more likely since the tsarist government compensated various other members of the *Oryol* expedition after the recovery of Astrakhan. Satisfying the request of the Dutch mariners was another trifle useful to further placate van Klenk.

Struys found more than one smouldering coal in the ashes, for all members of the ambassador's retinue were given some precious animal fur (probably sable) upon their departure from Moscow in 1676.[19] The embassy departed the Russian capital on 21 June 1676, and travelling much faster (which was necessary to avoid the onset of frost preventing seafaring) on its way home, left Arkhangel'sk on 9 September 1676, and landed in Holland on 23 October 1676.[20]

If Struys was indeed an agent for van Klenk during the embassy's sojourn in Russia, he may have been paid above his official wage as constable-palfryman by the ambassador. While his trip to Russia thus yielded him a tidy sum in 1676 (all included, perhaps more than a thousand guilders), it was too small to retire comfortably. The royalties for the book added a few hundred guilders to his income in 1675 and 1676, but Struys now craved for more comforts than those enjoyed by the average craftsman. His participation in the book and in the embassy alerted him to the possibility that he might be able to escape a subsistence-level existence altogether.

Two years after his second return from Moscow, when his earnings from that trip and from *Reysen*'s publication were likely dwindling,

Struys resurfaced in Denmark. In October 1678, Jacob le Maire, the long-time Dutch representative in Copenhagen, wrote to the Estates-General in The Hague:

> There is a Hollander here named Jan Janse Struijs, who has been master of ceremonies with Mr. van Klenck in Muscovy, and is making a model of a ship for the King of Denmark that cannot be sunk by shot; he has made this model solidly, and shown [it] to the King, who likes it, and also believes it to be practicable; he [King Christian V] also has requested me to confirm it [...], which I have done [,] taking with me an experienced Hollander master ship's carpenter, who praises the work, and is of the opinion that it could work, because of which the King has resolved to have a large ship built according to this model, to see if it is practicable in large as well as in small size; he also claims that he intends to make it in such a way that the debris caused by balls cannot do damage to the ship.[21]

Thousands of Dutch sailors served on Christian V's ships in a Danish-Swedish war in 1675–79, and the Danes even recruited soldiers in the Republic, a complicated albeit ultimately successful endeavour in a country that was itself at war with France until 1678.[22] Danish records indicate that Struys joined these compatriots in the Danish navy, at the rank of captain-lieutenant, not long before he presented his ship's model to the king in the fall of 1678.[23]

Reysen confirms that Struys knew how to build ship's models.[24] When he was quagmired in Shemakha in the summer of 1671, he built a miniature galley and ship for his (by then former) master Haji Biram Ali.[25] Haji Biram Ali was so impressed that he presented the vessels as gifts to the Shemakha *khan*, who invited Struys to his court to report on Western-European seafaring. The governor, however, showed no interest in having ships built to scale.[26]

The innovations Struys presented to King Christian V (r. 1670–99) can be approximated from an outline in the second edition of Nicolaas Witsen's book on shipbuilding.[27] Struys used a combination of methods, such as applying water-resistant substances between the ship's ribs over the planks, using a double layer of planks to construct the ship's hull, drying nails and boiling them in oil (to prevent rusting), applying moss or similar material on the inside of the hull to smother cannon-ball hits, and making Titanic-like compartments in the hold to avoid having it fill up with water in its entirety if springing a leak. Experiments of the sort were also conducted under the auspices of the Royal

Society in England at the time, but few found application in shipbuilding.[28]

Witsen did not believe in the feasibility of these ideas by a 'certain Jan Jansz. Struis', but was more positive about another of Struys's inventions, that of small folding ferry-type boats made of leather stretched on a wooden frame, which could be easily transported overland on a cart, and thus used effectively on interior waterways.[29] They were capable of carrying a heavier load than the vessels commonly used for light water transport at the time. It may be that Struys adopted here the nautical principles he had seen applied in central Russia. Its frequent portages made imperative the use of light-weight boats that could be moved easily across land.

As captain-lieutenant, Struys's wage in Danish service was about three times the average sailor's.[30] But by late 1679 he had left the Danish navy and probably Denmark itself, indicative that his efforts to build an unsinkable ship had not convinced the Danes.[31] Although other causes may have played a role, such as the Swedish-Danish peace and the ensuing anti-Dutch turn in Denmark of that year, the Danes may have grown suspicious of his honesty. They may have discovered that he could not write, which should have disqualified him to serve in their navy as a ship's officer. Additionally, Dutch sources may have informed the Danes that Struys had inflated his importance for the van Klenk embassy (of which he had not been master of ceremonies, as he had apparently pretended).[32] In any case, Witsen's dismissal of Struys's ship design seems to imply that the grandiose plan to build an unsinkable ship was a chimera, which the Danes eventually realized.

Nonetheless, there is a remarkable parallel between Struys's activities on behalf of the Danish monarch and his former companion Karsten Brandt's involvement in the creation of a Russian seafaring fleet.[33] In 1688, when Brandt worked in the capital's *sloboda* as furniture maker, he emulated Struys's efforts in Denmark a decade earlier.[34] In a warehouse once owned by his uncle Nikita Romanov, the young Tsar Peter encountered a decrepit vessel, which Brandt restored to sailworthiness on the suggestion of Peter's Dutch-born mathematics' teacher, Frans Timmerman. Brandt proceeded to build several other small ships for Peter during the few remaining years of his life (Brandt died in 1693). In 1720, Peter the Great recalled how he found inspiration to rekindle his father's project for a modern Russian seafaring merchant fleet and navy by this felicitous introduction to Karsten Brandt.[35] In his Naval Statute, Peter recognized beyond Brandt's pivotal role the impetus provided by his father's pioneering effort to develop a Caspian fleet.[36]

Different from his father, Peter believed in the quintessential necessity of a Russian navy, and was to pursue its creation tenaciously.

Whereas Jan Janszoon Struys's achievements as one of the pioneers of Russian seafaring are overshadowed by those of his comrade Brandt, the publication of *Reysen* in 1676 rescued Struys's name from the almost absolute anonymity that was the norm for the thousands of sailors, artisans and workers of the Dutch Golden Age. After his departure from Danish royal service, Struys's life seems lost in the mist of time but for his Rotterdam visit of 1694; we further know Friedrichstadt, a small town in Schlesvig-Holstein, as the place where he spent his last years and was buried.[37] Friedrichstadt was located in Schleswig-Holstein, duchies ruled by the house of Holstein-Gottorp, but gradually brought under Danish royal control in the second half of the seventeenth century (even if remaining officially part of the Empire).[38] Here Struys died in the 1690s, remaining famous enough to have the location of his grave noted in a chorography of Denmark published a century later.[39]

Ironically, the duchies had been ruled until 1659 by Duke Frederick III of Holstein-Gottorp, Olearius's sponsor as well as Tsar Mikhail's partner in the *Friedrich* project. In 1619, Duke Frederick allowed Arminians who fled Holland after the Synod of Dordt (1617–18) to found Friedrichstadt.[40] The duke gave Friedrichstadt a charter which offered religious freedom to the Arminians. Frederick hoped thereby to attract Dutch capitalists to use the port as a lauching pad for their overseas ventures. Although no such efforts bore fruit during the 1620s, Duke Frederick persisted in his plan to make Holstein into a middling maritime power, dispatching embassies to Moscow in the 1630s (of which Olearius was secretary!). Friedrichstadt was designated as the port to receive the Persian silk from Muscovy, but the loss of the ship *Friedrich* on the Caspian Sea halted this effort. The king of Denmark eventually reduced Holstein-Gottorp to a dependent status, putting paid to any grandiose schemes of its duke (who died besieged by Danish forces) and chasing Frederick's son and heir Christian Albrecht (r. 1659–95) for a long while from his realm.

Friedrichstadt instead remained a small-sized haven of tolerance in the Empire, boasting Remonstrant, Catholic, Mennonite, and Lutheran churches as well as a synagogue by 1700.[41] It remained mainly Dutch-speaking far into the eighteenth century. We can only speculate what informed Struys's move to this small town in his later years. Perhaps the plans to transform Friedrichstadt into a significant maritime port were revived in the 1680s, when more silk began to be shipped from

Swedish Baltic ports (in no small measure thanks to Lodewijk Faber's efforts).[42] Struys may have offered the Holsteiners his knowledge of Iran and Muscovy and his skills as a sailor and shipwright. Amsterdam's baptismal records seem to suggest that Struys became a Lutheran on the eve of his second marriage in 1668.[43] Perhaps residence in a Dutch-speaking haven, in which Lutherans enjoyed the same civic rights as their fellow townsmen, was preferable over a more discriminating Holland. Tantalizing as these suggestions are, none can be be supported by conclusive evidence.

In Friedrichstadt's unusually tolerant community, Struys apparently kept in rather close touch with his native country. He was well off in his last years, able to dress and behave in a way that had Rabus take him for a sort of 'gentleman' (*heer*) (he was likely a '*poorter*,' that is, a burgher with civic rights) in 1694.[44] During the last years of his life, the humble sailmaker Jan Struys had therefore not only become a celebrity, but was also able to enjoy in a tranquil and enlightened setting the material comforts that he had begun to pursue in earnest in December 1647.

Conclusion

In *Reysen*'s third part, Jan Struys's adventures and mishaps spring to life when the text's dry chorographic descriptions recede before the seafaring hero's epic. Besides *Reysen*'s comparatively modest price, the individual quality of Struys's observations made the work far more exciting to read than cosmographies such as Dapper's earlier books. Struys's account of the Razin rebellion and his enslavement by Muslims enlivened *Reysen*'s more standard exploration of foreign customs, while even its chorographic sections were strategically animated by instances of incest, infanticide, the legend of Noah's Ark, people with tails, marauding elephants, torture, and appalling murders. These latter vignettes saturated the text in a baroque manner that conformed to contemporary aesthetic sensibilities. Treated to another epic following Linschoten's template and borrowing from a host of other popular texts in form and content, *Reysen*'s intended audience in the United Provinces was also enticed by its subtext. It emphasized the unique and superior blend of Dutch resolve, inventiveness, and tolerance that sustained their country in a hostile world. Dutch readers, shaken by the assault on their country of the Franco-English-German alliance in 1672, could fortify themselves through reading about their countrymen's continued global prominence in *Reysen*, as it charted the omnipresence and great power of the Dutch East India Company in Asia, with its operations stretching from Persia to Batavia, Japan, and South-Africa. *Reysen*'s three parts present the short-lived Dutch hegemony within the European-dominated 'world economy', emphasizing several key components of the Republic's strength: the Dutch economic (and often political) presence in utterly remote places, seafaring prowess, and technological lead.[1] Hence the success of *Reysen*.

Meanwhile, other sources than *Reysen* indicate how the harrowing series of adventures recounted in the third part of the book exerted a cathartic effect on the historical sailmaker Jan Struys. In 1673, he arrived as a much more confident man in Amsterdam, as demonstrated in his moves during the following years. The successful pursuit of lucrative opportunities enabled him to leave behind the modest circumstances to which he seems to have resigned himself before 1668. It appears that the adventures that befell Struys between the summer of 1668 and the autumn of 1673 led him to shed his pre-modern, pre-capitalist mindset for a modern, capitalist outlook. In a sense, Struys's personal metamorphosis reflected at an individual level his native country's (or at least its coastal provinces') transformation toward a precocious modernity during its 'long seventeenth century' (*c.*1580–1713).

Or, to phrase it differently, Struys's actions indicate how the capitalist outlook of the Republic's mercantile elite had began to percolate further down the social hierarchy and began to affect even some of those born to the lower levels of Dutch (or Hollandish) society. As we saw in Chapter 3, social historians rank Dutch sailors and sailmakers barely above the *lumpenproletariat,* living an existence at the mercy of natural and human forces which they barely understood and before which they were powerless. But the behaviour of Struys and his *Oryol* comrades show how not all of these subalterns resigned themselves to their fate. No longer accepting the traditional conviction that their station in life was determined by birth, they began to perceive a possibility of making their fortune irrespective of their background. Struys's evident transformation (which can also be discerned in the movements of Karsten Brandt or Lodewijk Faber) is a stark example of the process in which many of his compatriots appear to adopt this more modern (or capitalist) attitude during the seventeenth century. This widespread diffusion of a capitalist disposition in the Dutch Republic is also readily apparent in *Reysen*'s production and marketing. The publishers' feigning as if *Reysen* was written by its humble protagonist as well as its elaborate packaging eloquently bespeak capitalist expertise as sophisticated as those informing modern advertisements.

Beyond their demonstration of Dutch baroque and capitalist sensibilities and customs, *Reysen* and Struys's biography allow us some other startling insights into the seventeenth-century world. Five further points may be singled out to indicate how *Reysen* and Struys's story aid our understanding of the age.

First of all, they attest to a Dutch collective mindset which, by its diffusion among broad layers of society and by the intensity of its chauvinism, seems a nationalism that foreshadows its nineteenth-century incarnation across Western and Central Europe. Beyond self-help-type capitalist convictions, it combined elements of a mutually reinforcing patriotic loyalty and Christian solidarity, a peculiar sense of being God's chosen people (as Vondel's or Cats's poetry reiterated), a growing secular outlook, and a pervasive misogyny, and was reinforced by widespread literacy (albeit not exclusively dependent on the ability to read). In this latter respect, *Reysen* and writings like it resemble the sort of folk tale collected by the Grimm brothers or an early version of a Wagnerian libretto, both of which would galvanize early German nationalism during the nineteenth century.

The support provided by expatriate Dutchmen to their compatriots and the aid supplied by the VOC to the *Oryol*'s sailors in their miserable predicament in the Caucasus reveal these nationalist convictions. While the Republic's nationalism crystallized in a struggle against a succession of foreign foes (Spain, England, France) and profiled itself against foreign others (such as *Reysen*'s lecherous Italians, barbarian Asians, superstitious Russians), it is harder to identify a Bismarckian 'enemy within' that underscored most nineteenth- and twentieth-century nationalisms.[2] Perhaps, as Schama suggests, this was caused by its prosperity linked to its welcoming reception of others who helped the United Provinces to flourish, while the tenacious Erasmian humanist tradition continued to emphasize Christian compassion.[3] Nationalism seems at its most inclusive in affluent times. This tolerance and openness declined together with economic stagnation and the rapid disappearance of the Republic as a European Power after 1713.

Reysen obliquely portrays its protagonist as a faithful member of the Dutch Reformed Church. The modest attention given to the Christian religion in *Reysen* suggests the prevalence of a tolerant and increasingly secular worldview rather than the competing image of seventeenth-century Dutchmen as strict, God-fearing Calvinists. The entire book is infused with a secular spirit and expresses the tradition of the Dutch 'liberal' (*rekkelijke*) or humanist tradition, rather than that of the opposing Calvinist or Puritan (*precieze*) strain. *Reysen* seems to reflect the circles of certain Dutch intellectuals of its day who ever more incisively questioned the nature of Christian theology and practice.[4] Not long after *Reysen* was published, Balthasar Bekker wrote his condemnation of witchcraft, and Bekker and Witsen, *Reysen*'s sponsor, were well acquainted.[5] Although neither Witsen nor Bekker was an atheist (Bekker was a Protestant minister), in their minds this world and the world-to-come had become largely separate issues. *Reysen* in this sense seems to embody the age of Spinoza and Locke, and reflects 1670s Holland as the safe-haven of freethinkers rather than as a bastion of puritanical Dutch Calvinism.

Despite women's occasional role as stand-in business managers, Dutch society was dominated by men. Women sporadically adopted male roles, but were seen as members of the weaker sex, men's passive objects or helpless victims. While *Reysen*'s Struys exhibits in his sufferings feminine traits as a member of a subaltern class, this is complicated by other depictions of him as master of his own fate, as required by his Dutch masculinity. I suggested in Chapter 9 that Dutch slave traders involved in the Triangular Trade might have come to terms with the enslavement of their compatriots portrayed in *Reysen*. The slavehaulers and investors in the human trade perhaps saw the 'great unwashed' at home as creatures equally inferior as their human commodities, passive feminized objects only fit to serve their betters. But, given the Dutch national pride that *Reysen* appeals to, it is equally possible that many readers read with indignation about the brutalities inflicted upon Struys and his fellows, because they happened to fellow Dutchmen, rather than anonymous Africans. In contemporary European texts, most non-Westerners were caricaturized or demonized. *Reysen* followed this tradition by presenting inflammatory stories such as the flaying of the Iranian-Polish woman or the alleged infanticide of the Malagasy. In a sort of progression leading from Struys to Buffon to Gobineau, these racist views were to develop into the Orientalization of the Muslim world after 1800, and into systematic racism toward non-Westerners.

Outside of their continent, seventeenth-century Europeans (such as the monks of Iran) seem to have been inclined to provide assistance out of a long-standing traditional sense of duty and compassion for fellow Christians who shared the common cultural heritage of the Classics and Bible. Although most information extant about the Polish envoy Gurdziecki testifies to his utterly opportunistic character, he nevertheless bailed out some of the desperate Dutchmen in 1670 and 1671. Not merely mercenary reasons, but also Christian compassion informed his behaviour, even if his hospitality and treatment of Struys was poor after his liberation. Apparently, the few (Western) Europeans in Muscovy and Iran often banded together before a hostile local community or government, ignoring all the intense internecine religious and political quarrels which tore apart their continent. This indicates how Dutch nationalism coexisted with older forms of European or Western Christian solidarity.

Secondly, *Reysen* and Struys's fate underscore how the Christian and Islamic worlds in the seventeenth century still met on equal terms, in stark contrast to modern imperialism's asymmetrical relations. Safavid power had been challenged by the VOC in the 1640s, but the spectre of capture by Muslims, which befell Struys three times, remained one of the most pervasive fears haunting the maritime communities of Western Europe far into the eighteenth century. The strength of this fear was so intense that it survives to this very day in the Netherlands in the evergreen story of St Nicholas's Moorish servants hauling off badly behaving children to Spain. Struys's adventures and *Reysen* demonstrate how the meeting between Islamic and Christian cultures in Eastern Europe and West Asia alternated between the poles of peaceful exchanges and violent quarrels.

Smouldering mutual animosity fed on religious disagreement, an inability to understand foreign languages, and self-serving justifications that made others less than human. The weaker party in a cross-cultural encounter could be exploited as an inferior being, as charted in *Reysen* in the enslavement of Struys on three occasions, or the gang-raping of the Aceh woman and Maria Brak. But Safavid Iran usually treated most outsiders (those residing and those visiting) with superior benevolence, as did the tsar's or the sultan's states. Based on a sense of superiority, tolerance did not equate acceptance, and, as *Reysen* shows in a great variety of places, could give way to intense hostility. Lack of curiosity in Western inventions was a natural consequence of this sense of supremacy, but this underestimation of the accelerating Western technological advance ultimately informed Islam's subjugation by the West in the Imperialist Age. *Reysen* hints at this error of judgement. Different from the Danish king in 1678, the Shemakha *khan* in 1670 never demonstrated interest in having Struys build standard-size rather than miniature ships.

Even though each culture viewed the other as intrinsically inferior, often demonizing it, mutually beneficial collaboration was frequently established in practice. This latter state of affairs allowed for the recognition of a common humanity, of which Struys's relationship with his benevolent (former) master Haji Biram Ali (who seems more than just a trope) is probably *Reysen*'s most commanding example.

Another countervailing trend is discernible in *Reysen* that undermines overly simple binary oppositions. Its compassion with Struys's trials and tribulations suggests that even a humble commoner has a right to human dignity and should be spared the inhumanity of slavery. Of course, in the worldview of the seventeenth-century Dutch not all people were entitled to such fair treatment: women and most male foreigners were usually excluded. But *Reysen*, in its lamentations about slavery in the Mediterranean and the Caucasus, in its frequent alarm at the *abject* subordination of women to men, and in its nuanced and positive depiction of certain Others as human beings, such as Haji Biram Ali, presents the merest hints of the future embrace of human equality, a concept that was only beginning to emerge in 1676.

Thirdly, *Reysen* reflects how Muscovy was entering the European orbit. More vulnerable than the Islamic empires, the tsar's realm had discarded its version of superior 'splendid isolation', and was beginning to accept the necessity of compromise with Westerners and Western ways rather than reject them contemptuously, even if this trend was in its infancy in the 1660s. In their turn, Western Europeans cautiously began to include Muscovy as part of Europe. By the time Struys died, Peter the Great was accelerating the Westernization of Russia's state and its elite. By Russia's partial embrace of things Western, she gained an ability to fend off Western attempts at subjugation and colonization. *Reysen*'s chorographic sections represent a sort of final recitation of the long-standing stereotype of coarse, primitive, and superstitious Orthodox Muscovites, showing how they traditionally were not considered to belong to this Europe. But the very project of the *Oryol* heralded Muscovy's arrival on the European stage. The Dutch role in this project and in other aspects of a modernizing or Westernizing seventeenth-century Russia was considerable.

Indeed, fourthly, Dutch influence in Muscovy and Iran was more crucial before Peter the Great's Grand Embassy than after its return in 1698. This is evident from the story of Jan Struys, his fellow sailors, and the Dutch merchants and entrepreneurs he encountered in Amsterdam and Muscovy. The height of Dutch prosperity and dominance within the world system was reached by the middle decades of the seventeenth century, and the Dutch role in the *Oryol* project in Russia is testimony to this importance. Many of the Dutch participants in the *Oryol* venture epitomize the skills, talents, and spirit that were the secret of Dutch success in the Republic's Golden Age. Much of this was due to literally workmanlike efforts. Dutch workmen were a remarkable crowd, who boldly faced, and often overcame in a determined fashion, ghastly challenges. And, while the shipbuilding project failed (as did many other Dutch-led projects in the seventeenth century, from the colonization of Formosa, Brazil and New Amsterdam to the drainage of the Haarlemmermeer,[6] or the creation of a viable Russian manufacturing sector), its initiative did lead ultimately to the creation of the modern Russian navy because of Karsten Brandt's fateful decision to return from Iran to Moscow. The *Oryol* may not have sailed on the Caspian Sea, but two decades later Brandt's *botik* would navigate a lake near Moscow, with long-lasting consequences. Peter's navy contributed to the definitive inclusion of the former borderlands of the Muscovite empire into the Concert of Europe after 1700.

Seventeenth-century Dutch organizational, technological, and entrepreneurial skills in Muscovy were remarkably prominent, from the activities of the army officers such as the van Bockhovens, Beem and Faber, the generous loans to the tsar's agent Hebdon by the firm of de Vogelaar and van Klenk, the series of weapon deliveries by a host of Amsterdam-based merchants, the armament works at Tula, the postal system managed by van Sweeden, Marselis, and A.D. Vinius, the shipbuilding craft of the Helt group and Karsten Brandt, the fortress builders such as Cosimo de Moucheron, the high quality sailcloth accompanying the sailmaker Struys, the map-making and seafaring regulations of David Butler and A.A. Vinius, to the surgeon's skills of Jan Termundt. In their activities in Muscovy, the Dutch seem to combine the entrepreneurial role they played in Denmark and Sweden and the trading role they performed in Asia. Their sustained contribution to the modernization of Muscovy relied on tight-knit kinship and patronage systems, demonstrating how social networks reinforced and sustained Dutch capitalist development at home and abroad.

Finally, some insights into the personality of Jan Struys may be gained from the uneven traces of evidence in *Reysen* and other documents. These are all the more important because we still know little about the mindset of the underprivileged commoners of Europe's Early Modern age. In the matter of religion, usually held to be the most important component of Early Modern European identity, Struys's spiritual views seem to have been modest and personal (indeed, the *Oryol* crew was not accompanied by a minister, a standard figure on VOC ships, an absence never remarked upon in *Reysen*). In his actions, he seems guided by a kind of stoicism, fatalism, or belief in predestination. Calvinists of course were supposed to believe that all was preordained, and nothing could be done to alter God's plan. Calvinistic or not, this sense of fatalism helped sailors cope with the stark dangers threatening their existence. Such spiritual convictions allowed Struys to cope with the death of wife and children, or abandon them for four years when enticed by steady and lucrative employment in faraway Muscovy, with uncertain odds of returning hale and wealthy. Struys did not precisely know the horrendous statistics on the minimal return of VOC servants to Western Europe. But from witnessing fellow sailors' death and from sharing stories with other seafarers, he would have surmised that most went missing in action during their service. He must have believed that he was among God's chosen ones after having survived so many voyages.[7]

Signs of Struys's personal fear, despair, and condemnation of barbaric behaviour occasionally crop up in *Reysen*. It may nevertheless be assumed that a stoic attitude helped the real Jan Struys face his own suffering and that of others (perhaps rooted in a strong inner faith), exposing great powers of perseverance and ingenuity in overcoming utter hardship.[8] Thus the fictional Struys blends with the historical Struys in representing the 'ideal Netherlander', exemplifying 'virtue, bravery, and zeal', and his constancy on *Reysen*'s pages in representing these traits was undoubtedly one of the secrets of its success.[9]

Despite his fatalistic streak, Struys increasingly showed a conviction that he did not have to surrender to the brutish lot traditionally seen as the standard for humble folk. His repeated departure from Holland in search of riches demonstrates a growing belief in the possibility of human agency to come out ahead, which became utterly manifest when he left for Moscow in 1668. After he recovered some of the wages lost in Russia, he proceeded to pursue a series of ventures that helped him reach a measure of affluence in his old age. He also was a man of versatile skills, another sign of a belief in the possibility of personal improvement. On various occasions in his life, he showed that he was capable of baking hard tack, making sails, sailing and building ships, handling shotguns and cannons, setting broken bones, telling a good yarn, and communicating in various languages.

Thus a crucial conclusion can be drawn from the documents (including *Reysen*) on Struys: after 1650, the modern, capitalist mindset diffused far beyond Holland's mercantile elite in the Republic. Struys's later years provide seem to prove how he abandoned a pre-modern mentality for a modern capitalist outlook. Forty years old, after a life that had known baffling hardship, he began to pursue projects that would yield him celebrity and sufficient money to retire a man of some fortune, rather than resigning himself to end his days in an old mariners' home. Ceaseless effort, some luck, a certain unscrupulousness, and a willingness to probe the boundaries allowed him to improve his lot in life. Dutch extraordinary preoccupation with Mammon is reflected in the English humour that seems to originate from the days of the Anglo-Dutch rivalry in the seventeenth century ('Dutch treat'). Jan Struys the narrator in *Reysen* and Jan Struys the historical figure, in deciding to return to Muscovy in 1675 or to travel to Denmark in 1678, undoubtedly exhibits this obsessive streak for monetary gain. The sailmaker could neither read nor write, and knew little about reckoning time according to a calendar, but he undoubtedly knew how to count!

In *Reysen*, Struys is meticulous in reporting on the wages he received or was promised. It seems that the real sailmaker surfaces in the repeated mention of his pay as sailmaker for the Genoese, Venetians, VOC, and tsar (or his anger at not receiving pay for work rendered on the ship from Amsterdam to Riga in 1668). Apart from the mariner's own insistence on supplying this information, its inclusion in *Reysen* is otherwise hard to explain. The readers of *Reysen*, hailing from rather more affluent circles than Struys, were little interested in knowing how much artisans received for their toil (unless they were keen on comparing Dutch or VOC wages to those paid by others to sailmakers). For Struys, however, these wages were the prime mover behind his odyssey, and mention of his takings crops up regularly in the book.

We do not have to be Marxists to understand this materialist obsession in workmen or sailors such as Jan Struys. In the commercialized Dutch economy, the social safety net of a return to the ancestral village had disappeared, and as an adult Struys never seems to have contemplated a return to his native grounds. By 1673 Struys realized that there were more ways to make money than make sail. He first cashed in on the captivating stories of his calamitous overseas adventures, and after 1674 amply provided for his old age by successfully exploring a variety of other lucrative opportunities for monetary gain.

Notes

Introduction: From Sailmaker to Celebrity

1 A.M. Loviagin, ed., *Posol'stvo Koenraada fan'-Klenka k' tsariam Alekseiu Mikhailovichu i Feodoru Alekseevichu* (Saint-Petersburg, 1900), 3, originally published as [Balthasar Coyett], *Historisch Verhael of Beschryving van de Voyagie gedaan onder de Suite van den Heere Koenraad van Klenck* (Amsterdam: Jan ten Hoorn, 1677).

2 See Bibliography; the book's first English edition of 1683 bore the title as rendered.

3 As a more concise version of the book's lengthy title I will use '*Reysen*' in the texts and notes. For its full Dutch title, see the Bibliography. Its lengthy title was the seventeenth-century norm, as was the separation of title page and frontispiece, possibly a tradition started by Peter-Paul Rubens (see L. Febvre, H.J. Martin, *The Coming of the Book* [London, 1976], 85–6).

4 P. Iurchenko, 'O puteshchestviia po Rossii Golandtsa Striusa,' *Russkii arkhiv'* (St Petersburg) 2 (1879) 265–9: 266; see as well K.N. Begichev', *Kavkazskie puteshestvenniki proshlykh stoletii. Iogan' Ioganson Striuis' (1670 g.)* (Tiflis, 1900), 3–4, footnote. They remain there in its collection today.

5 The work's first extended Russian translation appeared in an archival journal in 1879; during the 1930s a complete translation was published in the Soviet Union. See 'Puteshestvye po Rossii Gollandtsa Striusa,' *Russkii arkhiv'* 1 (1880) 17–108; A. Morozov, ed., *Tri puteshestviia Ia.Ia. Streis*, trans. E. Borodina (Moskva, 1935); see also Chapter 13.

6 See Chapters 1 and 10.

7 See Chapters 1 and 12.

8 See further Chapter 14.

9 On his death in Friedrichstadt, see Chapter 14.

10 S.P. Orlenko, for example, lists several travel accounts of seventeenth-century Muscovy by foreigners (S.P. Orlenko, *Vykhodtsy iz Zapadnoi Evropy v Rossii XVII veka* [Moskva, 2004], 40). For some standard Western views of Muscovy, see M. Poe, *'A People Born to Slavery'* (Ithaca, NY, 2000); S. Mund, *Orbis Russiarum* (Genève, 2003); M. Mervaud, J.-C. Roberti, *Une infinie brutalité* (Paris, 1991); G. Scheidegger, *Perverses Abendland-barbarisches Rußland* (Zürich, 1993). Some have dismissed much of *Reysen* as derivative, for example, see Vinal Smith, Lach and Van Kley, and Adelung, and contrast them with the rather more positive Floor, or Lach and van Kley themselves (G. Vinal Smith, *The Dutch in Seventeenth-Century Thailand* [DeKalb, Il, 1977], 128–9; D.F. Lach and E. Van Kley, *Asia in the Making of Europe*, vol. 3: *A Century of Advance* [Chicago, 1993] 497–8, 1801–5; F. Adelung, *Kritisch-literärische Übersicht*, vol. 2 [St Petersburg, 1846] 344–5; W. Floor, 'Fact or Fiction: The Most Perilous Journeys of Jan Jansz. Struys,' in *Etudes Safavides*, ed. Jean Calmard [Paris, 1993] 57–68). See Chapter 1.

11 See Poe, *'A People'*, 5; see also S.B. Schwartz, 'Introduction,' in *Implicit Understandings*, ed. S.B. Schwarz (Cambridge, 1994) 1–19: 1–2.

12 See Feofan Prokopovich's remarks upon the Peter the Great's return from Europe in 1717 (see L. Hughes, *Russia in the Age of Peter the Great* [London, 1998], 432).
13 F. Liechtenhan, *Les trois christianismes et la Russie* (Paris, 2002), 9.
14 Ibid., 179.
15 Liechtenhan suggests that the emphasis on exclusivity and purity caused by Reformation and Counterreformation sharpened Western condemnation of Eastern Christianity (see Liechtenhan, *Les trois christianismes*, 177, 179).
16 Again, see ibid., 179.
17 Most vociferous about this is the criticism of Struys's account of Thailand by G. Vinal Smith (see Smith, *The Dutch*, 128–9, as well as Chapters 1, 2, 7–10).
18 M.G. Aune, 'Early Modern European Travel Writing after Orientalism,' *Journal for Early Modern Cultural Studies* 2 (2005) 120–38: 121.
19 E. Said, 'Orientalism Reconsidered,' *Cultural Critique* 1 (1985) 89–107: 97.
20 See J.D. Gurney, 'Pietro Della Valle: The Limits of Perception,' *Bulletin of the School of Oriental and African Studies* 1 (1986) 103–16: 103; Floor, 'Fact'; D. Kaiser, 'Whose Wife Will She Be at Resurrection? Marriage and Remarriage in Early Modern Russia,' *Slavic Review* 2 (2003) 302–23: 308–9; R.C. Davis, *Christian Slaves, Muslim Masters* (Basingstoke, 2004), 53; R. Barendse, *The Arabian Seas* (Armonk, NY, 2002), 108–9; P. Longworth, 'The Role of Westerners in Russia's Penetrations of Asia, 17th–18th Century,' in *Mesto Rossii v Evrazii*, ed. G. Szvak (Budapest, 2001) 207–13: 207. For a fruitful recent use of Struys's work by a literary scholar, see E. Brancaforte, *Visions of Persia* (Cambridge, MA, 2003), 102–6.
21 Thus in seventeenth-century texts on Asian empires, John Emerson found data not contained in indigenous sources (see J. Emerson, 'Sir John Chardin,' *Encyclopaedia Iranica*, ed. Ehsan Yarshater [available at http://www.iranica.com/newsite/accessed 12 January 2007]). It should be noted nevertheless that in their precise descriptions other texts appear superior to *Reysen* (see A. Olearius, *Moskowitische und Persische Reise*, ed. Detlef Haberland [Stuttgart, 1986]; Pietro della Valle, *Viaggi di Pietro della Valle, il Pellegrino*, 4 vols. [Rome, 1650]; E. Kaempfer, *Am Hofe des persischen Großkönigs, 1684–5*, [Leipzig, 1940]; John Chardin, *A New and Accurate Description of Persia*, 2 vols [London, 1724]; J. Chardin, *Le couronnement de Soleïmaan Troisième roy de Perse* [Paris, 1671]; Raphaël du Mans, *Estat de la Perse en 1660*, ed. Ch. Scheffer [Paris, 1960]; F. Bernier, *Histoire de la dernière révolution des états du Grand Mogol* [Paris, 1670]; J.B. Tavernier, *Les Six Voyages en Turquie et en Perse* [Paris, 1676]: In the first volume of the copy of this book in the Library of Congress, a front engraving indicates that Johannes van Someren printed an edition of it in Amsterdam in 1678).
22 These restrictions were only abolished by Peter (see Hughes, *Russia*, xiv).
23 Most of the Safavid's Empire's archives were destroyed in the eighteenth century (see for example S. Faroqhi, *The Ottoman Empire and the World Around It* (London, 2004), 36).
24 Omitting the preamble of the two letters, which are only tangentially linked to the rest of the book.
25 For this identification, see Chapter 12.
26 Giving support to Schama's observations about Dutch seventeenth-century descriptions of Amsterdam Jews: 'The tone ... is that of curiosity (...) rather

than fear and hatred' (S. Schama, *The Embarrassment of Riches* [Berkeley, CA, 1988] 589).

27 Said suggested that Orientalism only found its full form after 1800 (see E. Said, *Orientalism* [London, 2003], 81–8).

28 *Reysen*, 70–9.

29 Indeed, the inclusion of a description of Italy was perhaps even more attractive to *Reysen*'s publishers in anticipation of their (or van Meurs's widow's) publication of translations of *Reysen*, divining similar interest in the cradle of the Renaissance among German and French readers. On the Grand Tour and Italy's central place in it, see P. Rietbergen, *Europe: A Cultural History*, second edn. (London, 2006), 287–90.

1 Struys's Youth and *Reysen*'s First Journey

1 Leaving out autobiographical details and silence about one's youth was usual before 1700, according to Dekker (see R. Dekker, *Childhood, Memory and Autobiography in Holland* [Basingstoke, 2000], 11, 101, 105, 107, 109).

2 In his first marriage banns (of 6 July 1658) preserved in the city archive of Amsterdam, Struys's age of twenty-nine is noted (see *Gemeentearchief Amsterdam, Doop, Trouw- en Begrafenisregister* [Municipal Archive of Amsterdam, Baptism, Marriage, and Burial Register; from here indicated as *GAA DTB*] 478, p. 462.

3 *Chorography*: A work which attempts to systematically describe a country or region (countries, regions), following the method of the Greek geographer Ptolemeus (see Mund, *Orbis Russiarum*, 171. *Cosmographies* are texts intended to provide a comprehensive overview of all that can be found under the firmament (human government and society, physical geography, flora and fauna, and so on) that have a broader scope than chorograpies, but the two terms are often used interchangeably. For the popularity of cosmographies during the sixteenth and seventeenth centuries, see Poe, *'A People,'* 28, 36–7.

4 *Reysen*, 368. Baptismal records for the village during this era have been lost.

5 J. Honig Jr., *Geschiedenis der Zaanlanden*, vol. 1 (Haarlem, 1849), 93–5.

6 See Honig Jr., *Geschiedenis*, vol. 1, 108–14, 163–82, 187–8.

7 See J. Israel, *The Dutch Republic* (Oxford, 1998), 363–4.

8 Israel, *Dutch Republic*, 231, 244–5. *Reysen* seems to indicate that he is Calvinist (*'Apostolisch Catholijk'*; see *Reysen*, 255, and Chapter 7); of course, his true religious identity could have been camouflaged by the editor. On the strength of Catholicism in the region, see for instance H. van Nierop, 'Catholics and the Law in Holland,' in *Calvinism and Religious Toleration in the Dutch Golden Age*, eds R. Po-chia Hsia and H. Van Nierop (Cambridge, 2002) 102–11: 107.

9 In this sense, even the baptismal record of his children is inconclusive, for a great number of sympathizers was never confirmed in the church they attended or used for such services (see further Chapter 4).

10 In 1667 Johannes van Someren published the works by the dissenter Jean de Labadie (1610–74; before Labadie's exclusion from the Walloon Reformed Church) in translation (see J. de Labadie, *erheffingen des geestes tot Godt*

[Amsterdam: J. van Someren, 1667]). But three years later, after Labadie's expulsion, van Someren proceeded with other partners to publish several anti-Arminian and anti-Labadist Calvinist treatises (see for example Johannes van der Waeyen, *Ernstige betuiginge der gereformeerde kercke* [Amsterdam: J. Van Someren, D. Bakkamude 1670]). For Labadie, see Israel, *Dutch Republic*, 669–71.

11 Friedrichstadt in Schlesvig. See Chapter 14.

12 Israel, *Dutch Republic*, 211; K. Schulten, 'Ontstaan van de Republiek en het Staatse Leger,' in *Met Man en Macht*, eds J.R. Bruijn, C.B. Wels (Baarn, 2003) 13–43: 40.

13 Israel, *Dutch Republic*, 241, 307–21.

14 J.L. Price, *Dutch Society, 1588–1713* (Harlow, 2000), 93; Israel, *Dutch Republic*, 450–65.

15 B. Schmidt, *Innocence Abroad* (Cambridge, 2001), 213.

16 See H. den Haan, *Moedernegotie en grote vaart* (Amsterdam, 1977); K. Amsberg, P. de Buck, eds, *De Moedernegotie*, radio documentary in 8 parts (Hilversum, 1999); J.W. Veluwenkamp, *Archangel* (Amsterdam, 2000).

17 See Chapter 3 and Price, *Dutch Society*, 69, 73.

18 See R.W. Unger, *Dutch Shipbuilding before 1800: Ships and Guilds* (Assen, 1978) 5–7.

19 A. van Deursen, *Een dorp in de polder*, sixth edition (Amsterdam, 2003), 52; A.M. van der Woude, *Het Noorderkwartier*, 3 vols (Wageningen, 1972), 50. For a view of the village in 1648, see 'Feestelijke optocht in Wormer, ter ere van de vrede van Munster,' available at http://www.digitaleatlasgeschiedenis.nl; accessed 18 February 2007.

20 Honig, *Geschiedenis* vol. 1, 234.

21 Price, *Dutch Society*, 117–18; Honig, *Geschiedenis* vol. 1, 258, 266. See also F. Braudel, *Civilization and Capitalism* (New York, 1982) vol. 3, 186, 191. Two magificent historical studies are available for the seventeenth-century history of Struys's birth region (van der Woude, *Noorderkwartier* and van Deursen, *Een Dorp*).

22 Honig, *Geschiedenis* vol. 1, 240–1, 246–56.

23 van Deursen, *Dorp*, 68.

24 About half of the Amsterdam sailors were illiterate in 1660; Struys's illiteracy was thus not uncommon (see S. Hart, *Geschrift en Getal* [Dordrecht, 1976], 204). See also van Deursen, *Dorp*, 131–5

25 R. Prud'homme van Reine, *Schittering en schandaal* (Amsterdam, 2001) 125–6.

26 *Reysen*, 2; Prud'homme van Reine, *Schittering*, 23; for his age, see *GAA DTB* 478, p. 462. This was a frequently travelled route in 1647, see J. Israel, *Dutch Primacy in the World Trade* (Oxford, 1989), 203. For the Dutch predeliction to take to sea in search of fortunes, see Barendse, *Arabian Seas*, 104–5; the adolescent runaway, meanwhile, was a literary trope common to the travel genre (see Percy G. Adams, *Travel Literature and the Evolution of the Novel* [Lexington, KY, 1983], 152).

27 *Reysen*, 1–2.

28 Johann von der Behr, *Reise nach Java, Vorder-Indien, Persien und Ceylon, 1641–50*, ed. S.P. L'Honoré Naber (Den Haag, 1930; first edn., 1668), 127 and 127n2.

29 This winter departure indicates that the Indonesian islands were the likely destination for the Genoese from the outset in Amsterdam, for one of the

VOC (*Vereenigde Oostindische Compagnie* [United East-India Company]) fleets left for Batavia every year in December.

30 This ownership is all the more likely as the Genoese Doge from July 1646 to July 1648 was *Giovanni Batista* Lomellini.

31 *Reysen*, 3–4.

32 *Reysen*, 4–11. On the corsairs, see Davis, *Christian Slaves*. The fact that the people of Sierra Leone are called '*Caffers*,' the Dutch rendition of a Muslim term for infidels, hints at a ghostwriter familiar with O. Dapper's work, for it uses the same term (see O. Dapper, *Naukeurige Beschrijvinge der Afrikaensche gewesten* [Amsterdam: Van Meurs, 1668]; *Reysen*, 9). See also D. Lanni, 'Une cartographie ethnique,' *Africultures: Le site et la revue de référence des cultures africaines* [2004; available at http://www.africultures.com//popup_article.asp?no=4025&print=1, accessed 25 February 2007]. My thanks to Igor Osipov of Syktyvkar, Komi Republic, Russian Federation, for alerting me to this last source.

33 *Reysen*, 13. There is no account of what occurred between early August and mid-October.

34 *Reysen*, 13–14.

35 Prud'homme van Reine, *Rechterhand*, 135–6.

36 It was an area frequented by Dutch ships (see Étienne de Flacourt, *Histoire de la grande isle Madagascar*, ed. C. Allibert [Paris, 1995; orig.: Paris: G. Clouzier, 1661], 131–2).

37 *Reysen*, 15.

38 *Reysen*, 15–24.

39 Dapper himself mainly used two French sources for the sixty [!] folio-size pages he dedicated to Madagascar in his description of Africa (*van Flakourt* and *Kauche*), see O. Dapper, *Naukeurige Beschrijvinge der Afrikaensche gewesten*, second edn. (Amsterdam: van Meurs, 1676), 3; see also Dapper's letter to Vossius (C.M. Dozy, 'Olfert Dapper,' *Tijdschrift voor het Aardrijkskundig Genootschap* 3 [1887] 414–35: 435). The sources mentioned were Flacourt's *Histoire* and Claude-Barthélemy Morisot, ed., *Relations veritables et curieuses de l'isle de Madagascar et du Brésil ... par François Cauche (de Rouen) et al.* (Paris: Augustin Courbé), 1651.

40 See Chapter 12.

41 Geography: *Reysen*, 15–16; resources: ibid., 16; fauna: ibid., 16–17; flora: ibid., 16. Female sexual lasciviousness: ibid., 19; devil worship: ibid., 22; infanticide: ibid., 20–1. In Dapper, *Naukeurige Beschrijvinge*, second edn., vol. 2, similar descriptions may be found for devil worship on 5, infanticide on 17 and 47–8, fauna on 34, or women's sexual behaviour on 45.

42 '*... seer begeerig om van den Hollandtschen aart te fokken*' (*Reysen*, 14).

43 See Laura Brown, 'Reading Race and Gender: Jonathan Swift,' *Eighteenth-Century Studies* 4 (1990) 425–43: 440–2.

44 In his work on Africa, Dapper, as Lanni suggests, follows a standard pattern of description, but does not adhere to it religiously (Lanni, 'Une cartographie ethnique').

45 Flacourt, *Histoire*, 176; *Reysen*, 21.

46 Flacourt, *Histoire*, 177; *Reysen*, 21.

47 *Reysen*, 24. According to L'Honoré Naber, the ships had entered the Sunda Straits on 15 April 1649 (von der Behr, *Reise nach Java*, 127n2).

48 *Reysen*, 24–5: *'geyle en ongebonden Landtaart'*.
49 See Chapter 9.
50 See Rietbergen, *Europe*, 295.
51 S.P. L'Honoré Naber, 'Vorwort,' in Johann von der Behr, *Reise nach Java, Vorder-Indien, Persien und Ceylon, 1641–50*, ed. S.P. L'Honoré Naber (Den Haag, 1930), ix-xii: xi; Pieter van Dam, *Beschryvinge van de Oostindische Compagnie*, ed. F.W. Stapel, 7 vols (Den Haag, 1927–1954), vol. 1, part 1, 560–1.
52 L'Honoré Naber,'Vorwort,' xi.
53 Von der Behr, *Reise*, 127; *Reysen*, 2–3.
54 Von der Behr, *Reise*, 127n2.
55 There is some further evidence, such as Struys's recollection that he later served on the ship the *Zwarte Beer*, which was the name of a ship in VOC service from 1644 to 1655 (see *Reysen*, 26; www.vocsite.nl/schepen/detail.html?id=11474).
56 *Reysen*, 25–6.
57 *Reysen*, 26.
58 Von der Behr, *Reise*, 127–8. Struys identifies von der Behr's 'General' as VOC Governor-General van der Lijn (see *Reysen*, 26). Cornelis van der Lijn and his deputy François Caron were honourably (on van der Lijn's request) discharged in October 1650, but there had been accusations of malfeasance (see J.J. Merklein, *Reise nach Java, Vorder- und Hinter-Indien, China und Japan, 1644–53*, ed. S.P. L'Honoré Naber [Den Haag, 1930; reprint of second edition, 1672], 77 and 77n2; www.vocsite.nl/geschiedenis/personalia/vanderlijn.html).
59 For the VOC wage, see van Dam, *Beschryvinge* vol. 1, part 1, 557.
60 *Reysen*, 26.
61 See M.S., ed., 'Jan Struys, the Perillous and most Unhappy Voyages of John Struys ... ,' *Journal of the Siam Society* 94 (2006) 177–209: 177.
62 *Reysen*, 26–57. In some sources, his name is spelled van Nieuwenroode. Compare C. Van N(e)ijenrode, 'Vertoogh van de Gelegentheijd des Koninrijk van Siam,' *Kroniek van het Historisch Genootschap te Utrecht* 27 (1872) 279–318: 280–302, and *Reysen*, 27–44: It shows clearly how Struys's editor used much of van Nijenrode's 1622 account describing Siam. Van Nijenrode's work was published for the first time in the nineteenth century (C. van Nijenrode, 'Remonstrantie ende verthooninghe der gelegenheijt des conickrijck van Siam', *Kronijk van het Historisch Genootschap gevestigd te Utrecht* 10 [1854] 176–91; van N(e)ijenrode, 'Vertoogh'). It is not as evident what the source was for the subsequent thirteen pages on Thailand in *Reysen* (44–57). *Reysen*'s passage on Siam does not resemble other works on Thailand: See Jeremias van Vliet, *Historiael Verhael der Sieckte ende Doot van Pra Interra Tsia 22en Coninck in Siam & den Regherende Pra Onghsry 1640* (Tokyo, 1956); I. van Vliet, *Naukeurige Beschryvinge van het Koningryck Siam* (Leiden: F. Harinck, 1692); F. Caron, J. Schouten, *A True Description of the Kingdoms of Japan and Siam*, ed. C.R. Boxer (New York, 1971; origin., 1663).
63 See Smith, *The Dutch*, 124–5, 128–9. Van der Kraan suggests that the chiefs of the VOC in East Asia had available to them most of the manuscripts mentioned in the previous note (A. van der Kraan, 'On Company Business: The Rijckloff van Goens Mission to Siam, 1650,' *Itinerario* 2 [1998] 42–84: 48–50). Witsen joined the VOC Board (the all-powerful *Heeren XVII*) in the

1690s and remained a VOC Board member until his death in 1717, see for example van Dam, *Beschryvinge* vol. 1, part 1, 112; François Valentijn, *Oud en Nieuw Oost-Indië* (Dordrecht and Amsterdam: J. van der Braam and G. Onder de Linden, 1724–6) vol. 1, 304.

64 *Reysen*, 67; van der Kraan, 'On Company,' 42.

65 *Reysen*, 60. And the year of the murder is probably incorrect, see W. Campbell, *Formosa under the Dutch* (London, 1903) 253. For good recent accounts of the Dutch at Formosa, see T. Andrade, 'The Rise and Fall of Dutch Taiwan, 1624–1662,' *Journal of World History* 4 (2006) 429–50; J.R. Shepherd, *Statecraft and Political Economy on the Taiwan Frontier, 1600–1800* (Stanford, CA, 1993), 47–90.

66 Rumours about tailed people in South-East Asia were common and persisted until the twentieth century (see Lach, Van Kley, *Asia* vol. 3, 1805 n259). Lach and van Kley believe Struys's description of Formosa to be more authentic than his account of Siam, although noting similarities with Dapper's *Gedenkwaerdig bedryf* (see Lach, Van Kley, *Asia* vol. 3, 1801–5, 1807; O. Dapper, *Gedenkwaerdig Bedryf der Nederlandsche Oostindische Maatschappye* [Amsterdam: J. van Meurs, 1670]).

67 They provide an incorrect page number in their reference (Lach, Van Kley, *Asia* vol. 3, 1804; *Reysen*, 61).

68 Dapper, *Gedenkwaerdig Bedryf*, 10–122. For an appreciation of Dapper's work as a source, see C.R. Boxer, *Dutch Merchants and Mariners in Asia, 1602–1795* (London, 1988), 20–1, 46; Lach, Van Kley, *Asia* vol. 3, 1801–18.

69 See L. Blussé, 'No Boats to China,' *Modern Asia Studies* 1 (1996) 51–76: 68. Thus in May 1650, *Reysen* has Struys arrive on Formosa, where Pieter over 't Water [Overtwater] is governor, but van Veen notes that Overtwater was its governor from 1646 to 1649; by 1650 Nicolaes Verburg had become its governor, even if it is possible that Overtwater still was on Formosa in early 1650, as Valentijn implies (*Reysen*, 58; E. van Veen, 'How the Dutch Ran a Seventeenth-Century Colony: The Occupation and Loss of Formosa, 1624–1662,' *Itinerario* XX [1996] 59–77: 70; F. Valentijn, *Oud en Nieuw Oost-Indië* vol. 4, part 2, 73, 83 [on China]; Campbell, *Formosa*, 63, 75, 249). Likewise, Jan van Muijden, the *Opperkoopman* (Upper Merchant) for the VOC in Siam, was recalled to Batavia in early 1650; implausibly, Struys's ship picks him up only in January 1651 (*Reysen*, 67; van der Kraan, 'On Company,' 42).

70 See Chapter 10.

71 Frans Jansz. van der Heiden, *Vervarelyke schip-breuk* (Amsterdam: van Meurs and van Someren, 1675); Wouter Schouten, *Oost Indische Voyagie* (Amsterdam: van Meurs and van Someren, 1676). For full titles, see bibliography.

72 J. van Goor, *Nederlandse Koloniën* (Den Haag, 1997), 49.

73 Barendse, *Arabian Seas*, 383–4.

74 See also M.E. van Opstall, 'From Alkmaar to Ayudhya and Back,' in *All of One Company: The VOC in Biographical Perspective* (Utrecht, 1986) 108–20: 109.

75 See John E. Wills, Jr., Review of *Colbert, Mercantilism, and the French Quest for Asian Trade* by Glenn Ames, *Annals of the American Academy of Political and Social Science* March (1998) 206–7.

76 *Reysen*, 63; see van der Kraan, 'On Company,' 71; www.vocsite.nl/schepen/detail. html?id=11680. See Chapter 6 for more on *fluyts*. Similarly, while

Struys recalled the ship that subsequently carried him from Formosa to Siam as the *Postpaert*, there was a ship in VOC service in East Asia called *Post*, which may have been his ship (www.vocsite.nl/schepen/detail.html?id=11824).

77 *Reysen*, 67–8.

78 See Valentijn, *Oud en Nieuw Oost-Indië* vol. 4, part 1, 296, 368.

79 See www.vocsite.nl/schepen/detail.html?id=11971; Valentijn, *Oud en Nieuw Oost-Indië* vol. 1, 230–1. The first part of the fleet, consisting of three or four ships, left Batavia in late 1650 and brought back Cornelis van der Lijn and François Caron to Holland (see van der Kraan, 'On Company,' 71; J.R. Bruijn et al., eds, *Dutch Asiatic-Shipping in the Seventeenth Century* vol. 3 [The Hague, 1979], 58–9; Valentijn, *Oud en Nieuw Oost-Indië* vol. 1, 230–1).

80 The VOC, although eventually a formidable record keeper, does not seem to have collected lists of all its seafaring or land personnel in the time of Struys's travels.

81 The printing in Latin script rather than Gothic script of the two letters reinforces my suspicion here (see 'Extract Uyt een Brief [...],' in *Reysen*, 1–4 [from here, *Reysen* [Extract]; *Reysen*, 'Extract Uyt den Brief geschreven in de Stadt Ispahan, van David Butler,' in *Reysen*, 5–34 [from here: *Reysen*, 'Extract' [Butler]).

2 The Second Voyage

1 *Reysen*, 69–119.

2 *Reysen*, 69.

3 Paintings depicting commoners' daily life were likewise pregnant with symbolism and highly stylized (for this interest, see M. Prak, *The Dutch Republic in the Seventeenth Century* [Cambridge, 2005], 238–40).

4 R. Prud'homme van Reine, 'De Republiek als grote en kleine mogendheid ter zee (1648–1763),' in *Met Man en Macht*, 105–39: 113, 115–16. Peace was signed in March 1654. See also J. Glete, *War and the State in Early Modern Europe* (London, 2002), 166. The standard work for the seventeenth-century Anglo-Dutch Wars is J.R. Jones, *The Anglo-Dutch Wars of the Seventeenth Century* (London, 1996). See further R. Hinton, *The Eastland Trade and the Common Weal in the Seventeenth Century* (Cambridge, 1959), 10, 89; David Ormrod, *The Rise of Commercial Empires* (Cambridge, 2003), 32–3.

5 Glete, *War and the State*, 166; Prud'homme van Reine, 'De Republiek,' 108–9, 113; Jones, *Anglo-Dutch Wars*, 27–8, 46, 130; de Jonge, *Geschiedenis* vol. 1, 665n1, 703; R. Prud'homme van Reine, *Rechterhand van Nederland* (Amsterdam, 1996), 203, 223. Sailors received slightly better remuneration on ships used for naval purposes during wartime. Also note Unger, *Dutch Shipbuilding*, 85.

6 Prud'homme van Reine, 'De Republiek,' 108–9; A.van Deursen, *Plain Lives in a Golden Age* (Cambridge, 1991), 22.

7 See Chapter 14.

8 For example, Willem van de Velde II (1633–1707) specialized in paintings of ships at sea, see M. Westermann, *A Worldly Art* (New Haven, CT, 1996), 112–13.

9 See Jones, *Anglo-Dutch Wars*, 11, 29–30.

10 The publishers included a large section on a heroic escape by the Dutch East India fleet from an English pursuit in European waters during the Second Anglo-Dutch War in Schouten's book (see Breet, *Oost-Indische voyagie*, 472–521).
11 See Chapter 12.
12 Jones, *Anglo-Dutch Wars*, 46–7.
13 *Reysen*, 69.
14 At the end of the seventeenth century, Durgerdam had about 500 inhabitants (see van Deursen, *Dorp*, 113). The story of his recruitment by Klaas Ketel hints at this subterfuge (*Reysen*, 69): As a villager of Durgerdam in 1655, Struys must have heard of his neighbour Ketel's search for a sailmaker, but somehow only finds out about this when he comes across Ketel in Amsterdam.
15 Struys's name is nowhere to be found in city records that should normally have listed him once he had become a more experienced sailmaker, such as the *poorter*-books that recorded those becoming citizens of Amsterdam (for example, see P.C. Spierenburg, *Judicial Violence in the Dutch Republic* [Amsterdam, 1978], 30–1; Unger, *Dutch Shipbuilding*, 78, 91–2). Such status normally accompanied a senior guild member. The upstart *Zaankanters* did not organize their crafts in guilds (see Unger, *Dutch Shipbuilding*, 5).
16 Compare, for example, to W.Y. Bontekoe, *Memorable Description of the East Indian Voyage, 1618–25*, ed. P. Geyl (London, 1929; origin. Hoorn: Jan Jansz. Deutel, 1646), 20–1, 23, 51–3, 68–9. On its popularity, see *Iournael ofte Gedenckwaedige beschrijvinghe vande Oost-Indische Reyse van Willem Ysbrantsz. Bontekoe van Hoorn*, eds G. Verhoeven, P. Verkruijsse (Zutphen, 1996).
17 G. Verhoeven, P. Verkruijsse, 'De vele gezichten van Bontekoe,' in *Iournael ofte Gedenckwaedige beschrijvinghe*, 39–79: 40–1. See for some examples R. Tavernier, *Russia and the Low Countries* (Groningen, 2006), 98–102.
18 *Reysen*, 209–21; see map between 236 and 237 and Chapter 9. Compare to Joan Blaeu, *Atlas Maior van 1665* [Köln, 2005], 109. Olearius's work, the template for much of that section in *Reysen*, rendered the sea as ball-shaped rather than elongated from north to south (see D. Haberland, 'Einführung des Herausgebers,' in Olearius, *Moskowitische Reise*, ed. Haberland, 13–46: 30–7).
19 *Reysen*, 64. The incidence of ships sinking at sea was in fact rather low in the seventeenth century (see below in this chapter).
20 My translation of '*Sijnde kennelijck, dat de menschen, door malkander genomen, somtijts wel kunne tegens d'ongemacken van de reysen, sonder verversingen te genieten, voor den tijt van vier maanden, maar niet voor vijf maanden, alsoo het op die laatste maent voornamentlijck aankomt. En wanneer die menschen dan komen in te vallen by gebreck van sodanige verversingen, en dat het schorbut sigh alsdan begint te openbaeren, tot geen verhaal kunnen geraecken en voort wegsterven*' (van Dam, *Beschryvinge* vol. 1, part 1, 514–15).
21 van Deursen, *Plain Lives*, 24; van Deursen, *Een Dorp*, 102–3; A.P. van Vliet, 'De Staatse Vloot in de Tachtigjarige Oorlog,' in *Met Man en Macht*, 44–62: 53; J.C. de Jonge, *Geschiedenis van het Nederlandsche Zeewezen*, third edn. (Zwolle, 1869) vol. 1, 663n1.
22 P.C. van Royen, *Zeevarenden op de koopvaardijvloot omstreeks 1700* (Amsterdam, 1987), 73.
23 van Goor, *Nederlandse Koloniën*, 49, table; van Deursen, *Plain Lives*, 22–6; Prak, *Dutch Republic*, 144; P. Linebaugh and M. Rediker, *The Many-Headed*

Hydra (Boston, 2000), 160; Breet, *Oost-Indische Voyagie*, 31; Prud'homme van Reine, *Rechterhand*, 223; Spierenburg, *Judicial Violence*, 167–8.

24 See for instance van Goor, *Nederlandse Koloniën*, 50; van Deursen, *Een Dorp*, 297–8, 300.

25 Jones speaks of 'miserable conditions endured by seamen in both naval ships and merchant vessels' (Jones, *Anglo-Dutch Wars*, 6; see also ibid., 59).

26 J. de Hullu, 'De matrozen en soldaten op de schepen der Oost-Indische Compagnie,' *Bijdragen tot de Taal-, Land- en Volkenkunde van Nederlandsch-Indie* 69 (1914) 318–65: 329–30; see also C. Boxer, *The Dutch Seaborne Empire* (New York, 1965), 71.

27 Jones, *Anglo-Dutch Wars*, 47.

28 Ibid., 330; Breet, *Oost-Indische Voyagie*, 31; Boxer, *Dutch Seaborne Empire*, 73.

29 Van Deursen, *Plain Lives*, 24; Van Deursen, *Dorp*, 103.

30 P.E. Pérez-Mallaína, *Spain's Men of the Sea* (Baltimore, MD, 1998), 101.

31 De Hullu, 'De matrozen en soldaten,' 349–57.

32 The size of the trunk(s) was not very strictly circumscribed before 1656; by the end of the century the trunk a sailmaker could fill was ordained to be 1.26 metres in length, and 56 centimetres in height and width (see van Dam, *Beschryvinge* vol. 1, part 1, 656–7).

33 Van Deursen, *Plain Lives*, 22, 25.

34 Van Deursen, *Plain Lives*, 25.

35 Van Deursen, *Plain Lives*, 26.

36 Pérez-Mallaína, *Spain's Men*, 25.

37 *Reysen*, 69, 71–7; his kidnapping was a bit of a cliché (see Barendse, *Arabian Seas*, 114). In 1655, work was indeed scarce when Amsterdam's economy was mired in a recession and even in normal circumstances year's end was already a lean time for sailors, when most of the fishing and cargo fleets were in harbour (see H. Brugmans, *Geschiedenis van Amsterdam*, second edn., vol. 3 (Utrecht, 1972–3) 163; de Jonge, *Geschiedenis* vol. 1, 660; Schama, *Embarrassment*, 186). Livorno was the port frequented most by Dutch ships (see Israel, *Dutch Primacy*, 204; D.W. Davies, *A Primer of Dutch Seventeenth Century Overseas Trade* [The Hague, 1961], 40; Prud'homme van Reine, *Schittering*, 200).

38 Glete, *War and the State*, 166; J.C. de Jonge, *Nederland en Venetië* ('s-Gravenhage, 1852), 226; R.C. Anderson, *Naval Wars in the Levant, 1559–1853* (Princeton, NJ, 1952), 125, 137.

39 *Reysen*, 71–7.

40 *Reysen*, 71–7.

41 Rudolf Dekker discovered merely two travel accounts *in manuscript form* written by men in subaltern positions (and none by women) for the entire seventeenth century, one by a sailor and another by someone from similar social rank (R. Dekker, 'Van Grand Tour tot treur- en sukkelreis: Nederlandse reisverslagen van de 16e tot begin 19e eeuw,' *Opossum* 13–14 [1994] 8–24: 8).

42 A. Maczak, *Travel in Early Modern Europe* (Cambridge, 1995), 142; (Marie-Catherine) Madame d'Aulnoy, *Relation du voyage d'Espagne* (La Haye: H. van Bulderen, 1691).

43 *Reysen*, 78. Anderson lists among the Venetian sailing ships the *Gallo d'Oro* (Anderson, *Naval Wars*, 159).

44 A. Hamilton, 'Introduction,' in *Friends and Rivals in the East*, eds A. Hamilton et al. (Leiden, 2000) 1–9: 4; W.H. McNeill, *Venice: The Hinge of Europe, 1081–1797* (Chicago, 1974), 151–2; Anderson, *Naval Wars*, 147.
45 Faroqhi, *Ottoman Empire*, 98.
46 *Reysen*, 79; Anderson, *Naval Wars*, 145. Born in (Danish-ruled) Norway and by marriage related to the Dutch naval hero Maarten Harpertszoon Tromp (1598–1653), Adeler had received most of his sea training on ships carrying the Dutch Republic's flag (de Jonge, *Nederland*, 254–5; Prud'homme van Reine, *Rechterhand*, 215). In Dutch his name is rendered as '*Coert Siewertszoon Adelaar*'. Since the 1650s, the Witsen family knew Adeler (see N.C. Witsen, *Moscovische reyse 1664–5. Journaal en aentekeningen* ['s-Gravenhage, 1966], 8n6). Klenk joined Adeler as director of the Danish king's salt company (*GAA Notarieel Archief* [from here *NA*] 3410 [Notary P. Engelbrecht], 4 June 1666, p. 301). Adeler was also a friend of Admiral Michiel de Ruyter (see Prud'homme van Reine, *Rechterhand*, 215). After Adeler's return to North-Western Europe, he served another stint under Dutch flag (1661–3), but turned down an offer to become admiral of the Dutch navy; instead, he commanded the Danish navy from 1666 until his death in 1675 (de Jonge, *Nederland*, 255, 257, 260; Prud'homme van Reine, *Rechterhand*, 215).
47 On this education, see below in this chapter and Chapter 12.
48 Together with some of the earlier descriptions of the archipelago, this perhaps was the first fruit of Olfert Dapper's research for his later book on the Greek islands (see O. Dapper, *Naukeurige beschryving der eilanden, in de archipel der Middelantsche zee* [Amsterdam: Wolfgangh, wed. J. van Someren, van Waesberge et al., 1688]). See Chapter 12 for more on Dapper's role.
49 *Reysen*, 83–6.
50 It is not clear whether Struys saw a link between the lack of fresh fruit and vegetables and scurvy, although he lamented the short supply of 'refreshments' (*ververschinge; Reysen*, 85).
51 *Reysen*, 85. The Venetian fleet dispatched such excursions onto the mainland regularly when blockading the Dardanelles (Anderson, *Naval Wars*, 164).
52 Davis, *Christian Slaves*, 75; Linebaugh, Rediker, *Many-Headed Hydra*, 15; de Jonge, *Nederland*, 226–8; McNeill, *Venice*, 129–30.
53 Davis, *Christian Slaves*, 73–82, 171. Escape, however, was not unusual (see Faroqhi, *Ottoman Empire*, 127).
54 *Reysen*, 85–7.
55 On the use of this language in the Mediterranean, see Davis, *Christian Slaves*, 113–15. In *Don Quijote*, the captive, who reflects Cervantes's own experience as a Barbary slave, explains it as ' ... that lingua franca employed all across Barbary, and even in Constantinople, which neither is nor isn't Moorish or Spanish, or any other language for that matter, but a jumble of languages, thus allowing us to communicate' (M. de Cervantes, *Don Quijote*, trans. B. Raffel [New York, 1999], 279).
56 There indeed was a Dutch-made ship of this name among the Venetian navy; the name was commonly used, as is evident from the fact that the ship carrying Struys to Russia in 1668 had the same name (see de Jonge, *Nederland*, 249n2; Anderson, *Naval Wars*, 125n1, 137, 137n1, 143, 159; *Reysen*, 121).

57 *Reysen*, 87.
58 Anderson, *Naval Wars*, 159.
59 *Reysen*, 94; this agrees with de Jonge's and Anderson's account (see de Jonge, *Nederland*, 258; Anderson, *Naval Wars*, 161).
60 De Jonge, *Nederland*, 258; *Reysen*, 88–90. *Reysen* gives different first names from de Jonge for the commanders Mocenigo, Contarini, Riva, Marcello and Morosini, who carry names of families prominent among the Venetian patriciate.
61 The *Mercurius* is de Jonge's source (see de Jonge, *Nederland*, 258).
62 de Jonge, *Nederland*, 228.
63 See for its full text http://www.dbnl.org/tekst/vond001dewe08_01/vond001 dewe08_01_0016.htm (accessed 12 April 2007). Its fiftieth line also mentions Marcello's death.
64 Such as its report of the destruction of *het Wapen van Nassau*, the activities of *Proveditor* Barbaro Badoer and the flag officer Antonio Barbaro, the arrival of the Malteser detachment in June, and the explosion of the *David en Goliath*, which cost indeed 22 men their lives (compare Anderson, *Naval Wars*, 159, 161; *Reysen*, 88–95; and de Jonge, *Nederland*, 258–9). In the case of the *(David en) Goliath*, de Jonge claims that 22 out of 144 survive, which seems a typo.
65 *Reysen*, 99–100. This sort of practice explains why Venetian rule was not popular on the islands (Faroqhi, *Ottoman Empire*, 78).
66 *Reysen*, 101–2.
67 *Reysen*, 103.
68 N. Witsen, *Architectura navalis et Regimen Nauticum* (Amsterdam: Pieter and Joan Blaeu, 1690), 198–9.
69 See van Deursen, *Plain Lives*, 7–8.
70 *Reysen*, 103–7.
71 Again, this is also evident from Anderson (Anderson, *Naval Wars*, 162–3).
72 See de Jonge, *Nederland*, 230–1.
73 M.J. Maxwell, 'Afanasii Nikitin: An Orthodox Russian's Spiritual Voyage in the *Dar al-Islam*, 1468–75,' *Journal of World History* 3 (2006) 243–66: 247–9; see Chapters 9 and 10.
74 See van Deursen, *Plain Lives*, 235, 242, 268: Israel, *Dutch Republic*, 474–7, 637–45. Van Deursen suggests, nevertheless, that with the growth of the full membership of the Reformed Church the zeal of religious experience and behaviour began to erode (van Deursen, *Dorp*, 87).
75 Van Deursen, *Plain Lives*, 286–7; Israel, *Dutch Republic*, 460–5.
76 For example, see Davis, *Christian Slaves*, 125–7; he makes evident that part of that image was a proclivity to homosexuality. Van Deursen observes how this creation of an evil Other was undoubtedly effective: many sailors avoided sailing the Mediterranean especially because of the fear of capture by the Barbary corsairs (van Deursen, *Een Dorp*, 303). In Holland, villages organized regularly collections to pay the ransom for those enslaved by Muslims (van Deursen, *Dorp*, 303–6).
77 *Reysen*, 111; see de Jonge, *Nederland*, 226–7; Anderson, *Naval Wars*, 166–7.
78 W.H. McNeill, *Europe's Steppe Frontier, 1500–1800* (Chicago, 1964), 139.
79 McNeill, *Venice*, 153; McNeill, *Europe's*, 141.
80 *Reysen*, 117.

81 *Reysen*, 118.
82 *Reysen*, 119. Van Royen points out that many Dutch captains in the Mediterranean doubled as privateers, having acquired commissions for this purpose (van Royen, *Zeevarenden*, 109).
83 A. Pagden, 'Europe: Conceptualizing a Continent,' in A. Pagden, ed., *The Idea of Europe* (Washington, DC, 2002) 33–54: 51–2.
84 A. Pagden, 'Europe and the World Around,' in E. Cameron, ed., *Early Modern Europe* (Oxford, 2001) 1–28: 5–6, 21.
85 Peter J. Brenner, 'Der Mythos des Reisens. Idee und Wirklichkeit der europäischen Reisekultur in der Frühen Neuzeit,' in M. Maurer, ed., *Neue Impulse der Reiseforschung* (Berlin, 1999) 13–61: 14, 28; Rietbergen, *Europe*, 292.
86 Brenner, 'Mythos,' 26.
87 C. Ginzburg, *The Cheese and the Worms* (Baltimore, MD, 1980), xiii.
88 In the endeavour to retrieve worthwhile information, a constant awareness of the conventions of various genres and the expectations of what Elisabeth Eisenstein has called the public (as imagined by the publishers and ghost-writer) is of course crucial (see E. Eisenstein, *The Printing Press as an Agent of Change*, 2 vols in one [Cambridge, 2005], 64). I do not apply a distinction between reading 'audiences' and (hypothetical) 'publics', in the manner of Eisenstein here.
89 See further Chapters 11 and 12.
90 *Reysen*, 373; see Chapter 10.
91 See W. Dalrymple, 'Homer in India: The Oral Epics of Rajasthan,' *The New Yorker* (20 November 2006) 48–55: 54.
92 Van Deursen, *Dorp*, 298–9.
93 His name is absent in the city's surviving *verpondingscohieren* (real-estate tax records).

3 The Dutch Republic

1 A. Smith, *An Inquiry into the Nature and Causes of the Wealth of Nations* (Oxford, 1976; origin., 1776), 479–88; K. Marx, *The Capital*, ed. F. Engels vol. 1 (New York, 1967; origin., 1867), 751–5; K. Marx, *The Capital*, ed. F. Engels vol. 3 (New York, 1967), 332–3; M. Weber, 'Die protestantische Ethik und der Geist des Kapitalismus,' in M. Weber, *Gesammelte Aufsätze zur Religionssoziologie* vol. 1 (Tübingen, 1922) 17–206; H. Kohn, *The Idea of Nationalism* (New York, 1967; origin., 1944), 476, 479, 607n67, 637n78, 650–1n51; S.N. Eisenstadt, *Modernization: Protest and Change* (Englewood Cliffs, NJ, 1966), 1; I. Wallerstein, *The Modern World System* 3 vols (New York, 1974–89), see especially vol. 1, 165–221, and vol. 2, 236–71; J. de Vries and A. van der Woude, *The First Modern Economy* (Cambridge, 1997), see especially 160; Schama, *Embarrassment*; J. Israel, *Radical Enlightenment* (Oxford, 2001); Israel, *Dutch Primacy*; Israel, *Dutch Republic*; E.A. Wrigley, 'The Divergence of England,' *Transactions of the Royal Historical Society*, 6th series, vol. 10 (2002) 117–41; Ormrod, *The Rise*; Gorski, *Disciplinary Revolution*; Glete, *War and the State*; K. Davids and J. Lucassen, eds, *A Miracle Mirrored* (Cambridge, 1995). Less convinced of a Dutch advance toward modernity are P.W. Klein, 'Nederland de eerste moderne economie? Een kritiek,'

Tijdschrift voor Geschiedenis 109 (1996) 514–20; some of the essayists in F. Krantz and M. Hohenberg, eds, *Failed Transitions to Modern Industrial Society* (Montréal, 1975); and Fernand Braudel (see Braudel, *Civilization* vol. 3, 215).

2 G.M. Phipps, 'Britons in Seventeenth-Century Russia: A Study in the Origins of Modernization,' unpubl. Ph.D. Diss. (Philadelphia: University of Pennsylvania, 1971), 4–5. Kotilaine and Poe's definition of modernization is narrower, see J. Kotilaine, M. Poe, 'Modernization in the Early Modern Context: The Case of Muscovy,' in *Modernizing Muscovy*, eds J. Kotilaine and M. Poe (London, 2004) 1–8: 3–4. As Simon Dixon points out, modernization theory derives much of its outline from the writings of both Max Weber and Ferdinand Tönnies (see S. Dixon, *The Modernisation of Russia, 1676–1825* [Cambridge, 1999], 1–2, 112; see M. Weber, *Economy and Society*, eds G. Roth, C. Wittich [Berkeley, CA, 1978]; F. Tönnies, *Gemeinschaft und Gesellschaft* [Leipzig, 1887]).

3 Phipps, 'Britons,' 4–5.

4 P. Gorski, *The Disciplinary Revolution* (Chicago, IL, 2003), 39.

5 'As envoy in The Hague [Sir George] Downing laboured for the relocation to England of Dutch families whose manufacturing skills would contribute both to national economic independence and to royal taxes' (J. Scott, '"Good Night Amsterdam": Sir George Downing and Anglo-Dutch Statebuilding,' *English Historical Review* 118 [2003] 334–56: 351) In Northern and Eastern Europe, Louis de Geer, Elias Trip, the Marselis family, Lus Tielman Akkema, and Andries Vinius were hired to build iron foundries and arms factories in Sweden and Russia, and Dutch engineers fortified cities, such as Terki in the Northern Caucasus (see *Reysen*, 214; J. Scheltema, *Rusland en de Nederlanden* 4 vols [Amsterdam, 1817] vol. 1, 121; see Chapter 4).

6 Compare de Vries and van der Woude, *First*, 713.

7 See Israel, *Dutch Primacy*, 197–291.

8 See for example J.L. Price, *Dutch Society, 1588–1713* (Harlow, 2000), 74.

9 B.N. Mironov, *The Social History of Imperial Russia, 1700–1917*, ed. B. Eklof, vol. 1 (Boulder, CO, 2000), 4, Table 1.1; Israel, *Dutch Republic*, 620.

10 De Vries and van der Woude, *First*, 129.

11 Wrigley, 'Divergence of England,' 131–2.

12 Wrigley, 'Divergence of England,' 119–22, 138–9; Ormrod, *Rise*, 346–7; J. de Vries, 'Holland: Commentary,' in Krantz and Hohenberg, eds, *Failed Transitions*, 55–7: 57. On the importance of windmills (and a picture of one of them at Struys's native Wormer), see Prak, *Dutch Republic*, 101–3.

13 See F. Febvre, H.J. Martin, *Coming*, 67, 129–31; Eisenstein, *Printing Press*, 22, 32–5, 44. On the significance of printing for modernization, see Eisenstein, *Printing Press*, 6, and *passim*. '[T]he book was the first modern-style mass-produced industrial commodity' (B. Anderson, *Imagined Communities*, rev. edn. [London, 2006], 34).

14 M. van Delft, C. de Wolf, eds, *Bibliopolis* (Zwolle, 2003), 75–6; Prak, *Dutch Republic*, 226, 241; L. Blussé, 'Op zoek naar een verdwenen manuscript,' in L. Blussé, R. Falkenburg, *Johan Nieuhofs beelden van een Chinareis 1655–1657* (Middelburg, 1987) 9–20: 13. On Dutch print culture, see further van Delft and de Wolf, eds, *Bibliopolis*, 57–8, and C. Berckvens-Stevelinck, ed., *Le Magasin de l'Univers* (Leiden, 1992).

15 During the 1590s, Cornelis Cornelisz. van Uitgeest built the first sawmill not far from Wormer (see H. Kaptein, *Het Schermereiland* [Bergen, 1988], 149). See also van der Woude, *Het Noorderkwartier.*
16 See van Royen, *Zeevarenden,* 15, Table 1–3.
17 Note in this regard Linebaugh's and Rediker's remarks on the Royal Navy (Linebaugh and Rediker, *Many-Headed Hydra,* 148). In an apt metaphor they call ships 'engines of capitalism' (see ibid., 144).
18 Prak, *Dutch Republic,* 273. On the exceptionally lively print culture in the Netherlands, see R. Darnton, *The Literary Underground of the Old Regime* (Cambridge, MA, 1982), ix; Schmidt, *Innocence,* 75–83. On literacy, see de Vries and van der Woude, *First Modern Economy,* Figure 5.1, 171; E. de Booy, *De weldaet der scholen* (Utrecht, 1977); Israel, *Dutch Republic,* 686–90. Some of the relative prosperity, obviously, derived from the ruthless exploitation of people in faraway regions by Dutch merchants and entrepreneurs (see Barendse, *Arabian Seas,* 68; Gorski, *Disciplinary Revolution,* 40; Schmidt, *Innocence,* 138).
19 For example, the German soldier in VOC service Johann von der Behr enjoyed tea in Batavia in January 1650, which was (according to the editor Naber) one of the earliest occasions documented of Europeans drinking tea in East Asia (von der Behr, *Reise nach Java,* 129–30, 129n2).
20 See Chapter 12. See also Weber, 'Die protestantische Ethik,' 23, 178, 202.
21 As a skilled sailmaker, Struys may have enjoyed some advantages over mere deckhands, see A. McCants, *Civic Charity in a Golden Age* (Urbana, IL, 1997), 71–82. We know less of women, who have left us with far fewer ego-documents, see Dekker, *Childhood,* 14.
22 See V.I. Lenin, *Imperialism: The Highest Stage of Capitalism,* in V.I. Lenin, *Collected Works,* vol. 22 (Moscow, 1964) 185–304: 279, 281–5; J.A. Hobson, *The Evolution of Modern Capitalism* (London, 1894).
23 Compare J. Swift, *Gulliver's Travels,* ed. Louis Landa (Boston, 1960), 196–7, with Linebaugh, Rediker, *Many-Headed Hydra,* 143–73; see also van Deursen, *Plain Lives,* 21–6.
24 Glete, *War and the State,* 166.
25 For the social mobility of some sailors, including Tromp and de Ruyter, see van Deursen, *Plain Lives,* 26, 76–8; see the works by Prud'homme van Reine for recent biographies of the two admirals. Compare as well the motivation of sixteenth-century Spaniards to go to sea in Pérez-Mallaína, *Spain's Men,* 23–6.
26 See Israel, *Dutch Primacy,* 292–358.
27 Part of this development was a switch to commission trading instead of trading for their own account, as lucidly outlined by Jonker and Sluyterman (see J. Jonker, K. Sluyterman, *At Home in the World Markets* [The Hague, 2000], 83–91).
28 Price, *Dutch Society,* 105.
29 See Spierenburg, *Judicial Violence,* 55–6.
30 Price, *Dutch Society,* 258–60; Prak, *Dutch Republic,* 263–4, 268; Israel, *Dutch Republic,* 998–9; Israel, *Dutch Primacy,* 377, 397; Wallerstein, *Modern World System* vol. 2, 51, 70. Even so, the wars of the 1670s led to at least a short-term period of hardship (see Prak, *Dutch Republic,* 248–9).

31 See Prak, *Dutch Republic*, 139. Prak uses completely modern *class* rubrics for his breakdown of Dutch society. For the six strata constituting Dutch society, see for example I. Schöffer, 'De Republiek der Verenigde Nederlanden, 1609–1702,' in *De Lage Landen van 1500 tot 1780*, eds I. Schöffer et al. (Amsterdam, 1978) 167–267: 182. The breakdown of society into five or six groups mainly applies to the most populated and powerful, urbanized and maritime province of Holland from which Struys hailed, and it leaves aside the considerable regional variation in the Republic (see Price, *Dutch Society*, 127).

32 P. Burke, *Venice and Amsterdam*, second edn. (Cambridge, 1994), 16.

33 The frequent choice of orphans of the middling groups (*brede middenstand*) who were raised in Amsterdam's *Burgerweeshuis* (one of the city's largest orphanages) to apprentice as sailmakers indicates the profession's respectable status (see McCants, *Civic Charity*, 72, Table 10).

34 Van Deursen, *Plain Lives*, 3.

35 Boxer, *Dutch Seaborne Empire*, 69.

36 See Price, *Dutch Society*, 54–82.

37 D.J. Roorda, *Partij en factie* (Groningen, 1978), 38–9; P. Bourdieu, *Outline of a Theory of Practice* (Cambridge, 1977), 178–9. Julia Adams calls the Dutch state 'a patriarchal patrimonial formation' (see J. Adams, *The Familial State* (Ithaca, NY, 2005), 3).

38 Roorda, *Partij en factie*, 42–3; N. Elias, *The Civilizing Process*, rev. edn. (Malden, MA, 2000), 421–33; Sir William Temple, *Observations upon the United Provinces of the Netherlands*, ed. G. Clark (Oxford, 1972; origin., 1673), 85.

39 Israel, *Dutch Republic*, 818, 960; Prak, *Dutch Republic*, 129. It is tempting to equate these clans with political *facties* (factions), usually differentiated into an Orangist, an anti-Stadtholder, and a moderate group (see Roorda, *Partij en factie*, 1–59; G. Hosking, *Rulers and Victims* [London, 2006], 133, 372–3). Some clan leaders, however, stayed away from political office, whilst their political fall hardly affected the social and economic domination of their clans; still others were barred from holding political office because of their religion (as was a rich and influential Catholic such as Matthijs van Overbeke, see R. Dekker, *Humour in Dutch Culture of the Golden Age* [Basingstoke, 2001], 47). *Facties* were occasionally devastated (as in 1619, 1650 or 1672) by political purges, but kinship networks or clans – intermarried families who dominated the economy and society of the Republic's towns and provinces – tended to maintain their hegemony throughout such episodes. Only on a handful of occasions were leaders of a clan killed, as was the case with Oldebarnevelt in 1619 and the brothers de Witt in 1672.

40 Prak, *Dutch Republic*, 129. See also Schama, *Embarrassment*, 341.

41 Jones, *Anglo-Dutch Wars*, 54; Peters, 'Nepotisme,' 83; Prak, *Dutch Republic*, 125–9; I. Buruma, *Murder in Amsterdam* (New York, 2006), 49–51; Barendse, *Arabian Seas*, 400–2. Muscovy had similar clan and patronage networks: for example, see R. Crummey, *Aristocrats and Servitors* (Princeton, NJ, 1983), 82–106, 167; Hughes, *Russia*, 416–17; V. Solov'ev, *Anatomiia russkogo bunta* (Moskva, 1994), 19; and N.S. Kollmann, *Kinship and Politics* (Stanford, CA, 1987). For the importance of networks in non-Western societies, see for example, Bourdieu, *Outline*, 178–80, and for *guanxi*, the comparable Chinese

version of social capital that sustained Jesuit missionary efforts there, see L.M. Brockey, *Journey to the East* (Cambridge, MA, 2007), 34.

42 John Ralston Saul during a TV interview on C-SPAN (USA) live, 19 November 2005.

43 In the 1690s, Witsen became a board member of the VOC (see M. Peters, 'From the Study of Nicolaes Witsen (1641–1717): His Life with Books and Manuscripts,' *Lias* 1 [1994] 1–45: 1).

44 And this network mirrored the behaviour of the ruling elite in the Republic (see for example M. Prak, *Gezeten Burgers* [Leiden, 1985], 10–15).

45 The Russian historian Kovrigina believes that family ties superseded any national allegiance, which is probably correct, but much of the culture of a considerable part of the Western-European business circles in Muscovy seems to have originated in the Dutch Republic, as did the ancestry of many traders, entrepreneurs, and agents (see V.A. Kovrigina, 'Inozemnye kuptsy-predprinimateli Moskvy Petrovskogo vremeni,' in L.A. Timoshina, I.A. Tikhoniuk, eds, *Torgovlia i predprinimatel'stvo v feodal'noi Rossiii*, [Moskva, 1994] 190–213: 208–9).

46 See further Chapter 5.

47 See Elias, *Civilizing Process*, 427–33; Adams, *Familial*, 47–8.

48 Gorski, *Disciplinary Revolution*, 40; McCants, *Civic Charity*, 3–6, 11, 15.

49 M.C. 't Hart, *The Making of a Bourgeois State: War, Politics, and Finance during the Dutch Revolt* (Manchester, 1993), 4–5.

50 See Israel, *Dutch Republic*, 1098–102; S. Schama, *Patriots and Liberators* (New York, 1992), 56–7, 64–8; Kohn, *Idea*, 478–81.

51 Gorski, *Disciplinary Revolution*, 39–40; Glete, *War and the State*, 143; Schama, *Embarrassment*, 224.

52 See Geoffrey Parker, *Success Is Never Final* (New York, 2002), 39–66.

53 *Wapenhandelinghe van roers, musquetten ende spiessen* ('s-Gravenhage: n.p., 1607); see [Johann Jacobi von Wallhausen] *Uchenie i khitrost' ratnogo stroeniia pekhotnykh liudei* (Moscow: Moskovskii pechatnyi dvor, 1647); see J.J. Driessen-van het Reve, *De Kunstkamera van Peter de Grote* (Hilversum, 2006), 28; Iu. Kurskov, *Vedushchee napravlenie obshchestvennoi mysli i proekty gosudarstvennykh preobrazovanii Rossii 40–60-kh godov XVII veka* (Chita, 1973), 110; G. Hosking, *Russia and the Russians* (Cambridge, MA, 2001), 154; the first English translation was Jacob de Gheyn, *The Exercise of Armes for Caliures, Muskettes, and Pikes* (The Hague: n.p., 1608).

54 R. Wittram, *Peter I, Czar und Kaiser* vol. 1 (Göttingen, 1964), 153.

55 Israel, *Dutch Republic*, 818; Van Nimwegen, *Deser landen*, 25, 268. Compare with Louis XIV's army, which surpassed – at least on paper – the quarter million mark in the 1670s (see J. Lynn, *The Wars of Louis XIV, 1667–1714* [London, 1999], 50–1). By then, the Dutch drill had for three generations disciplined its soldiers' bodies into an early version of the mechanically docile units that would march for eighteenth-century Prussia. For a recent discussion which somewhat downplays the Dutch importance for the Military Revolution, see van Nimwegen, *Deser landen*, 17–26, 249.

56 For the size of the Russian army, see R. Hellie, *Enserfment and Military Change in Muscovy* (Chicago, 1971), 183.

57 See Barendse, *Arabian Seas*, 381–5.

58 See H. Kamen, *Empire* (New York, 2004), 487–9.
59 For some of the less attractive side to their eastward explorations and conquests, see Richard H. Grove, *Green Imperialism* (Cambridge, 1995), 44–5, 54, 135, 150–1. See also G. Milton, *Nathaniel's Nutmeg* (London, 1999), 343.
60 De Vries and van der Woude, *First Modern Economy*, 714; also see Glete, *War and the State*, 6, 145, 172.
61 Van Nierop, 'Catholics,' 109–11.
62 See M. Foucault, *Discipline and Punish: The Birth of the Prison* (New York, 1995), 77–8; Israel, *Dutch Republic*, 864–5. Spierenburg's dissertation shows how the penalty of imprisonment was still accompanied by various corporal punishments and torture was routinely used in pre-trial investigations in the Republic (see Spierenburg, *Judicial Violence*, 11, 74–8, 147, 152–3, 173–8). But there were already treatises diffused advocating torture's abolition since 1624, and torture was applied in less than one in a hundred trials in the period between 1651 and 1683 (ibid., 149, 157–8).
63 Spierenburg, *Judicial Violence*, 176. Spierenburg's words here summarize Dapper's point (see O. Dapper, *Historische beschryving der stadt Amsterdam* [Amsterdam: J. van Meurs, 1663], 425–32). Spierenburg notes how in the houses of correction (*tuchthuizen*) corporal punishment was applied as well.
64 Temple, *Observations*, 88. This sort of planned 'regularity' mightily impressed Peter the Great, who tried to copy it in building his new capital of St Petersburg (Hughes, *Russia*, 382).
65 Nicholas Thomas, *Colonialism's Culture* (Cambridge, 1994), 4; see as well Gorski, *Disciplinary Revolution*, 18–21, 25, 164; Dekker, *Humour*, 10–11; Spierenburg, *Judicial Violence*, 202–3.
66 For Struys's praise for the Dutch government, see *Reysen*, 141–2, 167–8.
67 See ibid., 148.
68 This acceptance of harsh retribution given to offenders seems to transpire from the castration episode, see *Reysen*, 357–8 and Chapter 10.
69 Especially *Reysen*'s description of Struys's wanderings through the Islamic world of the Caucasus and Persia illustrates this mindset (see *Reysen*, 215–368, and see below in Chapters 9 and 10; John Bunyan, *The Pilgrim's Progress from This World* [London: N. Ponder, 1678]).
70 Dekker, *Childhood*, 12–13; Gorski, *Disciplinary Revolution*, xv–xvi; Eisenstadt, *Modernization*, 5. Also see for the advance of individualism Elias, *Civilizing Process*. The Dutch case is exceptional for the widespread diffusion of its individualism, not for the emergence of the phenomenon as such; see for emerging contemporary Muscovite individualism, see V.M. Zhivov, 'The Emergence of the Individual,' in *Religion and Culture*, eds S. Baron and N. Shields Kollmann (DeKalb, IL, 1997) 184–98.
71 Van Deursen, *Dorp*, 31–46. '*Struys*' indicated someone strong, tall, or of heavy physical stature, or who was brave or courageous.
72 Schama, *Embarrassment*, 34–7.
73 On this, and especially the availability of the authorized Dutch-language Bible or *Statenbijbel* from 1637 onwards, see A. Hastings, *The Construction of Nationhood* (Cambridge, 1997), 8, 24.

74 '[E]ncounters of different peoples ... crystallize political ideologies and concepts of identity' (E. Foner, 'American Freedom in a Global Age,' *American Historical Review* 1 [2001] 1–16: 4).

75 Prak, *Dutch Republic*, 141; van Deursen, *Plain Lives*, 37; 't Hart, *The Making*, 5; B.J. Kaplan, ' "Dutch" Religious Tolerance: Celebration and Revision,' in R. Po-chia Hsia and van Nierop, eds, *Calvinism*, 8–26: 17–18; Gorski, *Disciplinary Revolution*, 163.

76 Orlenko, *Vykhodtsy*, prilozhenie 2, 287.

77 For Slavic Muscovites religion was clearly the first marker of identity; in Russian the word for Orthodox is *pravoslavnyi*, literally meaning the *right worship*. But loyalty to the tsar was at least part of the self-image of Russian speakers. The *Ulozhenie* (law code) of 1649 did distinguish baptized *inozemtsy* ('foreign residents') as different from native Orthodox (see Orlenko, *Vykhodtsy*, 80).

78 See Orlenko, *Vykhodtsy*, 102–3).

79 For example, see R. Tillyard, *The Elizabethan World Picture* (New York, 1944); H.J. Cook, *Matters of Exchange: Commerce, Medicine and Science in the Dutch Golden Age* (New Haven: Yale UP, 2007), and K. Thomas, *Man and the Natural World* (Oxford, 1996).

80 See *Reysen*, 176. E. Van Den Boogaart, *Civil and Corrupt Asia* (Chicago, IL, 2003), 10: 'Some morals and practices were better than others in terms of technology or religion. Not only were there differences in civility within and between societies, but there was a hierarchy of civility, differing degrees of socially enforced restraint of the passions and socially beneficial application of reason.' See also ibid., 29. For further evidence of racialist bigotry, see Breet, *Oost-Indische Voyagie*, 248, and, for a detailed analysis of Dutch supremacist and hostile view of Asians, see M.P. Vink, 'Images and Ideologies of Dutch-South Asian Contact,' *Itinerario* 2 (1997) 102–12. Thomas agrees that 'racism' needs to be differentiated in analysing colonialist attitudes (Thomas, *Colonialism's Culture*, 14).

81 *Reysen*, 242. On Muslims and Jews, see Breet, *Oost-Indische Voyagie*, 263–4; Scheidegger, *Perverses Abendland*, 23. Thomas emphasizes the importance of religion in justifying Christian superiority in colonialist projects (Thomas, *Colonialism's Culture*, 14). Europeans' positive estimation of the Chinese was exceptional (see van den Boogaart, *Civil and Corrupt*, 21–2).

82 *Reysen*, 175. For Schouten's contempt of Hindus, see Breet, *Oost-Indische Voyagie*, 257–8.

83 *Reysen*, 185. See also ibid., 220, where Dagestani are called 'horrible to look at'.

84 On Others' duplicity and sexual abandon, see *Reysen*, 176, 217, 233–5. See also Breet, *Oost-Indische Voyagie*, 250, 261–2, 445 (but note how the Dutch also engaged in rape; see ibid., 214).

85 See M. Habsmeier, 'Reisebeschreibungen als mentalitätsgeschichtliche Quellen: Überlegungen zu einer historisch-anthropologischen Untersuchung frühneuzeitlicher deutscher Reisebeschreibungen,' in *Reisebeschreibungen als Quellen europäischer Kulturgeschichte*, eds A. Maczak and H.J. Teuteberg (Wolfenbüttel, 1982) 1–32: 3–4.

86 Milton, *Nathaniel's Nutmeg*, 349–50; Breet, *Oost-Indische Voyagie*, 417.

87 See Chapters 9 and 10.
88 Compare to Schama, *Embarrassment*, 589.
89 Anderson, *Imagined Communities*, 3, 40.
90 Schama, *Embarrassment*, 6–7.
91 The role of printing in forming this national identity was considerable even if it exerted little influence on wholly illiterate minority, see Eisenstein, *Printing Press*, 117–18, 127. P.C. Hooft, *Neederlandsche histoorien* (Amsterdam: Elzevier, 1642).
92 Schmidt, *Innocence Abroad*, xxiii, 85, 228–39, 252–3. Further on seventeenth-century Dutch identity, see for instance van Deursen, *Plain Lives*, 42–3, 45–6. P.C. Hooft wrote a history of the Dutch revolt after the model of Tacitus's *Annales* (see Hooft, *Neederlandsche histoorien*). The pride in one's Dutch identity was combined with a more local pride in one's town, at least in Holland (see McCants, *Civic Charity*, 9). In the orphanage for children from middling groups in Amsterdam the teaching of 'fatherland' (*vaderlandsche*) history was part of the curriculum in the fourth form (see ibid., 38, 63).
93 McCants, *Civic Charity*, 9.
94 See de Jonge, *Nederland*, 259. In Dutch: '*Naest Godt, den grooten zegenaer, Heeft zich de Batavier gequeeten, Met raet en daet, op't zeealtaer.*' Vondel was not alone (see de Jonge, *Nederland*, 394–6).
95 For the pirates, see Chapter 2, and Prud'homme van Reine, *Rechterhand*, 117, which identifies in the 1660s the corsair activities of one Yusuf Reys, a Barbary pirate admiral stationed in Tunis, who had been born as Gerrit Jacobsz in Enkhuizen.
96 See Chapter 6. Roorda notes how Dutch critics believed patriotism to be commensurate with wealth: the richer one was, the more patriotic, for the more one had to lose (Roorda, *Partij en factie*, 41).
97 For Struys's repeated choice to serve a foreign master, see *Reysen*, *passim*, and for his service for the tsar, S. Soloviev, *A History of Russia from Earliest Times*, vol. 23, trans. and ed. M.L. Lahara (Gulf Breeze, FL, 1998), 81–4. For the modern quality of travel as such, see Thomas, *Colonialism's Culture*, 5.
98 I use the term 'culture' here in the straightforward sense of the 'intellectual, spiritual, and aesthetic' (see R. Williams, *Keywords*, rev. edn. [London, 1983], 90).
99 Meanwhile, in most of Western and Central Europe flourishing genres in literature or plastic arts entered a period of transition toward 1700 (see M.D. Knowles, 'Presidential Address: Great Historical Enterprises I. The Bollandists,' *Transactions of the Royal Historical Society*, fifth series 8 [1958] 147–66: 147).
100 M.C. Brands, rev. of *Op het Breukvlak van Twee Eeuwen* by Jan Romein, *History and Theory* 1 (1970) 116–21: 117; see also K.W. Swart, 'Holland's Bourgeoisie and the Retarded Industrialization of the Netherlands,' in *Failed Transitions*, eds Krantz and Hohenberg, 44–8: 47–8.
101 Brands, rev. of *Op het Breukvlak*, 117.
102 Brenner, 'Mythos,' 17–18, 47–8. For Peter the Great's fondness of travel accounts, one of which has young Russian sailors travel to Holland (!), see Hughes, *Russia*, 326.

103 A. Olearius, *Offt begehrte beschreibung der neuen orientalischen Reise* (Schleswig: n.p., 1647). An expanded edition followed in 1656: Adam Olearius, *Vermehrte Neue Beschreibung der Moskowitischen und Persischen reise* (Schleswig: J. Holwein, 1656).

104 See note 106 below for the 1651 Dutch editions of Olearius's work.

105 Olearius, *Offt begehrte beschreibung*; his work, too, was mined by Balthasar Coyett (see A.M. Loviagin, 'Vvedenie,' in *Posol'stvo Kunraada fan' Klenka*, ed. A.M. Loviagin, [St Petersburg, 1900] iv–clxxvi: cxx–cxxiii). Olearius was not quite original, for his work followed the model imposed by Herberstein in 1549 (see Herberstein, *Rerum*).

106 Olearius, *Offt begehrte beschreibung*; Olearius, *Vermehrte Neue Beschreibung*; Olearius, *The Voyages & Travels of the Ambassadors sent by Frederick Duke of Holstein*, trans. John Davies (London: Thomas Dring and John Starkey, 1662); *The Voyages and Travells of the Ambassadors*, second edn., trans. John Davies (London: Starkey and Basset, 1669). Three Dutch editions of Olearius thus saw the light in 1651: A. Olearius, *Beschrijvingh vande nieuwe Parciaensche, ofte orientaelsche reyse*, trans. Dirck van Wageninge (Amsterdam: Jacob Benjamyn and Adriaen Roest, 1651); A. Olearius, *Beschrijvingh van de nieuwe Parciaensche Oste Orientaelsche reyse*, trans. Dirck Van Wageninge (Utrecht: L. Roeck, 1651); A. Olearius, *Persiaensche Reyse uyt Holsteyn, door Lijflandt, Moscovien, Tartarien in Persien*, second edn. (Amsterdam: Jan Jansz, 1651).

107 Brenner, 'Mythos,' 49.

108 For an overview, see for instance Prak, *Dutch Republic*, 222–33; Cook, *Matters, passim.*

109 On Spinoza's importance for the beginnings of the Enlightenment, see Israel, *Radical Enlightenment*, 704–11. A crucial book compiled and published in Holland that prepared the ground for the Enlightenment was P. Bayle, *Projet et fragment d'un dictionnaire critique* (Rotterdam: Reinier Leers, 1692).

110 On respect for foreign doctors among Russian elite, see for instance Philip Longworth, *Alexis* (New York, 1984), 134; Breet, *Oost-Indische Voyagie*, 211.

111 'Extract' [Butler], 17, 32. Termund, about whom more below, would later be a boon companion of Tsar Peter the Great.

112 The friendship with the entomologist Swammerdam can be found in W.J. Schiphouwer, 'Jan Jansen Struys,' *De Zaende* 6 (1951) 199–205: 202, but the evidence may be little more than *Reysen*'s reference to it (see *Reysen*, 191; see also [W.C. Ackerdijck], 'Verhandeling over den Nederlandschen reiziger Jan Jansen Struijs,' *De Fakkel* [Rotterdam], 1826, 117). For Swammerdam, see A. Schierbeek, *Jan Swammerdam, 1637–1680* (Amsterdam, 1967). It is equally likely that *Reysen* refers here to his father, Jan Jacobsz Swammerdam (d. 1678), a famous collector of naturalia (see Cook, *Matters*, 30, 141–2, 277–8, 281).

113 See M. Spufford, 'Literacy, Trade and Religion in the Commercial Centres of Europe,' in *Miracle Mirrored*, eds Davids and Lucassen, 229–83: 263.

114 His first scientific work was on shipbuilding: see N. Witsen, *Aeloude en Hedendaegsche Scheeps-Bouw en Bestier* (Amsterdam: C. Commelijn, and Broer en Jan Appelaer, 1671).

115 N. Witsen, *Noord en Oost Tartarye* (Amsterdam, 1705; first edn., 1692). About the impact of this enumeration, Mikhail Bakhtin noted how '[g]eneralization, empirical abstraction, and typification acquired a leading role in the world picture [in the seventeenth century],' see M. Bakhtin, *Rabelais and His World* (Cambridge, MA, 1968), 115. Rietbergen presents Witsen as a prime example of someone immersed in such efforts (Rietbergen, *Europe*, 310–11).

116 Grove ascribes the pioneering Dutch interest in 'redescribing and revaluing the natural world' to a Calvinist-infused 'anxiety about society and its discontents' (Grove, *Green Imperialism*, 14–15).

117 For the spread of a secular mindset, see Israel, *Dutch Republic*, 361–98, 581–91, 637–76, 690–9, 889–9, and Israel, *Radical Enlightenment*, 14–22.

118 For Dutch religious tolerance, see Schama, *Embarrassment*, 59–62, 122, 339–43; Israel, *Dutch Republic*, 361–98, 637–45; for some of its limits, see Prak, *Dutch Republic*, 219–20; C. Parker, 'Paying for the Privilege,' *Journal of World History* 3 (2006) 267–96: 287–95; and R. Po-Chia Hsia, 'Introduction,' in *Calvinism*, eds R. Po-chia Hsia and van Nierop, 1–7: 5.

119 See Cook, *Matters*, 158–9.

120 B. Bekker, *De betoverde weereld, zijnde een grondig ondersoek van't gemeen gevoelen aangaande de geesten* (Amsterdam: n.p., 1691–3).

121 Price, *Dutch Society*, 128–9. Ginzburg suggests that there may have been across Europe a 'popular current ... favoring toleration' (Ginzburg, *Cheese*, 49).

122 See Schama, *Embarrassment*, 148.

123 Especially if he was indeed Dapper (see Chapter 12),

124 *Reysen*, 158–9, 160–1, 163–5.

125 For Wouter Schouten's much more zealous Calvinism, see Breet, *Oost-Indische Voyagie*, 278, 319, 371, 463, 484, 521; Schouten, *Oost Indische Voyagie*.

126 The 1670s was the decade when Spinoza's pantheism was spreading among Dutch intellectuals, and accusations of atheism had been levelled in the 1660s at religious moderates, the heirs to the Arminian Remonstrants. The historical Jan Struys chose to live in his final days in the originally Remonstrant enclave of Friedrichstadt, a haven of tolerance (see Israel, *Dutch Republic*, 463, 654, 692, 697, 1047–9). We already saw how van Someren had published work by Labadie, a notorious (albeit zealously Christian) schismatic, in 1667, before his teachings were condemned by the Dutch Reformed Church (see Israel, *Dutch Republic*, 669–71; Schama, *Embarrassment*, 411; de Labadie, *erheffingen des geestes*). Olfert Dapper, van Meurs's favourite writer and the probable ghostwriter of *Reysen*, possibly had the reputation of an atheist (see Chapter 12 and Dozy, 'Olfert,' 423–4).

4 The Dutch in Muscovy

1 On the transfer of technology from the Republic, see de Vries and van der Woude, *First*, 345, 348–9.

2 He was incapacitated in 1665 and died in 1668 (see G. Kotoshikhin, P. Gordon, J. Streis, Tsar' Aleksei Mikhailovich, *Moskoviia i Evropa* [Moskva,

2000], 196; R. O. Crummey, *Aristocrats and Servitors* [Princeton, NJ, 1983], 85; P. Longworth, *Alexis* [New York, 1984], 190).

3 For the Dutch trade with the Baltic ports, see den Haan, *Moedernegotie*; R. Daalder, ed., *Goud uit graan* (Zwolle, 1998); N. Davies, *God's Playground*, vol. 1 (New York, 2005), 98. For a while in the mid-sixteenth century, Narva had been a Russian port, on which the Dutch traded; after Narva's loss, this trade moved to Arkhangel'sk in the 1580s, although some Russian goods continued to be shipped in Riga (see S. Baron, 'Shipbuilding and Seafaring in Sixteenth-Century Russia,' in *Essays in Honor of A.A. Zimin*, ed. D. Waugh [Columbus, OH, 1983] 102–29: 109; H. Nolte, 'The Netherlands and Russia in the 17th Century,' *Review of the Fernand Braudel Center* 2 [1986] 230–44: 233n5; Veluwenkamp, *Archangel*; E.H. Wijnroks, *De handel tussen Rusland en de Nederlanden 1560–1640* [Hilversum, 2003]; T.S. Jansma, 'Olivier Brunel te Dordrecht,' *Tijdschrift voor Geschiedenis* 59 [1946] 337–62; C.C. Uhlenbeck, *Verslag aangaande een onderzoek van de archieven van Rusland* ['s Gravenhage, 1891], 247–8). For Dutch grain shipments from Arkhangel'sk, see Veluwenkamp, *Archangel*, 100, 125–7, 132; A.V. Demkin, A.V. Demkin, *Zapadnoevropeiskoe kupechestvo v Rossii v xvii v.* 2 vols (Moskva, 2004) vol. 1, 20; Loviagin, 'Vvedenie', lxxviii–lxxix.

4 For example, see Veluwenkamp, *Archangel*, 27–34.

5 L. Khakkebord, 'Arkheologicheskie i kartograficheskie aspekty rannikh russko-gollandskikh torgovikh sviazei na Russkom Severe,' in *Niderlandy i Severnaia Rossiia*, eds Iu.N. Besplatnikh et al. (St Petersburg, 2003) 7–13: 8–9; Scheltema, *Rusland* vol. 1, 38–9, 42–4.

6 For some of the losses sustained by Dutch merchants during the Time of Troubles, see J. Kotilaine, *Russia's Foreign Trade and Economic Expansion in the Seventeenth Century* (Leiden, 2005), 71.

7 And like the Marselis family, the de Moucherons branched out across Europe, so that by 1632 one of them was the representative of the Duke of Schleswig-Holstein in Moscow (see Adam Olearius, *The Travels of Olearius in Seventeenth-Century Russia*, ed. S. Baron [Stanford, CA, 1967], 59; E. Amburger, *Die Familie Marselis* [Giessen, 1957]). On the de Moucherons, see Veluwenkamp, *Archangel*, 46–9. They were related to the de la Dale and Ruts families.

8 Veluwenkamp, *Archangel*, 27–34, 46–62; N.N. Bantysh-Kamenskii, *Obzor' vneshnikh snoshenii Rossiia* vol. 1 (Moskva, 1894), 174–6; N.N. Repin, 'Gollandskie kuptsy v Arkhangel'ske vo vtoroi polovine xviii v.,' in *Niderlandy i Severnaia Rossiia*, eds Besplatnikh et al., 14–36: 20; Scheltema, *Rusland* vol. 1, 60, 102–15; J. Keuning, 'Isaac Massa, 1586–1643,' *Imago Mundi* X (1953) 64–79.

9 They also all helped to found the VOC (see Scheltema, *Rusland* vol. 1, 62–3).

10 Ia. V. Veluvenkamp, 'Kompaniia "de Vogelar i Klenk" v gollandsko-russkikh kommercheskikh otnosheniiakh xvii v,' in *Niderlandy i Severnaia Rossiia*, eds Besplatnikh et al., 37–73: 37. Adams notes how these familial compacts among the mercantile elite were strongly developed in the Republic, and also monopolized the various political offices (see Adams, *Familial*, 33–4).

11 Maria Van Sweeden-Ruts managed to persuade the tsar to confirm her ownership of many of her husband's enterprises and to pay her for many of the costs her husband had recently incurred in hiring artisans, including the sailors, to work for the tsar – although she was forced to petition later again

for full remittance (Rossiiskaia Gosudarstvennaia Arkhiv Drevnikh Aktov [Russian State Archive for Ancient Acts] in Moscow [from here abbreviated as *RGADA*] *fond* 141 *delo* 371 (1668), *listy* 1–8 [from here abbreviated as 141/371 (1668), ll.1–8] and 159/*opis'*2/1363 [no later than 1675], ll.1–3). See as well Demkin, *kupechestvo* vol. 2, 12 (he errs by a year). The agency of Dutch women is also apparent from the 1670s petition by *Alenitsa Ivanova dochershika* (Aaltje Jansdochter?), the widow of the *Oryol* carpenter Dirk Pietersz. for her husband's wages (see *Krest'ianskaia voina pod predvoditel'stvom Stepana Razina*, vol. 4 [Moscow, 1954], 135). The model of female assertiveness may have been the widow of 'founding father' Marcus de Vogelaer, Margriet van Balkenburg, who took over his business in 1610, including his seat among the VOC's *Heeren XVII* (Veluvenkamp, 'Kompaniia,' 44–5).

12 Kotoshikhin et al., *Moskovia*, 224; M.A. Obolenskii, M.E. Possel't, *Dnevniki generala Patrika Gordona* (Moskva, 1892), 18n21. On Gordon, see also G.P. Herd, 'General Patrick Gordon of Auchleuchries: A Scot in Seventeeth Century Russian Service,' Unpubl. Ph.D. Diss. (Aberdeen: University of Aberdeen, 1994). Van Sweeden may have enlisted the aid of Colonel Cornelis van Bockhoven to persuade the tsar and Ordin to choose Holland as recruiting ground for shipwrights and crew (Arkheograficheskaia kommissia, *Dopolneniia k aktam istoricheskim*, 12 vols (St Petersburg, 1846–77) [from here indicated as *DAK*] vol. 5, 218).

13 Veluwenkamp, *Archangel*, 108–10. The ban was caused by Muscovite anger at the beheading of Charles I.

14 See Demkin, *kupechestvo* vol. 1, 26, 33; Orlenko, *Vykhodtsy*, 45, 50.

15 Scheltema, *Rusland* vol. 1, 69.

16 For example, see the efforts of 1658 and 1659 involving the '*goosten*' (from Russian *gosti*) de Vogelaer and Klen(c)k, with John Hebdon representing the tsar, in *GAA NA* 2205 (Notary A. Lock), p. 786 (26 November 1658), and *GAA NA* 2206 (Notary A. Lock), pp. 19–20 (3 January 1659). Klenk in 1676 also purchased grain on behalf of the Republic (Loviagin, 'Vvedenie,' xliii–iv). See I.M. Kulisher, *Ocherk istorii russkoi torgovli* (Peterburg, 1923), 140; Kotilaine, *Russia's*, 76–7, 81.

17 Scheltema, *Rusland* vol. 1, 268. Gurliand' lists about ten major arms deals between denizens of the Republic and the tsarist government between 1618 and 1660 (see I.Ia. Gurliand', *Ivan' Gebdon' kommissarius' i rezident'* (Iaroslavl', 1903), 1, 9–10). The first recorded sale of arms by Dutch merchants to the tsar was sanctioned by the Estates-General in 1618 (see Keuning, 'Isaac Massa,' 71). At the beginning of the war with Poland in 1654, Vinius had been buying carbines in Amsterdam (see J.G. van Dillen, ed., *Bronnen tot de geschiedenis van het bedrijfsleven en het gildewezen van Amsterdam* vol. 3 [Den Haag, 1974], 623–4). In 1660, the companionship of de Vogelaar and Klenk provided the tsar with a crucial loan to finance some weapons purchases (Gurliand', *Ivan' Gebdon'*, 21). The sums involved were enormous, as the total transaction was to amount to 90,000 rubles, or 450,000 guilders, according to Giurland'; the Dutch merchants were paid back mainly in hemp (ibid., 21).

18 In 1630 about one hundred Dutch ships arrived and only a handful of English ones, but this was a peak year (Kulisher, *Ocherk istorii*, 128–9).

19 J. Kotilaine, 'When the Twain Did Meet: Foreign Merchants and Russia's Economic Expansion in the Seventeenth Century,' Unpubl. Ph.D. Diss. (Harvard University, 2000), 1055, 1057.

20 The Hals is property of the Rijksmuseum in Amsterdam, while the Ruts portrait is in New York's Frick Collection.

21 Glete, *War and the State*, 173; Hinton, *Eastland Trade*, 10; G.V. Scammell, *The First Imperial Age* (London, 1989), 101–5; Jonker, Sluyterman, *At Home*, 63–4, 67, 73.

22 Veluvenkamp, 'Kompaniia,' 41–2. Marcus Joostsz. de Vogelaer was one of the original directors of the VOC in 1602 (Veluvenkamp, 'Kompaniia,' 44). For the entrepôt function of Amsterdam, see de Vries, *Economy*, 120–2.

23 As we saw, the rate of intermarriage between the main Dutch trading families involved in the trade on Muscovy during the seventeenth century was high (see Veluwenkamp, *Archangel*, 203–8).

24 Wittram, *Czar* vol. 1, 146; M.M. Bogoslovskii, *Petr I. Materialy dlia biografii* [Moskva, 1940], 124n1. The Brandenburg Elector was after 1700 known as King Frederick I of Prussia, whose mother's father was Frederick Henry of Orange (1583–1647; ibid., 124n1). As curious is the evidence of a letter written in Dutch in 1618 by John Merrick, a Russian-based English merchant and agent, to the Swedish general de la Gardie (see Uhlenbeck, *Verslag*, 6–7). See also Burke, *Toward*, 14.

25 See S.T. Christensen, 'Introduction, ' *Scandinavian Journal of History* 3–4 (2003) 151–64: 157; Janis Kreslins, 'Linguistic Landscapes in the Baltic,' *Scandinavian Journal of History* 3–4 (2003) 165–74: 169. On the diffusion of Dutch cultural influence, which also spread to the Rzeczpospolita and Brandenburg-Prussia, see K.A. Ottenheym, 'Dutch Contributions to the Classicist Tradition in Northern Europe,' *Scandinavian Journal of History* 3–4 (2003) 227–42: 241.

26 Kreslins, 'Linguistic Landscapes,' 169. See also P. Burke, *Toward a Social History of Early Modern Dutch* (Amsterdam, 2005), 14.

27 Burke, *Toward*, 14.

28 S. Collins, *The Present State of Russia*, ed. Marshall Poe (London: Dorman Newman, 1671), 128–9.

29 Ibid. For more on the publication history and editions of *The Present State*, see I. Osipov, *Obzor svedenii ob izdaniiakh i perevodakh v xvii stoletii sochineniia Semiuelia Kollinsa 'The Present State of Russia'* (Komi, Russian Federation, 2007).

30 V. Barbour, *Capitalism in Amsterdam in the Seventeenth Century* (Baltimore, 1950), 115.

31 Adams situates the formation of the Dutch mercantile elite in the half century before 1620, after which it closed itself off to newcomers (see Adams, *Familial*, 38).

32 See de Vries and van der Woude, *First*, 362.

33 See I.P. Kozlovskii, *Andrei Vinius', sotrudnik Petra Velikogo (1641–1717 g.)* (Sankt-Peterburkh, 1911), 5; D.S. van Zuiden, 'Nieuwe bijdrage tot de kennis van de Hollandsch-Russische relaties in de 16e-18e eeuw,' *Economisch-Historisch Jaarboek* 2 (1916) 258–95: 276–80; Olearius, *The Travels*, 120 and 120n26.

34 Van Zuiden, 'Nieuwe bijdrage,' 276.

35 See Orlenko, *Vykhodtsy*, prilozhenie 2, 287.

36 Kozlovskii, *Andrei Vinius'*, 8; Wittram, *Czar*, vol. 1, 73.

37 Amsterdam was an unrivalled collecting point of information, one of several reasons for the Dutch advance in the seventeenth century (W.D. Smith, 'The Function of Commercial Centers in the Modernization of European Capitalism,' *Journal of Economic History* 4 [1984] 985–1005: 987).

38 See Lach and Van Kley, *Asia* vol. 3, Book 1, 89. Dutch merchants had a hand in the French *Compagnie des Indes*, and both the Ostend and Swedish East India Companies (see Scammell, *First Imperial Age*, 236).

39 For a while, they headed together the Danish king's salt company: for example, see *GAA NA* 3410 (Notary Philips Engelbrecht), 301 (4 June 1666).

40 Amburger, *Familie*, 7, 24–5, 94–5, 97–8; Wittram, *Czar* vol. 1, 73; R. Hellie, *The Economy and Material Culture of Russia, 1600–1725* (Chicago, 1999), 133, 211n7. On the Trip and de Geer enterprises, see Prak, *Dutch Republic*, 122–4; H. Nusteling, *Welvaart en Werkgelegenheid in Amsterdam, 1540–1860* (Amsterdam, 1985), 188; Jonker, Sluyterman, *At Home*, 58–61. Most detailed and systematic on foreign merchants is the recent work by A.V. Demkin (Demkin, *kupechestvo* vol. 1, 16, and *passim*). On the Dutch in Sweden, see Braudel, *Civilization* vol. 3, 252.

41 See for example U. Birgegard, ed., *J.G. Sparwenfeld's Diary of a Journey to Russia 1684–7* (Stockholm, 2002), 304–5n482. Hellie disparages the quality of the production at Tula, which forced Russians to continue to import better quality iron from Sweden (see Hellie, *Economy*, 148–9, 401). On the Coyetts, see Reger, 'In the Service,' 20; Phipps, 'Britons,' 258; Muliukin, *Priezd*, 100–4.

42 Amburger, *Familie*, 7; Barbour, *Capitalism*, 119.

43 Jonker, Sluyterman, *At Home*, 83. See for instance some of the notarial records where as Ruts's representative in Muscovy he hires apprentices in Amsterdam (*GAA NA* 1100, p. 123, 6 June 1652 [Notary J. v.d. Ven]; *GAA NA* 1100, p. 147, 13 June 1652 [Notary J. v.d. Ven]; *GAA NA* 1100, p. 203, 27 June 1652 [Notary J. v.d. Ven]; *GAA NA* 1100, p. 220, 4 July 1652 [Notary J. v.d. Ven]).

44 *GAA NA* 1079 (Notary J. v.d. Ven), p. 169 (25 June 1646); Veluwenkamp, *Archangel*, 149.

45 Demkin, *kupechestvo* 2, 30–1; Bantysh-Kamenskii, *Obzor'*, 185; Kotilaine, *Russia's*, 81–2, 276, 291–2, Table 5.5. In one of the Amsterdam notarial records, the import of pistols and carbines to Arkhangel'sk by seven ships hired by van Sweeden is described as 'contraband' (*GAA NA* 3015 [Notary H. Venkel], p. 795 [15 November 1660]). Some of his arms imports were stolen in 1661 (see Demkin, *kupechestvo* vol. 1, 116). Likewise, Struys's later benefactor Klenk was involved in the arms trade (Locher, de Buck, 'Inleiding,' xxv–xxviii). The number of arms in such transactions was enormous: van Sweeden and his partner Hendrick Swellengrebel imported at least 100,000 firearms into Muscovy around 1660 (Veluwenkamp, *Archangel*, 149). Despite considerable domestic production, Muscovite manufacturing produced insufficient firearms for its troops in the Thirteen Years' War (see Hellie, *Enserfment*, 182–3).

46 I. Zabelin, *Domashnii byt' Russkikh' tsarei*, vol. 1, second edn. (Moskva, 1872), 212–13. The bed cost the tsar a fortune, 2800 rubles, more than a quarter of the cost of the building of the *Oryol*!

47 His contract as postmaster started in 1664 (*Russkaia istoricheskaia biblioteka*, vol. 21, *Dela tainogo prikaza* [St Petersburg, 1907], 1066). On 25 May 1668, van Sweeden's contract with the government to run the postal system expired; it was then transferred to Peter Marselis's son Leonard (Kurskov, *Vedushchee napravlenie*, 53–4; Wittram, *Czar*, vol. 1, 74). The Dutch savvy regarding the regular exchange of information, which included a well-functioning postal system, was unrivalled in the seventeenth century (see Smith, 'The Function,' 991). See further Longworth, *Alexis*, 160; Veluwenkamp, *Archangel*, 149–51; Locher, de Buck, 'Inleiding,' xxv–xxviii; Amburger, *Die Familie*, 118, 155–6; Uhlenbeck, *Verslag*, 43; *DAK* vol. 5, 216–18; Demkin, *kupechestvo* vol. 2, 11; Hellie, *Economy*, 260. Van Sweeden, too, owned a non-Orthodox slave, while the textile mills were worked by serfs he was given by the tsar who paid their fees (*obrok*) in the form of cloth (Demkin, *kupechestvo* vol. 1, 120 and vol. 2, 11).

48 Veluwenkamp, *Archangel*, 148–51; Demkin, *kupechestvo* vol. 2, 11–12.

49 It is noteworthy that Louis XIV enacted a protectionist tariff in 1667, the year of the Muscovite New Trade Statute (P. Goubert, *The Course of French History* [London, 1991], 126). A cogent argument regarding the role of mercantilist ideas in Russian economic policy has been presented by Jarmo Kotilaine (J. Kotilaine, 'Mercantilism in Pre-Petrine Russia,' in *Modernizing Muscovy*, eds Kotilaine and Poe, 143–73). Gebhard linked the creation of van Sweeden's enterprises with a mercantilistic policy of import substitution, see J.F. Gebhard, *Het leven van Mr Nicolaas Cornelisz. Witsen* (Utrecht, 1881), vol. 1, 39n2. Mercantilism was of course never a fully coherent set of ideas about economic policy, and only received its name in the eighteenth century, when the popularity of its key principles were waning. See also Wittram, *Czar*, vol. 2, 515n44.

50 On him, see the next chapter and Kurskov, *Vedushchee napravlenie*; V.O. Kliuchevskii, *Istoricheskie portrety* (Moskva, 1990), 131–5; Amburger, *Familie*, 136, 139–40; Longworth, *Alexis*, 182.

51 For Witsen's meeting with van Sweeden, see Locher, de Buck, 'Inleiding,' xlix; Witsen, *Moscovische reyse*, 373–83; Witsen, *Noord en Oost Tartarye*. For a certain resemblance between van Sweeden and other Dutch 'early modern cultural and political agent[s] or broker[s]' elsewhere, see M. Klebusek, 'The Business of News,' *Scandinavian Journal of History* 3–4 (2003) 205–13: 213.

52 See Witsen, *Moscovische Reyse*, 373–88.

53 See Uhlenbeck, *Verslag*, 21–2.

54 On the A(c)kema family from Harlingen (Friesland province), see W. Eekhoff, 'Friezen in Rusland vóór en onder Keizer Peter den Groote,' *Nieuwe Friesche Volks-Almanak* (1859) 29–39: 31–2. They were apparently Mennonites, who began trading with Russia in the first decade of the seventeenth century.

55 Orlenko, *Vykhodtsy*, 129–30.

56 Orlenko, *Vykhodtsy*, 52–8, 137–8; I. Wladimiroff, 'Andries Winius and Nicolaas Witsen, Tsar Peter's Dutch Connection,' in *Around Peter the Great*, eds C. Horstmeier et al. (Groningen, 1997) 5–23: 7. There were a few other foreigners who converted (see A. Olearius, *Podrobnoe opisanie puteshestviia golshtinskago posol'stva* [Moskva, 1870], 316–23; Orlenko, *Vykhodtsy*, 146–7).

57 See for instance M.S. Arel, 'The Archangelsk Trade,' in *Modernizing Muscovy*, eds Kotilaine and Poe, 175–201: 176, 182, 185. Arel draws attention to the

parallel between the trading policies of the tsar and various Asian rulers, such as the shah and the Mughal emperor (ibid., 177–8). See for example Barendse, *Arabian Seas*, 87, for Asian rulers' practice to grant *inam*s, which resemble the concessions enjoyed by Klenk and de Vogelaer and others.

58 Demkin, *kupechestvo* vol. 1, 32. Demkin traces this term back to a decree by tsar Boris Godunov of 1599 (ibid., 55). If of special service to the tsar, they were sometimes given the honourific title of *gost'*. See also Kotilaine, 'Mercantilism,' 165.

59 See R. White, *The Middle Ground* (Cambridge, 1991).

60 Demkin, *kupechestvo* vol. 2, 34–5.

61 For example, see Belov, 'Rossiia i Gollandiia,' 66–7.

62 For some of Patrick Gordon's correspondence from Muscovy to London indicating that he was an informant for Charles II's government, see National Archives, UK (from here abbreviated as *PRO*) 91/3 (State Papers Russia), 126–7; and for the former(?) tsarist agent John Hebdon's profession of allegiance to Charles, *PRO* 91/3, 83.

63 For the various special privileges enjoyed by European *natios* in Asia, see Barendse, *Arabian Seas*, 92–3.

64 B. Naarden, 'Dutchmen and the European Image of Russia before 1917,' in *Russians and Dutchmen*, eds J. Braat et al. (Groningen, 1993) 1–19: 4.

65 See Chapter 13.

66 See for instance M.I. Belov, 'Rossiia i Gollandiia v poslednei chetverti XVII v.,' in *Mezhdunarodnye sviazi Rossii v XVII-XVIII vv.*, ed. L. Beskrovnyi (Moskva, 1966) 58–83: 63.

67 M.I. Belov, 'Rossiia i Gollandiia v poslednei chetverti XVII v.,' in *Mezhdunarodnye sviazi Rossii v XVII–XVIII vv.*, ed. L. Beskrovnyi (Moskva, 1966), 58–83: 73.

68 Repin, 'Gollandskie kuptsy,' 21–2.

69 *RGADA* 50/9 (1676), ll.190–190ob.; Belov, 'Rossiia i Gollandiia,' 67; Houtman was a business partner of Klenk (see A.V. Demkin, A.A. Preobrazhenskii, eds, *Zapadnoevropeiskie kuptsy i ikh tovary v Rossii xvii veka* [Moskva, 1992], 72).

70 Scheltema, *Rusland* vol. 1, 309.

71 Phipps, 'Britons,' 31. The status of ownership was not clear in the case of the Dutch-founded enterprises. As Richard Pipes has argued, the tsar considered all of Muscovy his personal property, and the rather whimsical fate of the Tula armament works shows how private property was not protected from the intrusions of the state (see R. Pipes, *Russia under the Old Regime*, second edn. [London, 1995], 94).

72 Uhlenbeck, *Verslag*, 252–3; Scheltema, *Rusland* vol. 1, 43–4. Liubimenko provided an utterly positive assessment of foreigners' significance on Muscovite history (see I. Liubimenko, 'Trud inozemtsev v Moskvoskom gosudarstve,' *Arkhiv Istorii Truda v Rossii* 6–7 [1923] 52–74: 52). But Soviet historians downplayed the strength of this influence; after the Second World War, they portrayed foreigners in Muscovy as debauched spies (see S.K. Bogoiavlenskii, 'Moskovskaia nemetskaia sloboda,' *Izvestiia AN SSSR, seriia istorii i filosofii* 3 [1947] 220–32: 222–3).

73 See for example J.A. de Moor, 'Experience and Experiment: Some Reflections upon the Military Developments in 16th and 17th Century Western

Europe," in *Exercise of Arms*, ed. M. van der Hoeven (Leiden, 1997) 17–32: 26–7.

74 Hellie, *Enserfment*, 187–9.

75 See Uhlenbeck, *Verslag*, 265.

76 Hellie, *Enserfment*, 190; Longworth, *Alexis*, 34–5; Bantysh-Kamenskii, *Obzor'* vol. 1, 181; Scheltema, *Rusland* vol. 1, 182, 196, 200–1.

77 See Chapter 3.

78 Bantysh-Kamenskii, *Obzor'* vol. 1, 184–6; Scheltema, *Rusland* vol. 1, 245–7; A.N. Popov, *O postroenii korablia Orla' v gosudarstvovanie tsaria Alekseia Mikhailovicha* [Moscow, 1858], 2; Gurliand', *Ivan' Gebdon'*, 1, 9–10, 14; Barbour, *Capitalism*, 39–40; Orlenko, *Vykhodtsy*, 47, 54–5; Demkin, Preobrazhenskii, *Zapadnoevropeiskie kuptsy*, 71.

79 See S. Konovalov, 'Ludwig Fabritius's Account of the Razin Rebellion,' *Oxford Slavonic Papers* 6 (1955), 72–101: 76; A.G. Man'kov, ed., *Zapiski inostrantsev o vosstanii Stepana Razina* (Leningrad, 1968), 14.

80 Perhaps one in five mercenaries was Dutch by the early 1670s (Orlenko, *Vykhodtsy*, 52).

81 Numerous Dutchmen can be encountered among all legal or professional categories used to distinguish foreign workers by A.S. Muliukin and I. Liubimenko (A.S. Muliukin, *Priezd inostrantsev v Moskovskoe gosudarstvo* [St Petersurg, 1909], 3; Liubimenko, 'Trud inozemtsev,' 57).

82 Hellie, *Enserfment*, 178; Uhlenbeck, *Verslag*, 21, 23.

83 Hellie, *Enserfment*, 181.

84 Bantysh-Kamenskii, *Obzor'* vol. 1, 175–9; Scheltema, *Rusland* vol. 1, 121–2.

85 Muliukin, *Priezd*, 50–1.

86 A key reason for underestimating the numbers of Muscovite Dutch in seventeenth century is that they often still called *themselves* often *Nederduytsch* (Lower '*Deutsch*' [German], therefore) in the seventeenth century and used a language which neither had a standard grammar nor a uniform spelling; this situation sometimes has historians mistake Dutchmen for German speakers.

87 *RGADA* 50/42, ll. 2, 3, and 13; see also Orlenko, *Vykhodtsy*, 50–1.

88 On the expulsion of the foreigners from Moscow proper in 1652, see Chapter 6.

89 Orlenko, *Vykhodtsy*, 49; A.V. Kovrigina, *Nemetskaia sloboda Moskvy* (Moskva, 1998), 36. The 1665 census identified as house-owners in the Moscow *sloboda* 142 officers, four arms manufacturers, twenty court artisans such as goldsmiths, four apothecaries and surgeons, three translators, twenty-three merchants, three church ministers, one Jew (distinguished by religion rather than profession), three without a clear occupation, and one solicitor (or notary). No independent artisans were listed, but they may have rented from any of these 200-odd owners (Muliukin, *Priezd inostrantsev*, 106–7). By 1676, the *sloboda* was the size of the small town of Muiden (of some 1,500 inhabitants) near Amsterdam, according to Coyett, who counted three Lutheran and one Reformed Church, indicating that most inhabitants (likely mercenaries) hailed from northern Germany, the Baltic littoral, and Scandinavia (see Loviagin, ed., *Posol'stvo*, 219; this was about one per cent of Moscow's total population, see Bogoiavlenskii, 'Moskovskaia,' 222; D. Tverskaia, *Moskva vtoroi poloviny XVII veka* [Moskva, 1959], 7–8).

90 Orlenko, *Vykhodtsy*, 49–50; Phipps, 'Britons,' 31.

91 As noted above, Demkin counts a total of 664 Dutch merchants and factotums residing in Muscovy during the seventeenth century (see Demkin, *kupechestvo* vol. 1, 26). I have estimated here a presence in Muscovy of one hundred Dutch people involved in trade, about sixty army officers, and about fifty artisans and specialists employed by the manufactories of Marselis, Ackema, Coyett, van Sweeden, the tsar's armament shop in Moscow as well as shopkeepers, tavern owners and innkeepers, apothecaries and doctors, and all of their wives and children. There were additionally a fair number of Dutch servants, since Orthodox believers were not permitted to work for heretics (some were of Tatars and other background, who sometimes converted to Reformed Christianity, such as Cornelis Brak's wife Maria). Of course, as elsewhere, several Dutch residents of Russia married exogamously and may have begun to be counted as Muscovite by the authorities or to consider themselves as such. Muscovy had no significant settlement by Dutch farmers, although several such settlements were founded in northern Germany, Poland, Denmark and elsewhere in Europe (see P. de la Court, *The True Interest and Political Maxims of the Republic of Holland* [New York, 1972, origin. 1662], 130). I am not counting some of the sailors who wintered at Arkhangel'sk if failing to depart before the ice build-up.

92 N.C. Landsman, 'The Middle Colonies: New Opportunities for Settlement, 1660–1700,' in N. Canny, ed., *The Origins of Empire* (Oxford, 1998) 351–74: 353–4.

93 A.V. Viskovatov, *Kratkii istoricheskii obzor morskikh pokhodov* (Sankt-Peterburg, 1994), 88.

94 Kotilaine, *Russia's*, 504.

5 Muscovy

1 Orlenko, *Vykhodtsy*, 100–1; see also Kotilaine, 'Mercantilism in Russia,' 171. At the court, Western influence spread sometimes unnoticed (see Wittram, *Czar* vol. 1, 62).

2 Kliuchevskii, *Istoricheskie portrety*, 121–2. See also Kotilaine, Poe, 'Modernization,' 1; Wittram, *Czar* vol. 1, 43, 45; Crummey, *Aristocrats*, 29; Kurskov, *Vedushchee napravlenie*, 101; Zhivov, 'The Emergence,' 187. Zhivov, however, emphasizes 'a far-reaching transformation of the cultural system' rather than 'Westernization', expressing itself in a separation of the secular and spiritual in literature (ibid., 188–9).

3 Among others, see N.V. Riasanovsky, *A History of Russia*, sixth edn. (Oxford, 2000), 147–52, 157–82; M. Khodarkovsky, *Russia's Steppe Frontier* (Bloomington, IN, 2002), 77–125; A. Kappeler, *The Russian Empire* (Harlow, 2001), 21–48. Hellie notes that even with foreign equipment, training, and men the results could be poor, as in the Smolensk War (1632–1634, see Hellie, *Enserfment*, 172–3). See also Wittram, *Czar* vol. 1, 48.

4 Orlenko, *Vykhodtsy*, 169–70; A.L. Gol'dberg, A.G. Man'kov, S.Ia. Marlinskii, 'Izvestiia o vosstanii S. Razina v zapadnoevropeiskikh periodicheskikh izdaniiakh i khronikakh XVII v.,' in *Inostrannye izvestiia o vosstanii Stepana Razina*, ed. A.G. Man'kov (Leningrad, 1975) 80–91: 85–6; V.V. Pundani,

'Evropeizatsiia Rossii v XVII v.,' in *Mezhdunarodnaia nauchnaia konferentsiia Rossiia i Zapadnaia Evropa* (Kurgan, 1993) 37–8: 38. In this regard, see Ordin's letter of 1659 to the tsar from the border area where he then served (S.M. Solov'ev, *Istoriia Rossii Istoriia Rossii*, vol. 3: parts 11–15, Third. edn. [St Petersburg, 1911], 43).

5 Khodarkovsky, *Russia's*, 131; Riasanovsky, *History*, 159.

6 For the defensive motive, see A. Toynbee, *Civilization on Trial* (New York, 1948), 167–9. Recently John LeDonne has argued that offensive motives should not be underestimated, see John P. LeDonne, *The Grand Strategy of the Russian Empire, 1650–1831* (Oxford, 2004), 4.

7 The classic work on this is G. Parker, *The Military Revolution*, second edn. (Cambridge, 2000). For Russia, see Riasanovsky, *History*, 146–7.

8 Hellie, *Enserfment*, 151–63.

9 R. Hellie, 'The Expanding Role of the State,' in *Modernizing Muscovy*, eds Kotilaine and Poe, 29–55: 41.

10 See for example Hellie, *Enserfment*, 163–4; C. Stevens, *Soldiers of the Steppe* (DeKalb, IL, 1995), 7–8. Even foreign mercenary officers were often partially paid in land allotments in order to keep more coin in the war chest (see Orlenko, *Vykhodtsy*, 76–8). See also P.B. Brown, 'Bureaucratic Administration in Seventeenth-Century Russia,' in *Modernizing Muscovy*, eds Kotilaine and Poe, 57–78: 57; and Kotilaine, Poe, 'Modernization,' 3–4.

11 Kotilaine, 'When the Twain,' 41–2; Wittram, *Czar* vol. 1, 31–2; LeDonne, *Grand Strategy*, 3, 6; V. Klyuchevsky, *Peter the Great*, trans. L. Archibald (London, 1958), 59–60; Kotilaine, *Russia's*, 323–8.

12 See Khodarkovsky, *Russia's*, 132.

13 Ibid., 133–5.

14 Ibid., *Russia's*, 54–6.

15 See Iu. Tushin, *Russkoe moreplavanie na Kaspiiskom, Azovskom i Chernom moriakh (XVII vek)* (Moscow, 1978), 11. For Black Sea piracy, see Khodarkovsky, *Russia's*, 131; Viskovatov, *obzor*, 157–62n43.

16 See for instance *DAK* vol. 6, 13; P. Avrich, *Russian Rebels* (New York, 1972), 73.

17 For its size, see Mironov, *Social History*, vol. 1, 2.

18 Baron, 'Shipbuilding,' 115, 123.

19 See Baron, 'Shipbuilding,' 120–1; Hellie, *Economy*, 483; Kotilaine, *Russia's*, 79–80; S.H. Baron, 'Osip Nepea and the Opening of Anglo-Russian Commercial Relations,' *Oxford Slavonic Papers*, New Series 11 (1978): 42–63. In the late sixteenth century several Muscovite merchants had tried to trade in Dutch Dordrecht, but had been relieved by local tricksters of their sable and other pelts without payment (see Jansma, 'Olivier Brunel,' 337, 354–6). It was a long-standing tradition, going back to the days of Hanseatic trade on Novgorod, to have European foreigners ship their own goods to the Russian 'ports' (see Pipes, *Russia*, 36).

20 Sporadically, Russian merchants managed to carry their goods to Swedish, northern German ports, or even Amsterdam by way of Riga, Narva or Reval (see Kotilaine, *Russia's*, 29, 43, 316).

21 Baron, 'Shipbuilding,' 104–5, 122; Kotilaine, 'When the Twain,' 1055–6, 1058; Kotilaine, *Russia's*, 448–50.

22 Wittram, *Czar* vol. 1, 23; Kotilaine, 'When the Twain,' 1066.

23 See Baron, 'Shipbuilding'.
24 Khodarkovsky, *Russia's*, 132. On the *Friedrich*, see E.J. Philips, *The Founding of Russia's Navy, 1688–1714* (Westport, CT, 1995), 14–18; Viskovatov, *obzor*, 79–84, 168–9n48, 169–72n49, 172–5n51, 176–8n52; Olearius, *The Travels*, 74n24, 78. Short but extraordinarily high waves of the Caspian Sea caused the *Friedrich* to break (Tushin, *Russkoe moreplavanie*, 44).
25 P. Bushkovitch, *Peter the Great* (Cambridge, 2001), 43.
26 Although ominous rumblings could be heard on the Don, see Riasanovsky, *History*, 179–82; Hellie, *Enserfment*, 135–7; S.H. Baron, 'A.L. Ordin-Nashchokin and the Orel Affair,' in S.H. Baron, *Explorations in Muscovite History* (Aldershot, 1991) 1–22: 2.
27 Riasanovsky, *History*, 181–2, 199; Longworth, *Alexis*, 177–86.
28 It should be noted that the northern littoral of the Black Sea and the Crimean peninsula remained in the hands of Tatar vassals of the Turkish Sultan.
29 Kurskov, *Vedushchee napravlenie*, 97.
30 See V.F. Starkov, V.L. Derzhavin, *Ekspeditsiia Villema Barentsa na Novoi Zemle, 1596–1597 gg.* (Moskva, 2003), 16–17. Aleksei concluded that a sea passage to China was not feasible, and sent in 1675 an embassy to China overland (ibid., 20).
31 Zabelin, *Domashnii byt'* vol. 1, 200.
32 For its full text, *Polnoe sobranie zakonov' Rossiiskoi imperii* vol. 1 (St Petersburg, 1830), 677–91. See also Amburger, *Familie*, 146; K.V. Basilevich, 'Novotorgoyyi ustav, 1667 g.,' *Izvestiia Akademii Nauk SSSR. Otdelenie obshchestvennykh nauk* 2 (1932) 589–622; B.G. Kurts, *Sochinenie Kil'burgera o 'russkoi torgovle v tsarstvovanie Alekseia Mikhailovicha'* (Kiev, 1915), 182; M.L. Lahana, 'Novaia Nemetskaia Sloboda: Seventeenth Century Moscow's Foreign Suburb,' unpubl. Ph.D. Diss. (University of North Carolina at Chapel Hill, 1983), 103. Kotilaine, 'When the Twain,' 46, notes how the Code discouraged foreign settlement as well.
33 At a minimum, as Amburger suggests, the Code promised to fill the treasury's coffers with more bullion (Amburger, *Familie*, 145).
34 See Kotilaine, 'When the Twain,' 4; Orlenko, *Vykhodtsy*, 170. Kotilaine emphasizes the risk of the trade on Arkhangel'sk even for the Western traders, which may have contributed to Russian reluctance to venture out (Kotilaine, 'When the Twain,' 29). Russians had no reliable network of trading partners abroad, and there was little Russian capital available for Muscovite long-term and expensive trading ventures (ibid., 36–7).
35 Lahana, 'Novaia Nemetskaia,' 103.
36 Orlenko, *Vykhodtsy*, 95.
37 See Kurskov, *Vedushchee napravlenie*, 125 and *passim*; Crummey, *Aristocrats*, 42; Longworth, *Alexis*, 189–92; Solov'ev, *Istoriia* vol. 3 (parts 11–15), 727.
38 VOC naval expeditions had cowed the shah in the 1640s, and some of India's rulers in the 1650s and 1660s, into granting advantageous trading privileges (see W. Floor, 'The Dutch and the Persian Silk Trade,' in *Safavid Persia*, ed. C. Melville [London, 1996] 323–68: 353; Lach, van Kley, *Asia* vol. 3, book 1, 57–9). On the small size of the port at Arkhangel'sk, see Kotilaine, 'When the Twain,' 17–18.

39 See M. Roberts, 'Introduction,' in *Swedish Diplomats at Cromwell's Court, 1655–1656* (London, 1988) 1–46: 19, 21; Kotilaine, *Russia's*, 312–45.

40 Jones, *Anglo-Dutch Wars*, 4. Mercantilists disagreed whether to advocate naval warfare; additionally, Western European governments built roads and canals and attempted to stimulate manufacturing.

41 Kurskov, *Vedushchee napravlenie*, 17–19, 61–2.

42 See Wittram, *Czar*, vol. 1, 67–8.

43 Prak, *Dutch Republic*, 123.

44 See Amburger, *Familie*, 25, 143–4; Kotilaine, 'When the Twain,' 42–3; Kurskov, *Vedushchee napravlenie*, 19, 36; Wittram, *Czar* vol. 2, 515n43. The tsarist government mined a variety of sources for information regarding the unfolding of the English Civil War and the developments in the Protectorate (Kurskov, *Vedushchee napravlenie*, 81–3). Baron suggests that Jean de Gron/Groen kindled Russian interest in mercantilism (Olearius, *The Travels*, 141n21; see below in this chapter for more on de Gron).

45 In this regard it may be instructive to remember how the protectionism of the *Zollverein* (and advocated by theorists such as Friedrich Liszt) contributed to the take-off of nineteenth-century German industry.

46 M.J. Braddick, 'Government, War, Trade, and Settlement,' in *The Origins of Empire*, ed. Canny, 286–308: 306; N. Zahedieh, 'Overseas Expansion and Trade,' in *The Origins of Empire*, ed. Canny, 398–422: 408, 418.

47 See Zahedieh, 'Overseas Expansion,' 418; J. Cracraft, *The Revolution of Peter the Great* (Cambridge, MA, 2003) 38–40.

48 For his support of 'mercantilism,' see Crummey, *Aristocrats*, 160.

49 Muliukin, *Priezd*, 58.

50 Among the Russian population at large, xenophobia varied in intensity; many Russians served without objection under Western officers in the seventeenth-century Muscovite army (Orlenko, *Vykhodtsy*, 185–9; see also Muliukin, *Priezd*, 58). At the same time, Razin's Cossacks were strongly hostile to foreigners, partially because they saw them as instruments and henchmen of the evil boyars' regime (see for instance N. Kostomarov, 'Bunt' Sten'ki Razina,' in N.I. Kostomarov, *Sobranie sochinenii* vol. 1 [Sankt-Peterburg', 1903] 405–505: 459; Konovalov, 'Ludwig Fabritius's Account,' 88).

51 Hellie, *Enserfment*, 56, 167. Some anti-foreigner measures of Aleksei's rule were directed against Tatars, as a decree of 1646 prohibiting the employment of non-Orthodox interpreters in the *Posol'skii prikaz* (even if they were transferred to the *Razriadnyi prikaz*, an army department; see T.A. Oparina, *Inozemtsy v Rossii XVI-XVII vv.* [Moskva, 2007]).

52 Orlenko, *Vykhodtsy*, 142–3, 149.

53 Lahana, 'Novaia,' 84–7; Orlenko, *Vykhodtsy*, 65–6, 149–51. There had been in the sixteenth century a residentially segregated foreigners' suburb, while, for example, non-Eastern Slavic Orthodox believers resided in their own separate community before 1672 as well (see Oparina, *Inozemtsy*, 330–1).

54 There is some doubt about this claim because only one copy of the decree expropriating Westerners is known (it applied to the Arzamas region; Orlenko, *Vykhodtsy*, 81–2, 152). Meanwhile, some foreigners, including Marselis, Ackema, and van Sweeden, continued to be rewarded with manors worked by serfs, albeit only for finite terms of usufruct (ibid., 84).

55 Wittram, *Czar* vol. 1, 59.

56 Although Catholics and Protestants sometimes preferred to stay apart, see Barendse, *Arabian Seas*, 87–8.
57 Orlenko, *Vykhodtsy*, 133–5; Longworth, *Alexis*, 210–11, 219–20, 223–5.
58 Orlenko, *Vykhodtsy*, 134; Loviagin, ed., *Posol'stvo*, 10.
59 Crummey, *Aristocrats*, 160. Kotilaine remarks that '[a]rguably the single most important failure of Muscovite mercantilism was its inability to promote capital accumulation and credit at home' (Kotilaine, 'When the Twain,' 40).
60 One of Ordin-Nashchokin's key insights was the urgent Muscovite need to learn everything useful from Western foreigners (Kliuchevskii, *Istoricheskie portrety*, 126). See also Kurskov, *Vedushchee napravlenie*, 42.
61 Collins believed that, in the latter half of the 1660s, Ordin-Nashchokin, a zealous monarchist, was inclined to prefer the English over the Dutch, while the tsar's other favourite, Bogdan Matveevich Khitrovo, was in the pocket of the Dutch trading interest, which bribed him handsomely (Collins, *Present State*, 107–9, 120–1; see also Kurskov, *Vedushchee napravlenie*, 3, 83). As main advisors of Aleksei, they had stepped into the shoes of the tsar's father-in-law, I.D. Miloslavskii, who in 1665 suffered a stroke (Collins, *Present State*, 106–7; Kotoshikhin et al., *Moskovia*, 196). Ordin was by no means a wholehearted Westernizer like Peter the Great (see for example Crummey, *Aristocrats*, 159–60).
62 These Armenians had been resettled in Isfahan's Julfa suburb by Shah Abbas I in 1603 and had been assigned the monopoly of the export trade to Muscovy, which was primarily in raw silk (see N.G. Kukanova, *Ocherki po istorii Russko-Iranskikh torgovykh otnoshenii v XVII-pervoi polovine XIX veka* [Saransk, 1977], 67–9). See also R. Matthee, *The Politics of Trade in Safavid Iran* (Cambridge, 1999), 193. On the omnipresence and importance of Armenian merchant communities in Europe and West-Asia in the seventeenth century, see Barendse, *Arabian Seas*, 70.
63 Longworth, *Alexis*, 192. Kurskov found a number of petitions by Muscovite merchants asking to limit foreigners' trade (Kurskov, *Vedushchee napravlenie*, 57). Klenk, by far the most experienced Dutch ambassador sent to Moscow in the seventeenth century, failed to persuade his hosts to rescind the Code in 1675–6 (Loviagin, 'Vvedenie,' cxlviii). When the leading Russian merchants were consulted by the Boyars M. Iu. Dologorukii and A.S. Matveev and *d'iak* (secretary) G. Bogdanov and others on behalf of the government in February 1676 (Old Style calendar), the merchants' response was a plea to categorically refuse to lift any restrictions imposed on Dutch, Hamburgers, or English traders included in the Commercial Code or in terms of the silk trade on Persia (Loviagin, 'Vvedenie,' cxlix–cli).
64 *Sobranie gosudarstvennykh gramot' i dogovorov'* vol. 4 (Moscow, 1828), 204–8; Matthee, *The Politics*, 193–4; Uhlenbeck, *Verslag*, 265: Viskovatov, *obzor*, 86; Loviagin, 'Vvedenie,' cxlvi-ii; Kukanova, *Ocherki*, 67; Faroqhi, *Ottoman Empire*, 54; Kotilaine, *Russia's*, 458. The Russians exchanged mainly hides for Persian silk (Kotilaine, 'When the Twain,' 1066). This agreement only became possible because of Shah Abbas II's death in 1666, as the Iranians had attacked Russian positions in the Caucasus in the early 1650s (see M. Kemper, *Herrschaft, Recht und Islam in Daghestan* [Wiesbaden, 2005], 131–2).
65 Viskovatov, *obzor*, 86–7; Kurts, *Sochinenie Kil'burgera*, 149–50.

66 Kotilaine, 'When the Twain,' 1066. Silk was mainly produced in northern Persian regions (Floor, 'The Dutch,' 336).
67 W. Floor, P. Clawson, 'Safavid Iran's Search for Silver and Gold,' *International Journal of Middle East Studies* 3 (2000) 345–68: 346, 357.
68 Ibid., 347; Braudel, *Civilization* vol. 3, 217–18.
69 Viskovatov, *obzor*, 60, 87–8. Veselago suggests that it was the Armenians who requested the tsar to construct warships to police their silk fleets across the Caspian Sea (F. Veselago, *Kratkaia istoriia russkago flota* vol. 1 [St Petersburg, 1893], 6; F. Veselago, *Kratkaia istoriia russkogo flota* [Moskva, 1939], 10).
70 Kukanova, *Ocherki*, 37; Kotilaine, 'When the Twain,' 1054.
71 See further Chapter 6. In *Reysen* the purpose of the ship(s) is expressed ambiguously: in the Dutch version Struys notes how sailors were recruited in Amsterdam in 1668 to sail the Persia–Astrakhan route and carry goods on Muscovite vessels, which in the English translation became 'to promote the Trade' (see *Reysen*, 120; Struys, *The Voyages*, 114).
72 See G.V. Scammell, 'European Exiles, Renegades and Outlaws and the Maritime Economy of Asia *c*.1500–1750,' *Modern Asia Studies* 4 (1992) 641–61: 645–6. See Chapter 14.
73 Officially ('by your great sovereign's decree'), Van Sweeden was ordered by Aleksei's inner council, the Chancellery of Secret Affairs (*Prikaz tainykh del'*), to recruit the shipwrights and sailors in Holland in 1667 (*DAK* vol. 5, 233). Van Sweeden may have already been assigned the recruitment of the Dutch specialists on 16 June 1666 (see Iurchenko, 'Predislovie', 8). The contract with Butler was signed in Amsterdam in February 1667 (*DAK*, vol. 5, 211–12).
74 Uhlenbeck, *Verslag*, 265. Tushin suggests that originally the ship's construction was to take place at Astrakhan, but that such plans were shelved because of worries about the increasing Don Cossack brigandage in 1666 and 1667 (Tushin, *Russkoe moreplavanie*, 45).
75 Viskovatov, *obzor*, 113; *DAK* vol. 5, 268, 283–4.
76 See Hughes, *Russia*, 466; Klyuchevsky, *Peter*, 83. At this time, the total government annual revenue from taxation was no more than 1.5 million rubles (as it was in 1680, see Hughes, *Russia*, 140).
77 Viskovatov, *obzor*, 89; *DAK* vol. 5, 211.
78 See N.M. Rogozhin, ed., *Boyarskaia kniga 1658 goda* (Moskva, 2004), 17; Longworth, *Alexis*, 182.
79 Rogozhin, ed., *Boyarskaia*, 20; Kliuchevskii, *Istoricheskie portrety*, 122.
80 Kliuchevskii, *Istoricheskie portrety*, 121.
81 See M. Poe, *The Russian Elite in the Seventeenth Century* (Helsinki, 2004) vol. 1, 210–11, 214, 216, 401, 405; Crummey, *Aristocrats*, 195.
82 Crummey, *Aristocrats*, 158.
83 See Kurskov, *Vedushchee napravlenie*, 46; see Chapters 6 and 7.
84 *DAK* vol. 5, 211; Philips, *Founding*, 20; B. Uroff, 'Grigory Karpovich Kotoshikhin, "On Russia in the Reign of Alexis Mikhailovich": An Annotated Translation,' Unpubl. Ph.D. Diss. (New York: Columbia University, 1970), 202.
85 *PRO* 91/3, 139v.
86 Viskovatov, *obzor*, 104.
87 For its full text *Polnoe sobranie zakonov'* vol. 1, 916–23. Further see Loviagin, 'Vvedenie,' cxlvii; Kotilaine, *Russia's*, 460–1; Kukanova, *Ocherki*, 83–4; Kurts, *Sochinenie Kil'burgera*, 150.

88 Loviagin, 'Vvedenie,' cxlvii–i; Matthee, *The Politics*, 149, 197; see also Faroqhi, *Ottoman Empire*, 100. On the various futile Dutch attempts to acquire a share in the transit of silk across Russia, see also Z.R. Dittrich, 'Illusies, misverstanden, wanklanken,' in *Rusland in Nederlandse ogen*, eds J. Driessen et al. (Amsterdam, 1986) 33–50: 40–2.
89 Kotiliane, *Russia's*, 462–5.

6 In the Tsar's Service

1 *GAA DTB* 478, p. 462. No other Jan Janszoon Struys, sailor, may be found in the Amsterdam archives for this period, and the age of the groom is almost exactly that of Struys the narrator. See *Reysen*, 120.
2 At one point in *Reysen*, Struys refers to two children at home in Holland (see *Reysen*, 233). See for Trijntje's death *GAA DTB* 1227, p. 105. In the burial record, Struys is named as 'Struis,' an identification omitted in the baptismal records.
3 *GAA DTB* 94, p. 491; 95, p. 104; 66, p. 84. The first two were baptized in the southern Reformed Church, the third child in the Reformed Chapel on the Nieuwezijds, both locations not far from their lodgings. In December 1666, Trijntje Pietersdochter was buried in the Heiligeweg and Leyden cemetery on the southside of the city (*GAA DTB* 1227, p. 105).
4 *GAA DTB* 1249, p. 145, notes the burial at the Heiligeweg and Leidsche cemetery (south of the city walls), of 'the child' of 'Jan Jans Struis' in the Reguliersdwarsstraat (the St Jorissteeg/straat was a sidestreet of this street) in January 1666. The fact that they had two daughters named Teuntje indicates that the first one died before the last one was born in the fall of 1666. On epidemics and mortality in Amsterdam, see Nusteling, *Welvaart*, 39; C.G.A. Clay, *Economic Expansion and Social Change* vol. 1 (Cambridge, 1984), 21; Brugmans, *Geschiedenis* vol. 3, 151–2; Jan de Vries, *The Economy of Europe in an Age of Crisis* (Cambridge, 1976), 8.
5 See de Vries, *Economy*, 7–12, 156–7.
6 *GAA DTB* 151, p. 40; 153, p. 6. The uncertain number of children is caused by the mention of the father as Jan Jansen, rather than Jan Janszoon (or Janse) Struys, in all baptismal records. The *names* of all three boys seem to underscore the parenthood of Jan Struys and Trijntje Pietersdochter: the oldest son, Jan, was named after his father and grandfather, Hendrick after Jan's brother Dirk, and Pieter after Trijntje's father. But unless the children were rebaptized, it is odd that Pieter and Hendrick were christened so long after their birth, which must have happened before December 1666 (and one baptism even occurred in the father's absence).
7 The St Jorisstraat address is given in several records related to Struys in the Amsterdam municipal archive (for example, see *GAA DTB* 478, p. 462; 1227, p. 105). See A. van Gelder, *Amsterdamsche Straatnamen* (Amsterdam, 1915); the area had been annexed by the city in 1593.
8 van Deursen, *Plain Lives*, 84, 87, 89; Clay, *Economic Expansion* vol. 1, 13, 23; A. Rowlands, 'Conditions of the Life of the Masses,' in *Early Modern Europe*, ed. Cameron, 31–62: 37.
9 de Jonge, *Geschiedenis* vol. 1, 640–1, 660–1; Jones, *Anglo-Dutch Wars*, 47.

10 Amsterdam shipbuilding was of course a massively thriving industry (the Dutch possessed more than 10,000 ships in the course of the seventeenth century), and, after the slump of 1654–55, the Dutch economy prospered again (see Brugmans, *Geschiedenis* vol. 3, 203; I. Schöffer, 'De Republiek,' 199, 201).
11 *GAA DTB* 491, p. 11.
12 *GAA DTB* 478, p. 462 and *GAA Collectie Kwijtschelding* 5062–52, p. 103 (Jansdr Marritje).
13 This expression reads in Dutch as '*mans goed is ingebracht*' (*GAA DTB* 491, p. 11). On wives' property rights, see Schama, *Embarrassment*, 79, 405–7.
14 His second wedding actually occurred a few days after the book has him depart (compare *GAA DTB* 491, p. 11, with *Reysen*, 121), a further instance of *Reysen*'s inaccurate dates.
15 *Reysen*, 121.
16 *Reysen*, 122–3.
17 See *RGADA* 50/42 (1668), l.46; bad weather had plagued their trip; in good weather the journey from Amsterdam to Riga took often less than two weeks (see *Reysen*, 121–2; the trip from Danzig to Amsterdam sometimes took no more than a week, see Davies, *God's Playground* vol. 1, 200). The trip's length in *Reysen* agrees with Butler's in his cost declaration to the tsar (*RGADA* 50/42 [1668]).
18 *Reysen*, 114–15.
19 *Reysen* mentions eighteen people, of whom six had already travelled to Moscow earlier; the list omits Struys, Karsten Brand(t) and Dani[e]l Cornelisz, but the text notes accurately how fifteen left Amsterdam in September 1668 (see *Reysen*, 121). Although *Reysen* omits any mention of him, one of the sailors, Gabriel Pietersz, fell ill on the trip and died soon after arrival in Livonia (*RGADA* 50/42 [1668], l.44]; see also *RGADA* 50/42 (1668), l.2 and *Reysen*, 121, 136). See also Viskovatov, *obzor*, 96.
20 *RGADA* 50/42 (1668), ll.1–4, 41–6, which file is dated as covering the period from 14 November 1668 to 7 March 1669.
21 R. Latham, ed., *Samuel Pepys and the Second Dutch War* (Aldershot, 1995), 29, 31.
22 Butler was a cousin of van Sweeden's wife Maria Ruts (see Veluwenkamp, *Archangel*, 207). In Russian sources he is called *David Ivanov syn*, indicating correctly that his father's name was Jan (see Viskovatov, *obzor*, 183n62; see also *GAA NA* 1842 [Notary N. Kruys], p. 601).
23 Butler was hired for four years, starting 31 March 1667 (Viskovatov, *obzor*, 184n62; *DAK* vol. 5, 211–12). Butler recruited the rest of the crew for a term of equal length to his own (see Viskovatov, *obzor*, 185–6n62).
24 *RGADA* 50/42 (1668). l.2. Trappen, who had been in Russia since 1667, joined the crew only when they neared Moscow, replacing Gabriel Pietersz in the record; later, Butler may have tried to pass off his servant Jan Vasselij (probably a Slav named Ivan Vasil'evich) as a member of his crew (in Pietersz's stead) to collect his wage (see van der Gaeten's accusations in Chapter 7; see *Reysen*, 121, 136). Struys notes that the Russian border was crossed about October 20, but his recollection of dates is poor; it is possible that news of the Dutch arriving on Muscovite territory reached Moscow on November 14 (see *Reysen*, 126–7). By November 20, the sailors were

registered in Moscow at the *Posol'skii prikaz* (see Viskovatov, *obzor*, 105–6). In the Russian record Struys appears also as 'Ian Iagans master of ship's sailing affairs' (*'Ian Iagans master karabelnogo parusnogo dela'*), and in the Dutch bill of the expenses Butler submitted to the tsar he is identified as *'Jan Jansen Struijss, zeijlemaecker'* (*RGADA* 50/42 (1668), l.3 and l.43).

25 *RGADA* 50/42 (1668), l.4
26 *Reysen*, 121.
27 Van Royen, *Zeevarenden*, 108–9.
28 Honig, *Geschiedenis*, vol. 2, 97; J. Scheltema, *Peter de Groote* (Amsterdam, 1814), vol. 1, 41.
29 Ibid., 128–9 (but see as well ibid., 133, where he depicts an abysmally poor hamlet in Russia). In the Aristotelian view of the world, advanced use of *techne* was a sign of civilized superiority (see Pagden, 'Europe and the World Around,' 15).
30 *Reysen*, 126–35.
31 The Dutch began to smoke tobacco in pipes around the time of Struys's birth (see Schmidt, *Innocence Abroad*, 266). Olearius also noted the Russian tobacco lust, and recalled a tsarist prohibition on smoking of 1634 (Olearius, *The Travels*, 146; perhaps the year was 1633, see Hellie, *Economy*, 106). This tobacco hunger is remarkable given the horrendous punishments for offenders (see Hellie, *Economy*, 106–7). In 1697, Peter the Great finally allowed smoking (Hughes, *Russia*, 147). Smoking at the time was likewise prohibited by Islamic scholars in the Caucasus (see M. Kemper, A. Sixsaidov, eds, *Die Islamgelehrten Daghestans und ihre arabischen Werke* [Berlin, 2004], 47).
32 Ibid., 133–6. That the journey could have been worse was evident when near Torzhok they passed by the graves of eight unnamed Dutch merchants who had been murdered while travelling the same route (*Reysen*, 133).
33 *Reysen*, 127.
34 Ibid.
35 Ibid., 132.
36 Orlenko, *Vykhodsty*, 134.
37 See Maczak, *Travel*, 71. It seems no coincidence that the greatest of Dutch historians, Johan Huizinga, wrote a pioneering work on human play in history (see J. Huizinga, *Homo Ludens* [Haarlem, 1938]).
38 Ice skating in Russia did not become popular then, but in 1698 members of Peter's Embassy in Holland enthusiastically tried to learn how to skate on the frozen waters of the Zaan and its tributaries (V. Kordt, ed., *Zapiski Ia.K. Nomena o prebyvanii Petra Velikago v Niderlandakh v 1697/98 i 1716/17 gg.* [Kiev, 1904], 58–9).
39 *Reysen*, 135.
40 See *GAA NA* 1100 (Notary J.van der Ven), p. 123 (13 June 1652; contract with a cooper); *GAA NA* 1100 (Notary J.van der Ven), p. 147 (6 June 1652; contract with a boxmaker).
41 Even if impressment was not practised in the Republic (see Chapter 2). H.B. Wheatley, ed., *The Diary of Samuel Pepys*, 2 vols (New York, 1942), [1667], 230–2, 231n1, 334n1. At this time, some Dutch merchants on Muscovy unsuccessfully sought dispensation from the Dutch embargo on outbound ships (Jones, *Anglo-Dutch Wars*, 46–7).
42 See *DAK* vol. 5, 230–3; Uhlenbeck, *Verslag*, 47; Bantysh, *Obzor'*, vol. 1, 187.

43 Viskovatov, *obzor*, 96; *DAK* vol. 5, 218–20, 233. Patrick Gordon had heard in early July 1667 about the arrival of skippers and shipwrights in Moscow: 'To encourage trading His Imp. [Majesty] diminished five of the ... of the Toll [examples follow] ... , wee are projecting to draw the Persian & Armenian Traffique through this Countreij Skippers & Ship Carpenters being sent for, & some come alreadij, wee are to build & prepare vessels to saile the Caspian Sea' (PRO 91/3, 126 [letter by Patrick Gordon from Moscow of 9 July 1667 to Joseph Williamson in Whitehall]); see also P. Gordon, *Dnevnik 1659–1667*, trans. and ed. D.G. Fedosov (Moskva, 2003), 211.

44 *DAK* vol. 5, 216. In December 1666, van Sweeden hired the patternmaker Pieter Jansz Veenhuijsen and three glassblowers, who would be accompanied by their wives and children, to work in his own factories near Moscow (*GAA NA 3426* [Notary W. Banning], p. 504 [7 December 1666], and p. 520 [9 December 1666]).

45 *Reysen*, 121; Viskovatov, *obzor*, 90; *DAK* vol. 5, 212–13. Helt, Klopper and Pietersz hailed from the *Streek*, the hinterland of the ports of Enkhuizen and Hoorn; Tolk-Schram was from Scheven[l]ingen, the coastal village near The Hague; and Trappen hailed from Vienna. Munster was from Amsterdam: while *Reysen* gives Munster's profession merely as diamond cutter, in Russian archival records Munster is listed as a shipwright, sailor and diamond cutter (see Viskovatov, *obzor*, 90–1; *DAK* vol. 5, 215–16, 231). Perhaps van Sweeden thought that Munster might also be employed in one of his other Muscovite enterprises or for the tsar's workshops, as the Russian monarch had been interested in the diamond trade with Persia (see Barendse, *Arabian Seas*, 106). The contract with Lambert Jacobsz Helt was signed in Amsterdam on 5 July 1667 (*DAK* vol. 5, 212–13).

46 Kotoshikhin et al., *Moskovia*, 570. Van Bockhoven was released by the *Reitarskii prikaz* (the cavalry's chancellery) on 6 July 1667, while his deputy Stark was transferred from the *Pushkarskii prikaz* (the artillery chancellery) on 22 July 1667 (Popov, *O postroenii*, 5n2). The Russian record indicates that van Bockhoven had extensive experience at sea, for which there is no other evidence: he was by trade a field officer (see Viskovatov, *obzor*, 91; *DAK* vol. 5, 218).

47 See Uhlenbeck, *Verslag*, 28–35, 265; Scheltema, *Rusland*, vol. 1, 196. There were at least three van Bockhovens serving tsar Alexis: in 1647 Ilya Miloslavskii recruited Isaac and his sons Cornelis, and (after a spell in the service of the English king Charles I) Philips Albrecht (see Viskovatov, *obzor*, 181–2n60; Kotoshikhin et al., *Moskovia*, 192–4, 200, 215).

48 De Jonge, *Nederland*, 232–6. On the dismantling of the Dutch armed forces in 1646–1647, see Prud'homme van Reine, *Schittering*, 125–6 and van Nimwegen, *Deser landen*, 237–8. See also Chapter 1.

49 Viskovatov, *obzor*, 181–2n60.

50 See Witsen, *Moscovische reyse*, 147–8n10; N. Vitsen, *Puteshestvie v Moskoviiu 1664–1665* (Sankt Peterburg, 1996), 241n194; see for the bodyguard also Orlenko, *Vykhodtsy*, 132.

51 Orlenko, *Vykhodtsy*, 184–5; Muliukin, *Priezd*, 59.

52 See W.M. Reger, 'In the Service of the Tsar,' unpublished Ph.D. Diss. (U. of Illinois at Urbana-Champaign, 1997), 285.

53 *DAK* vol. 5, 221, 228, 248. Schak and Schram are noted in *Reysen*, 121. Coyett noted with apparent surprise how the tsar's '*Duitsche*' colonels were paid high wages and bonuses, while Russian officers received far less (Loviagin, ed., *Posol'stvo*, 194–5). But Russian officers were partially remunerated by manner of land grants (which included serfs). Van Bockhoven's remuneration amounted to about four times Struys's wage, but was at the going rate for foreign infantry colonels (see Orlenko, *Vykhodtsy*, 194).
54 Viskovatov, *obzor*, 91–3, 95–7; *DAK* vol. 5, 220–1, 225–6; Philips, *Founding*, 21–2. Solov'ev (in a recent translation) and Philips provide both solid detailed accounts in English (see Philips, *Founding*, 19–26; S.M. Solov'ev, *A History of Russia from Earliest Times*, vol. 23 [Gulf Breeze, FL], 1998, 81–4).
55 Viskovatov, *obzor*, 93–6; *DAK* vol. 5, 225, 228–9, 236.
56 Viskovatov, *obzor*, 97–100; *DAK* vol. 5, 221–4, 235–6; Philips, *Founding*, 22–4.
57 Viskovatov, *obzor*, 98–9.
58 Viskovatov, *obzor*, 100–1; *DAK* vol. 5 237–9, 241–2. This sort of problem still plagued Peter the Great's shipbuilding projects in the eighteenth century (see Hughes, *Russia*, 170).
59 Viskovatov, *obzor*, 101; *DAK* vol. 5, 256.
60 *DAK* vol. 5, 245.
61 *DAK* vol. 5 231–4.
62 *DAK* vol. 5, 239–41, 245. He also suggested that a Dutchman by the name of Jan, who was in the employment of Philimon Philimonovich (Tieleman Ackema), could carve a figurehead for the ship. This knowledge shows how familiar van Bockhoven was with the Dutch expatriate community.
63 Viskovatov, *obzor*, 101.
64 Viskovatov, *obzor*, 101–2.
65 *DAK* vol. 5, 257–62, 267; Viskovatov, *Kratkii istoricheskii obzor*, 103–6. Cornelis van Bockhoven was temperamental (see Olearius, *The Travels*, 141).
66 Their tasks were not precisely circumscribed, but, as the tsar's deputies, the *voevody* enjoyed in principle absolute power to commandeer the population of their realm: Thus they could oblige the local population to build forts, roads and bridges, and to staff garrisons and other armed forces (see Pipes, *Russia*, 99).
67 *DAK* vol. 8, 250–1; Tushin, *Russkoe moreplavanie*, 50. This is the only elaborate description of the vessels that has survived. The lion seems to have been the result of a misunderstanding about the figurehead during the Dedinovo construction (the ship was named *Oryol* late in the process). On an engraving, *Reysen* shows a sketchy *Oryol* and yacht (with Dutch flag) in the water before Astrakhan, but the picture lacks detail and may be fabulous (*Reysen*, between 188 and 189).
68 No sources extant furnish an unequivocal description of the specific task(s) that were to be performed by the *Oryol*. However, because Dutch shipwrights built few outright ships-of-the-line at this time yet, a combination of a trading and a military role seems to have been intended for the ship.
69 Viskovatov, *obzor*, 112; *Spisok' russkikh voennykh' sudov s' 1668 po 1860 god'* (St Petersburg, 1872), 2–3; Unger, *Dutch Shipbuilding*, 37. See van Dam, *Beschryvinge* vol. 1, part 1, 480–2, 734; see for example the dimensions of

the *'fluytje'* the *Marcken* built in 1670 on the Amsterdam wharves of the VOC: length, 28 metres; width, seven metres; hold, three metres (ibid., 482, see also ibid., 469, 471, for the description of a *'kleyne'* [small] *fluyt* of the 1650s and 1660s of almost exactly the *Oryol*'s proportions). In van Dam's text the 'Amsterdam' measurements then current are likely used: one foot was 0.28 metres (ibid., 741).

70 When Cornelis de Bruyn set sail for Russia in 1701, he still embarked on a *fluyt* at the island of Texel manned by a crew of eighteen, carrying eight guns (C. le Bruyn, *Travels into Muscovy, Persia and Part of the East-Indies* [London: Bettesworth and company, 1737] vol. 1, 2). See also Unger, *Dutch Shipbuilding*, 26.

71 Viskovatov, *obzor*, 112; *Spisok'*, 2–3.

72 See *Reysen*, 121, 171. See Unger, *Dutch Shipbuilding*, 60.

73 See Jones, *Anglo-Dutch Wars*, 51–2; Unger, *Dutch Shipbuilding*, 38.

74 Viskovatov, *obzor*, 112; *Spisok'*, 2–3. *'Jacht* even in the seventeenth century was a generic term, but then for vessels of medium size with one or two masts and some form of fore-and-aft rig. The word may have originally meant a vessel which chased and so was to be recommended for its speed' (Unger, *Dutch Shipbuilding*, 50).

75 Ibid., 135–6. His death is noted as well in *RGADA* 141/371 (1668), l.2. See also Viskovatov, *obzor*, 78.

76 Longworth, *Alexis*, 191; Davies, *God's Playground* vol. 1, 255–6, 355–7.

7 *Reysen*'s Muscovy and Struys's Muscovy

1 See the evocative description of this ordering process in Rohinton Mistry, *A Fine Balance* (New York, 1997) 554–5. See also W. Benjamin, *Illuminations* (New York, 1969) 91.

2 Benjamin, *Illuminations*, 85–6, 89.

3 *Reysen*, 140–68. For the commonality of such descriptions in contemporary texts, see for instance Poe, *'A People'*; Mund, *Orbis Russiarum*; Mervaud, Roberti, *Une infinie brutalité*; Scheidegger, *Perverses Abendland*; Wittram, *Czar* vol. 1, 33–6; for a brief listing, see ibid., 33, as well as Poe, *'A People'*.

4 *Reysen*, 120; nevertheless, his curiosity was not wholly insignificant, as Justin Stagl makes evident (J. Stagl, *A History of Curiosity* [Chur, 1995], 2).

5 *Reysen*, 120.

6 *DAK* vol. 5, 231; Uhlenbeck, *Verslag*, 47.

7 '[O]*ngemeene gelegentheyt*,' see *Reysen*, 121. Exceptionally, the shipwrights and crew were hired for a fixed period of four years: mercenaries were usually expected to serve the tsar all their lives (see *DAK* vol. 5, 212, 214; Olearius, *Travels*, 175n4). The fixed term may indicate uncertainty regarding the project (and a bargaining chip used by van Sweeden to sell his project to Alexis?): a finite service term released the tsar from long-term commitments to these expensive foreign servitors.

8 De Vries and van der Woude, *First*, 172. It is possible that the brevity of his stint as officer in the Danish navy (see Chapter 14) was caused by his continued inability to cope with written material.

9 *DAK* vol. 5, 263, 268, 270–1.

10 This was an extraordinary sum of money, ten times the usual sum (see de Vries and van der Woude, *First*, 645).
11 *RGADA* 50/42 (1668), l.46.
12 *Reysen*, 121–2. Normally, the captain took care of provisions on Dutch warships, see de Jonge, *Geschiedenis* vol. 1, 665n1, 703. Butler's bill is mostly written in Dutch, and the currency used is the Dutch guilder, which had an exchange rate of 0.20 rubles or 20 kopecks (for this rate, see *Nationaal Archief* of the Netherlands [Den Haag], *Archief der Staten-Generaal* [from here:] *NA SG* 8586, p. 361; Vitsen, *Puteshestvie*, 240n183). *Reysen*'s Struys is usually meticulous in remembering wages, even if fifty-seven guilders translates into an untidy sum of 11 rubles and 40 copecks, whereas archival records show payment to the crew in rounded ruble figures (compare *Reysen*, 121; *DAK* vol. 5, 263, 268, 270–2; and *RGADA* 50/42 [1668], l.43). *Reysen* does correctly list the wages of the common sailors (compare *Reysen*, 121, to *RGADA* 50/42 [1668], l.44). Struys, it seems, told the ghostwriter his average monthly remuneration calculated from the total sum of his first year's wages, during the first two-thirds of which he was paid twelve rubles, and the last one-third 50 guilders (ten rubles) per month. His annual earnings were thus 680 guilders, and his average monthly wage would then amount to 56.67, or about 57 guilders. In the spring of 1669, his and other crew members's wages were lowered (see *RGADA* 50/42 [1668], l.43; *DAK* vol. 5, 263).
13 In the Muscovy enterprises of Marselis and Ackema, foreign master craftsmen annually earned 50–100 rubles (250–500 guilders) and foreign apprentices 30 to 72 rubles (150–360 guilders), substantially more than the usually less skilled Russians they employed (see Phipps, 'Britons,' 31).
14 Loviagin, ed., *Posol'stvo*, 195. Russian officers usually were also remunerated with manors they held in conditional tenure (*pomest'ia*). For further conformation of the exchange rate of 1674, see, too, P. Bushkovitch, *The Merchants of Moscow* (Cambridge, 1980), 54, Table 3.9.
15 Hellie, *Enserfment*, 163; Hellie, *Economy*, 431.
16 *RGADA* 50/42 (1668), l.13.
17 van Deursen, *Plain Lives*, 7–8; Dekker, *Humour*, 47; Prud'homme van Reine, *Rechterhand*, 219.
18 Nusteling, *Welvaart*, 129.
19 *GAA Collectie Kwijtschelding* 5062–52, 103 (Marritje Jansdr).
20 See for the foreigners in Moscow and for the reputation of sailors, for example, Bogoiavlenskii, 'Moskovskaia Nemetskaia sloboda,' 223; Pérez-Mallaína, *Spain's Men*, 23–35; van Goor, *Nederlandse Koloniën*, 24; Scammel, 'European Exiles,' 645. Having sometimes nothing to lose, sailors formed something akin to a maritime proletariat (see for example van Deursen, *Plain Lives*, 26; Barendse, *Arabian Seas*, 109–10). Struys, however, had something to lose by 1668.
21 On the highly developed Amsterdam business 'information exchange', see Smith, 'Function', *passim*.
22 Honig, *Geschiedenis* vol. 2, 97; J.W. Veluwenkamp, 'The Arkhangelsk business venture of the Amsterdam merchant David Leeuw, 1714–1724,' in *Around Peter the Great*, eds C. Horstmeier et al. (Groningen, 1997) 92–102: 94. Even 'news' about Muscovy began to be published by the 1650s in such magazines

as the *Hollandtsche Mercurius* (see Loviagin, 'Vvedenie,' xviii–xxii). Such magazines and books were not likely read by artisans and sailors; oral accounts were a much more common source of information.

23 According to Honig, the people from Struys's native Zaan region were attracted by the chances for profit and opportunities to ply a trade that was scarce in Muscovy (Honig, *Geschiedenis* vol. 2, 97).

24 Van Goor, *Nederlandse Koloniën*, 49, table, shows how only one in three sailors returned from journeys with the VOC to the Far East in the seventeenth and eighteenth centuries.

25 *Reysen*, 129.

26 Mankov, ed., *Zapiski*, 29.

27 Orlenko, *Vykhodtsy*, 193.

28 Orlenko, *Vykhodtsy*, 230.

29 On this issue, see Colley, 'Going Native,' 185–6.

30 Isaac Massa apparently taught himself to write, according to a dedication he wrote to Prince Maurice of Orange; he also managed to learn a fair bit of Russian (see A. van der Linde, *Isaac Massa van Haarlem* [Amsterdam, 1864], 72). But Struys, of course, still could not sign his full name in 1668, when he was nearing forty.

31 Struys, *Reysen*, 136–7; *DAK* vol. 5, 262; Orlenko, *Vykhodtsy*, 70; Viskovatov, *obzor*, 106.

32 Demkin, *kupechestvo* vol. 1, 111. Nowadays the area is called 'clean ponds' (*chistye prudy*). Van Sweeden is not on this list, but his house may have been formally owned by his brother-in-law Hendrik Swellengrebel, or may have been absent from the list since van Sweeden had the status of *Moskovskii inozemets*, a foreign-born Muscovite who had not converted to Orthodoxy, rather than *gollandets* or *nemets*, and was no longer classified as a pure foreigner. For the residences in Moscow at the pond, see also Demkin, Preobrazhenskii, eds, *Zapadnoevropeiskie kuptsy*, 72.

33 In 1675–6, members of van Klenk's retinue would go on to stay there (see Loviagin, 'Vvedenie,' cviii; Veluwenkamp, *Archangel*, 148).

34 Demkin, *kupechestvo* vol. 1, 114. Van Sweeden was not alone in enjoying this privilege, granted to a very small number of individual Western European merchants (see Orlenko, *Vykhodtsy*, 68–9).

35 They did stand out in the city since they were prohibited from dressing in Russian manner (Orlenko, *Vykhodtsy*, 70, 152; see also N.V. Sedov, 'Peremeny v odezhde praviashchikh verkhov Rossii v kontse xvii v,' in *Mesto Rossii v Evrazii*, ed. G. Szvak [Budapest, 2001] 173–81: 174).

36 Viskovatov, *obzor*, 106, 187n63.

37 *DAK* vol. 5, 404–5; Viskovatov, *obzor*, 106–7, 187–90n63. Vinius's proposal shows that he knew exactly who managed the shipbuilding project, as he addressed it to the tsar and to Dokhturov, Golosov, and Iur'ev (and not to the absent Ordin; Viskovatov, *obzor*, 187–8n63). Previously, *Jan Vegron*, the French native Jean de Gron, had submitted a plan for a fleet aiming to aid Russian economic development (Olearius, *Travels*, 141n21; Viskovatov, *obzor*, 113, 196–202n70). A.A. Vinius had started to work in the *Posol'skii prikaz* in 1664; he was promoted to *d'iak* in 1675 and would become one of Peter the Great's closest collaborators (see for instance U. Birgegard, ed., *Sparwenfeld's*, 312n531; Kozlovskii, *Andrei Vinius'*, 9).

38 Viskovatov, *obzor*, 187–8n63; *DAK* vol. 5, 404. On the Holstein project of the 1630s, see Phillips, *Founding*, 14–18.
39 Viskovatov, *obzor*, 188–9n63.
40 Viskovatov, *obzor*, 107, 190–3n64; Kurskov, *Vedushchee napravlenie*, 125.
41 Butler claimed to have purchased seamaps (*seekaerten*) in Amsterdam in preparation for the trip for which he asked restitution from the tsar (*RGADA* 50/42 (1668), l.43). A.A. Vinius, meanwhile, maintained a strong interest in cartography throughout his life, drawing several maps himself and being a key source for Witsen's writing and map making (see J. Keuning, 'Nicolaas Witsen as a Cartographer,' *Imago Mundi* 11 [1954] 95–110: 99; L. Bagrow, 'Semyon Remezov: A Siberian Cartographer,' *Imago Mundi* 11 [1954] 111–25: 125).
42 See map of Caspian Sea in *Reysen*, between 236 and 237.
43 Perhaps a hint may be found of this in the vague statement that 'we [were ordered] to turn to shipbuilding' ('*kregen wy last … ons na den Scheeps-bouw te begeven*', see *Reysen*, 137). But the men mainly resided in Moscow until the spring of 1669.
44 *Reysen*, 137.
45 Orlenko, *Vykhodtsy*, 71–2, 232–3, 235–7. Not infrequently in the countryside, peasants fled their villages upon spotting a contingent of foreign emissaries on their way to or from Moscow (Orlenko, *Vykhodtsy*, 260–2). Orlenko notes that by the 1660s strict social and cultural segregation between foreign and Russian army officers began to wane, as is evident from both Patrick Gordon's and Cornelis van Bockhoven's interaction with Russians (see Orlenko, *Vykhodtsy*, 192–3, and above). But *Reysen* shows very little evidence of any sustained peaceful interaction between Struys (or his comrades) and Russians. The Dutchmen limited their contact with the local population to business transactions and matters related to their service for the tsar. Usually, the ice between foreign-born servants of the tsar and Orthodox Muscovites only melted after the former lived a number of years in the tsar's realm (Orlenko, *Vykhodtsy*, 193). Instead, Westerners's life usually centred on the *sloboda* and similar foreigners' communities.
46 *Reysen*, 137–8.
47 *Reysen*, 138. It is unfortunately impossible to exactly trace this episode in archival records; the Dutch Reformed Church's registers were destroyed in a fire (Scheltema, *Rusland* vol. 1, 120). See D. Tsvetaev, *Protestantstvo i protestanty v Rossiiu epokhi preobrazovanii* (Moskva, 1890), 410. The first Calvinist Church in Moscow had been founded in 1629 (see Olearius, *Travels*, 98n17). Its extant records in Amsterdam's archive are meagre indeed.
48 See van Deursen, *Plain Lives*, 262–3.
49 Her historical existence is proven by VOC records (see Chapters 9 and 10).
50 Butler thought that the Brak's child was about six months old in August or September 1670, which would indicate that she may have become pregnant in June or July of 1669, when the crew was descending the Volga; she could have had a miscarriage before, however, and she may have once missed her menses (not always a good indicator of pregnancy in an era when people were generally malnourished) if she and Brak began to cohabitate in

December, leading to the marriage (see *Reysen*, 'Extract [Butler],' 30; *Reysen*, 208). The Viennese-born Trappen, too, married, albeit in a Lutheran service: his wife was a Lübeck-born servant of a medical doctor; both women accompanied the troupe to Astrakhan (*Reysen*, 121).

51 *Reysen*, 138.

52 Lahana, 'Novaia Nemetskaia,' 68–70.

53 *Reysen*, 138.

54 See Said, *Orientalism*, 15, 23, 51, 166. Certainly, Adam Olearius may be read as a pioneering Russianist and Orientalist.

55 M.S. Anderson, *Britain's Discovery of Russia, 1553–1815* (London, 1958), 40.

56 On the powerful influence of certain authoritative books in this construction (as in the construction of Muscovy), see Said, *Orientalism*, 92.

57 See for Buffon, Said, *Orientalism*, 87, 119, as well as Chapter 13. Said did not investigate Dutch contributions to the emergence of European Orientalism (see Said, *Orientalism*, 16–17, 24).

58 *Reysen*, 140–68. In addition, elite's burial customs were discussed in connection with the demise of tsaritsa Maria Miloslavskaia on 13 March 1669 (ibid., 138–9). The lavish Russian celebration of Holy Week and Easter were conveniently inserted at the moment the narrative of Struys's travels resumes (ibid., 169–70).

59 *Reysen*, 140–7. Olearius's account of these things is far more elaborate (see Olearius, *Travels*, 111–17).

60 Again, Olearius provides a more detailed discussion which seems summarized in Struys's book (Olearius, *Travels*, 165–8; *Reysen*, 153–6).

61 *Reysen*, 147–57.

62 *' ... sterk, dik, en gedrongen van lichaam, en bysonder grof van hooft, armen, en beenen'* (ibid., 147).

63 *'Sy hebben weynigh Huysraat, bestaande het selve in eenige smeerige Potten en Schotels, houte Bakken en Nappen, tinne Brandewijns-schalen en Meede-beekers, die sy ook selden reynigen. Sy bemoeien sich ook niet met de Wanden op te schikken, gelijk de Nederlanders, als alleen met een paar of meer geschilderde Heyligen, en voor al een van St. Nicolaas'* (ibid., 148). This resembles Olearius, *Travels*, 155.

64 *' ... , de mannen niet willende toegeven, mede sich niet ontsien hebben haar kleederen te pandt te laten, ja alle eere en schaamte ter zijden zettende in het openbaar als geyle beesten allerley ontuchtigheyt te bedrijven, welke ongeregeltheden by haar voorheene ook niet voor groote schande geacht wierden en slechts voor wat kluchtigs gehouden, maar het is heden ten dage wat beter sedert dat den Grootvorst op aanraden van den Patriarch het groot getal der kleyne en heymelijke Sluyp-kroeghjens vermindert heeft ... '* (*Reysen*, 150; see for seventeenth-century English translation [which often veers rather far from the Dutch original's text], Struys, *Voyages*, 137[9]–140). The last clause appears to have been adopted from Olearius (see Olearius, *Travels*, 144). Olearius states more discreetly that regarding dead-drunk women '[o]ne may easily imagine the peril to honor and modesty, and its frequent ruin, under such conditions of life' (Olearius, *Travels*, 145).

65 *' ... geeft het in Moscovien voor de Vrouwen 't soetste leven niet, daar sy weynigh minder als de Turkinnen blijven opgeslooten en op het minste afterdenken van haar mannen geslagen, beklaagt en verstooten werden ... '* (*Reysen*, 158). This remark is omitted in the English translation (see Struys, *Voyages*, 146–7).

66 Olearius, *Travels*, 172; *Reysen*, 158.

67 Israel, *Dutch Republic*, 679–80. In its suppression of sexuality, Dutch society resembled rather Elias's nineteenth-century bourgeois society than his seventeenth-century court society (see Elias, *Civilizing Process*, 156–7). In the course of the seventeenth century, Western Europe's elite adopted a more refined code of behaviour, as traced by Norbert Elias (Elias, *Civilizing Process*, 78–9, 85–6).

68 Olearius, *Travels*, 141; *Reysen*, 149.

69 Olearius, *Travels*, 142; Collins, *Present State*, 106.

70 Dekker, *Humour*, 109; Pérez-Mallaína, *Spain's Men*, 33–5, 170–6.

71 For the Western standards (or lack thereof) regarding bathing and bodily functions, see Scheidegger, *Perverses Abendland*, 52–7. Struys's ghostwriter here mechanically copied Olearius: compare *Reysen*, 159–60, and Olearius, *Travels*, 142, and especially ibid., 162, where some of the blatant plagiarism becomes apparent. Thus Olearius, '[i]n Narva I saw with amazement how Russian and Finnish boys eight, nine, or ten years of age ... walked and stood barefooted in the snow ... just like geese ... ' (Olearius, *Travels*, 162) and *Reysen*, 'Children of 8 or 9 years may trot barefoot across the ice, no different as if they were geese' (Struys, *Reysen*, 160). Again copying Olearius, *Reysen* claims that Struys saw baths resembling the Russian *banya* among resident *Duytschen* (see Olearius, *Travels*, 162–3; *Reysen*, 159–60).

72 *Reysen*, 160–3.

73 Ibid., 163–6.

74 Ibid., 166–8. Olearius in contrast engages in a lengthy exposé regarding Muscovy's political system (Olearius, *Travels*, 173–202)

75 *Reysen*, 166. This phrase erroneously suggests that Romanow means 'son of the Roman', and was a moniker adopted by Tsar Ivan Vasil'evich to claim descendance from the Roman emperor. It is unclear whether Ivan III Vasil'evich (r. 1462–1505) or Ivan IV Vasil'evich (r. 1533–84) is meant here.

76 *'Sy zijn uyt der nature slaafachtigh en gelijk tot dienstbaarheyt gebooren en sullen selden door een edelmoedige of heusche aansporinge hun werk doen; maar altijdt door slagen daar toe moeten werden gedreven; ja sy hebben soo weynigh mishagen of verdriet in haar slaverny dat sy door den doodt of goedertierenheyt van haar Heeren eenmaal vry geworden zijnde sich selven voorts weder aan een ander verkoopen; sulks dat die gene die wel van haar wil gedient zijn hoe barmhertigh en mededogende hy ook wesen mach gedwongen werdt sijn vuysten noch stok niet te sparen'* (*Reysen*, 148).

77 On this alleged unbridled quality (or anarchistic streak) of the Russians, see A.N. Sakharov, 'Demokratiia i volia v nashem otechestve,' *Svobodnaia mysl'* 17 (1991) 42–53. For its relevance to the great revolts of the seventeenth and eighteenth centuries, see Avrich, *Russian Rebels*, 1–7, 256–8, 262. See as well Olearius, *Travels*, 153.

78 *'sy werden echter doorgaans te schraal met spijse onderhouden en dit verwekt veel Dieven en Moordenaars, sulks dat ieder die wat te verliesen heeft wel snedigh mach oppassen: want onaangesien de sware straffen op kleyne dieverijen, soo konnen de Brandewijn- en Tabacq-gretige Slaven daar door niet werden in toom gehouden'* (*Reysen*, 148).

79 Olearius, *Travels*, 147, 151.

80 The Holsteiner repeatedly calls Russians crafty, treacherous, superstitious, arrogant, uncouth, untutored, rude, crude, lewd, or drunk (see for example Olearius, *Travels*, 131, 133–4, 137–9, 141–5). We might note that Pepys by his own account consumed alcohol every day, usually from early morning onwards; as van Deursen observes, the Dutch were notorious for their excessive drinking, too (for example, see *Diary of Samuel Pepys*, ed. Wheatley, 209 [16th September 1665], [1666–7], 336 [11 August 1667]; van Deursen, *Plain Lives*, 100–2; Dekker, *Humour*, 8.

81 Th. Barran, rev. of *Une infinie brutalité*, by M. Mervaud and J.-C. Roberti in *Slavic and East European Journal* 1 (1994) 205–7: 206.

82 Viskovatov, *obzor*, 107; *DAK* vol. 5, 263–4.

83 Viskovatov, *obzor*, 107–8; *DAK* vol. 5, 265. Van der Gaeten is named in notarial records about a dispute in 1664 involving Dutch merchants' trade on Muscovy, including van Sweeden (*GAA NA* 1722 [Notary P. de Bary] p. 421 [24 November 1664]).

84 Khitrovo's servant might be *Reysen*'s Jan Fassely, mentioned once as a servant (*'knecht'*) of David Butler; Fassely is not found among the Dutchmen in the *RGADA* record of their entry, and seems omitted from the crew's list – indicated by their jobs instead of their names – leaving Moscow around 1 May 1669 (see *Reysen*, 121; *RGADA* 50/42 [1668], l.2). This servant may have been forced to stay behind after his 'unmasking': 'Jan Fassely,' quite likely, was a Dutch version of the Russian name Ivan Vasil'evich.

85 *GAA NA* 1722 (Notary Pieter de Bary), p. 421 (24 November 1664).

86 Viskovatov, *obzor*, 108; *DAK* vol. 5, 268–9.

87 Viskovatov, *obzor*, 105.

88 *DAK* vol. 5 268 (at this point his wage was lowered to ten rubles per month). Under the new conditions of their employment by the tsar, those who had travelled with Butler from Amsterdam were considered to be in the tsar's service for the period from 1 November 1668 to 1 November 1672. Perhaps the advance wages paid in Amsterdam were merely to cover the months of travelling to Moscow and did not fall under the terms of this contract.

89 Viskovatov, *obzor*, 108; *DAK* vol. 5, 273. If Struys did participate in this ceremony, it is strange that *Reysen* leaves it unmentioned; perhaps only the ship's officers were invited.

90 Viskovatov, *obzor*, 109; *DAK* vol. 5, 274.

91 Struys states May 4, but his dating is highly unreliable (*Reysen*, 170).

92 A *strug* was one of the several sorts of vessels travelling on the Russian rivers, and was also used for trade across the Caspian Sea. It was a primitive albeit versatile boat, with a length between six and fifteen metres, and it could carry up to 500 tonnes (see Kukanova, *Ocherki*, 35). Usually rowed, it sometimes used a sail, and it had a cabin on deck. A crew of

twenty would man the largest *strugy*. It may be that a *strug* was the 'sloop' with which the Struys group fled from Astrakhan, but they may have used a *bus*, another small one-sail ship, which was incapable of sailing against the wind; the *bus* was the more usual traditional boat plying the Caspian Sea from Astrakhan to Baku or Derbent (ibid., 36; Tushin, *Russkoe moreplavanie*, 40–2). Besides *strugi* many other vessels travelled the Volga (Tushin, *Russkoe moreplavanie*, 37).

93 Struys does not identify them by name (*Reysen*, 171).

94 Viskovatov, *obzor*, 109; *DAK* vol. 5, 276–7.

95 *DAK* vol. 5, 277–8. They were in such a hurry (on Poluekhtov's instigation) that they did not await cannon and ammunition that was to arrive from the Tula works (Viskovatov, *obzor*, 110). *Reysen* notes the date of May 12, but the archival source is more reliable than its uncertain dating (*Reysen*, 170).

96 In a sense, Butler's commission was a remarkable break with tradition: no unsupervised foreigner ever commanded a military contingent before Peter the Great's time (Orlenko, *Vykhodtsy*, 120). But the largely Dutch crew was hardly the size of an army, and the *Oryol* served both mercantile and military purposes.

97 His task was to make and maintain the wooden spindles (*bloocken* or *blokken*) through which the ropage of the sails was manipulated.

98 Viskovatov, *obzor*, 105–6, 110; Struys, *Reysen*, 121.

99 See Viskovatov, *obzor*, 96, where S(c)hak is identified as *lekar'* who joins van Bockhoven's team at Dedinovo in the fall of 1667; Struys erroneously lists Schak as part of the crew who left Holland with him (*Reysen*, 121).

100 Viskovatov, *obzor*, 110.

101 Tushin, *Russkoe moreplavanie*, 36.

102 Viskovatov, *obzor*, 110–1; Tushin, *Russkoe moreplavanie*, 40.

103 He died at the Turkish siege of Chyhyryn [in Russian, Chigirin], where the Russian defence was led by (his then former in-law) Patrick Gordon, in April 1678, see Obolenskii, Posselt, eds, *Tagebuch*, 544.

104 Arkheograficheskaia kommissia, *Akty istoricheskie* vol. 4 (1645–76) (St Petersburg, 1842), 410; *Krest'ianskaia voina pod predvoditel'stvom Stepana Razina*, vol. 3 (Moscow, 1954), 212–13.

105 *Reysen*, 135, 173; *DAK* vol. 5, 263. Both Olearius and Struys remarked on the low cost of food in Nizhnii Novgorod, although they offer different examples, showing that Struys rather than Olearius was *Reysen*'s source (Olearius, *The Travels*, 193; *Reysen*, 173).

106 *DAK* vol. 5, 280. Struys remembered June 21 (*Reysen*, 173).

107 Viskovatov, *obzor*, 110; *Reysen*, 173.

108 Viskovatov, *obzor*, 110–1.

109 The Volga's water levels are at their height in May and June, and quickly begin to drop in July; many are the occasions mentioned when the fleet ran aground and had to be pulled from sand banks (*Reysen*, 174, 177, 182–4). The average time for a journey via from Moscow to Astrakhan was less than 90 days, and some trips took a mere 40 days (Dedinovo is well-nigh equidistant with Moscow from Astrakhan, some 3,000 kilometres by river; Kukanova, *Ocherki*, 35, Kotilaine, *Russia's*, 59).

110 Th.M. Barrett, 'Lines of Uncertainty: The Frontiers of the North Caucasus,' *Slavic Review* 3 (1995) 578–601: 581.

111 Olearius, *Travels*, 298–300; *Reysen*, 175–6. It should also be noted that *Reysen* has Struys at Vasil'gorod for no more than one full day (*Reysen*, 174, 177). For the suggestion that Chuvash in the hills on the 'right' (western) side of the Volga, rather than the Cheremiss, who lived on the plains of the 'left' bank, see 'Puteshestvie Iana Iansena Streisa po Rossiiu v 1668 gody,' *Severnyi arkhiv'* 5 (1824) 275–90: 284n.

112 Mund has observed an inclination among all Western authors to essentialize the Volga peoples as abject barbarians (Mund, *Orbis Russiarum*, 299). This seems to be linked to a general loathing of nomadic peoples as primitive savages.

113 Olearius, *Travels*, 300; *Reysen*, 176. Struys did briefly halt on Ceylon at least once (*Reysen*, 369).

114 A. Herport, *Reise nach Java, Formosa, Vorder-Indien und Ceylon, 1659–1668*, ed. S.P. L'Honoré Naber (Den Haag, 1930; origin. 1669), 133; *Reysen*, 176. See also Chapter 12.

115 '*und [the father] sagt, daß es ihme gebühre von dem Baum, den er gepflantzet, die erste Frucht zugeniessen*', became in Dutch '*[s]ou ik een Boomken planten, en daar af geen vrucht leesen voor ik die aan een ander overgeef, dat waar een groote slechtigheyt*' (Herport, *Reise*, 133; *Reysen*, 176). Schouten's text does depict the people of Ceylon, but speaks of unspecified '*bloedschande*' (incest, see Breet, *Oost-Indische Voyagie*, 293–9).

116 Olearius, *Travels*, 300–1; *Reysen*, 177.

117 *Reysen*, 178. Estimates for the amount of people captured by Tatars in the tsar's realm in the first half of the seventeenth century have been as high as 150,000 to 200,000 (see Tushin, *Russkoe moreplavanie*, 23).

118 *Reysen*, 179.

119 *Reysen*, 181; see Honig, *Geschiedenis* vol. 1, 231, 244–5.

120 Olearius, *Travels*, 308–9; *Reysen*, 183.

121 *Reysen*, 185. The actual print of the Kalmyks was placed only after page 215 (see *Reysen*, 'Bericht'). Olearius has little to say about them, possibly because they only became a regular fixture on the western steppe during the 1630s (Olearius, *Travels*, 315; Khodarkovsky, *Russia's*, 130, 134).

122 See Hughes, *Russia*, 259.

123 *Reysen*, 186.

124 *Reysen*, 186: Olearius uses the name indeed only to identify the river after which the fortress would be named later (Olearius, *Travels*, 316–17).

125 Most of the stops at the towns along the river also allowed for repairs, since almost all had a wharf (Tushin, *Russkoe moreplavanie*, 37).

126 *Reysen*, 187–93; Olearius, *Travels*, 322–30. They arrived in Astrakhan in late August (following the Gregorian Calendar); see the anonymous letter in *Reysen* which uses the Julian calendar and mentions August 13 (*Reysen* [Extract], 2).

127 As Willem Floor argues (see Floor, 'Fact or Fiction'; see as well Brancaforte, *Visions*, 209n93). For example, the descriptions of the ten-day commemoration in May of Husayn (*'Ashura*) in Olearius (who witnessed this in Ardabil) and *Reysen* (Struys observed it in Shemakha) are alike, but *Reysen* is lengthier and more detailed than Olearius (see Olearius, *Podrobnoe*,

572–4; *Reysen*, 270, 280–1; for 'Ashura, part of the *Muharram* commemoration, see J. Calmard, 'Shi'i Rituals and Power II. The Consolidation of Safavid Shi'ism: Folklore and Popular Religion,' in *Safavid Persia*, ed. Melville, 139–190, 141, 148–51, 159; for [Twelver] Shi'ism and its connections with the Safavids in Iran, see R. Savory, *Iran under the Safavids* [Cambridge], 1980, 27).

128 Compare for instance *Reysen*, 270, 275, 278–9, 284–5, 290, with A.L. Gol'dberg, A.G. Man'kov, S.Ia. Marlinskii, 'Izvestiia o vosstanii S. Razina v zapadnoevropeiskikh periodicheskikh izdaniiakh i khronikakh XVII v.,' in *Inostrannye izvestiia o vosstanii Stepana Razina*, ed. A.G. Man'kov (Leningrad, 1975), 80–91; *Kort waerachtigh Verhaal, van de bloedige Rebellye in Moscovien* (Haarlem, 1671); *Hollandtsche Mercurius* 1670, 113 (August 1670), 141 (November 1670); *Hollandtsche Mercurius* 1671, 78 (June 1671), 85–98 (July 1671).

8 The Volga Delta and the *Oryol*'s Demise

1 The northern sector of the Caspian is usually frozen solid from early December to early March (X. de Planhol, 'Caspian Sea,' in E. Yarshalter, ed., *Encyclopaedia Iranica* (available at http://www.iranica.com/newsite/, accessed on 12 January 2007).

2 Kotilaine, 'When the Twain,' 1065; Kotilaine, *Russia's*, 452. The *Oryol* was in principle of course much faster than such vessels.

3 Struys and Faber first met in September 1669 after a stand-off between L'vov's forces and Razin's cossacks had ended amicably, and Struys met many '*Duytsche*' officers who commanded the tsarist troops and paid a visit to the ship that had just arrived at the city (see *Reysen*, 194).

4 Faber (Fabricius) deposed his recollections half a century later in Sweden on the request of Peter the Great, see *Krest'ianskaia voina* vol. 4, 216–17. Viskovatov argued that at first the ship was not prepared and rigged properly (Viskovatov, *obzor*, 86).

5 See for similar problems Hughes, *Russia*, 466.

6 See de Planhol, 'Caspian Sea.'

7 Phillips, *Founding*, 14–18.

8 His brother had been an ambassador to England in the early 1660s (see Gurliand', *Ivan' Gebdon'*, 26–7).

9 *Krest'ianskaia voina* vol. 4, 216–17. The *Oryol* had participated in facing down the Cossacks by firing a few cannon volleys in August 1669 (*Reysen* [Extract], 3).

10 *Krest'ianskaia voina* vol. 4, 216–17; *Reysen*, 'Extract [Butler],' 5.

11 Compare Jones's remarks on the cost of warships (Jones, *Anglo-Dutch Wars*, 61–2). Anisimov compares shipbuilding costs to those of space programmes today (E. Anisimov, *The Reforms of Peter the Great* [Armonk, NY, 1993], 66).

12 See Orlenko, *Vykhodtsy*, prilozheniia, 304–6.

13 Tushin, *Russkoe moreplavanie*, 49–50; *DAK* vol. 8, 250–1; see Chapter 6.

14 Viskovatov notes that Ordin only returned in the spring of 1670 (*obzor*, 78).

15 *DAK* vol. 5, 268.
16 For his pro-Western sentiments, see 'Skazanie Adolfa Lizeka o posol'stve ot Imperatora Leopol'da k Tsarius Aleksiiu Mikhailovichu v 1675 godu,' *Zhurnal' ministerstva narodnago prosveshcheniia* (Noiabr' 1837) 327–94: 367–8; see also Solov'ev, *Istoriia Rossii*, vol. 3 (parts 11–15), 760, 768. Kurskov notes a different emphasis in Matveev's priorities in foreign policy as head of the *Posol'skii Prikaz* (1671–6; see Kurskov, *Vedushchee napravlenie*, 93, 130–1). Matveev fell from grace in 1676, as the new Dutch resident van Keller reported to The Hague (*NA SG* 7364 [21 December 1676] and *NA SG* 7364 [28 December 1676]).
17 The process had begun in November 1669 and lasted until May 1670 (Zabelin, *Domashnii byt'* vol. 2, 264–6).
18 Viskovatov, *obzor*, 104–5; Solov'ev, *Istoriia Rossii* vol. 3 (parts 11–15), 681.
19 Court intrigues against Ordin had likely gained momentum during his absence abroad, forcing him to take the tonsure: evidence of his earlier quarrels about precedence with the *stol'nik* Matvei Pushkin may be seen in Solov'ev, *Istoriia,* vol. 3 (parts 11–15), 681.
20 Kurskov, *Vedushchee napravlenie*, 46, 139; Viskovatov, *obzor*, 78; Crummey, *Aristocrats*, 42–3, 97–8, 100–1, 151; Longworth, *Alexis*, 207–8. The exact year according to our calendar of his taking the tonsure is not quite evident, but 1671 seems most likely; he died in 1680 (see Poe, *Russian Elite* vol. 1, 222). It is possible that he also succumbed to pressure on the part of the older boyar families, who were displeased with the powerful position of the upstart Ordin and his reformist policies (see Crummey, *Aristocrats*, 28, 87, 100–1; Solov'ev, *Istoriia Rossii*, vol. 3 [parts 11–15], 681).
21 After tsarist forces recovered Astrakhan in 1671, authorities made no attempt to refurbish the naval squadron, which survived, even if damaged (see Chapter 6 and *DAK* vol. 8, 250–1; Tushin, *Russkoe moreplavanie*, 50; Longworth, *Alexis*, 194, 200–1).
22 Even if its volume did increase after 1675 (see Matthee, *Politics*, 196–7; see also Kotilaine, 'When the Twain,' 1046, 1064, 1074, 1122). Meanwhile, the long-lasting peace of 1639 between Ottoman Turkey and Safavid Persia provided a relatively smooth caravan transport of the bulk of the silk trade from Iran to the Mediterranean ports (a smaller part was exported via Bandar-e-'Abbas), thus preempting any urgency to trade across Russia (see Man'kov, ed., *Zapiski* vol. 1, 154n2).
23 See Man'kov, ed., *Zapiski* vol. 1, 132.
24 Man'kov writes of some 50 texts written by 'Europeans' travelling through Muscovy and Ivan between 1475 and 1700 (Man'kov, *Zapiski* vol. 1, 133).
25 See *Reysen*, 237–8.
26 Loviagin, 'Vvedenie,' lxxviii.
27 *NA SG* 8586, 384–97; Loviagin, 'Vvedenie,' lxxvi; Scheltema, *Rusland* vol. 1, 335–6; *RGADA* 50/9 (1676), ll.373–373ob.
28 E.S. Zevakin, 'Persidskii vopros v russko-evropeiskikh otnosheniiakh XVII v.,' *Istoricheskie zapiski* 8 (1940) 129–62: 151.
29 Zevakin, 'Persidskii vopros,' 157; Kotilaine, *Russia's*, 462–6.
30 As Paul Bushkovitch has shown (see Bushkovitch, *Merchants*, 168–73).
31 Hellie, *Economy*, 403.

32 Davies, *God's Playground* vol. 1, 356.
33 *Polnoe sobranie zakonov'* vol. 1, 911–12.
34 See Chapter 14.
35 See *Reysen*, 190–1.
36 See *Reysen*, 205.
37 For van Klenk's proposals, see Loviagin, 'Vvedenie,' lxxvi–lxxviii.
38 See Coyett, *Historisch Verhael*.
39 *Krest'ianskaia voina* vol. 3, 379; Viskovatov, *obzor*, 110.
40 Eekhoff, 'Friezen,' 33; W.M. von Richter, *Geschichte der Medicin in Russland* vol. 2 (Moscow, 1815), 416.
41 *Reysen*, 194. He was one of the '*Duytsche*' (Germanic) officers of the tsarist army who came aboard the ship around 1 September 1669.
42 *Reysen*, 188–204.
43 Avrich, *Russian Rebels*, 83.
44 Kotilaine, 'When the Twain,' 1046; on the six ports used in 1667, see ibid., 1050. See as well Kotilaine, *Russia's*, 57–8.
45 Kotilaine, 'When the Twain,' 1051.
46 Ibid., 1050.
47 *Reysen*, 188.
48 Kotilaine, 'When the Twain,' 1052; *Reysen*, 189.
49 *Reysen*, 188–9.
50 Ibid., 190–1.
51 Ibid., 190, and see below.
52 The 'Tatar Lamb' had fascinated the Western Europeans for some time; already around 1600 an English ambassador to Tsar Boris Godunov had received a robe allegedly made of the fur of this 'creature', which was in reality a sort of cotton; in 1666 a presentation about it was made before the fledgling Royal Society (see J.H. Appleby, 'The Royal Society and the Tartar Lamb,' *Notes and Records of the Royal Society of London* 1 [1997] 23–34: 23–4).
53 *Reysen*, 'Extract [Butler],' 5. The English translation suggests the crew's adaptation of the sailship; the original Dutch is ambiguous here, but Shcherbatov's 1680 report indicates that the rowing vessel was built from scratch (see Struys, *Voyages*, ['Narrativ of Butler'], 364; see Chapter 4). Butler's letter provides no detail and leaves moot how far this project progressed before the Cossacks' encirclement of Astrakhan ended it.
54 Kostomarov, 'Bunt,' 412–25.
55 For the high incidence of flight by serfs around 1670, see Solov'ev, *Istoriia*, vol. 3 (parts 11–15), 733–4.
56 See Khodarkovsky, *Russia's*, 130–9.
57 Khodarkovsky, *Russia's*, 30–1. Struys notes that at the conjunction of the Usa and Volga one may find some of the best soil but that 'robbers and Cossacks', hiding in the forest made it too dangerous to stay (*Reysen*, 184). Olearius's group was also warned about Cossacks making the lands along the Volga unsafe in the 1630s (Olearius, *Travels*, 307, 311–13).
58 The term 'Cossack' appears Turkic in origin, meaning something such as freeman or itinerant person (see Man'kov, ed., *Zapiski* vol.1, 120n2); for the sale of Muslim captives (*iasyri*), see for instance B. Tikhomirov, *Razinshchina* (Moskva, 1930), 49.

59 See for this and his early days Avrich, *Russian Rebels*, 66–8.
60 Kostomarov, 'Bunt',' 427–8.
61 Khodarkovsky, *Russia's*, 136–9.
62 See among others Avrich, *Russian Rebels*, 64.
63 Tikhomirov, *Razinshchina*, 50–1; Solov'ev, *Anatomiia*, 37.
64 Khodarkovsky, *Russia's*, 137; Tikhomirov, *Razinshchina*, 52. At the time and after the rebellion's suppression, Cossacks themselves explained that the Azov fortifications caused this shift (see A.N. Popov, 'Istoriia vozmushcheniia Sten'ki Razina,' *Russkaia beseda* 1 [1857] 47–104: 99; *DAK* vol. 6, 16).
65 It is uncertain why Razin, a registered Cossack himself, defied the official Don Cossack host by leading the restless unregistered frontiersmen; *Reysen* dismisses the claim that Razin sought revenge for the death of his brother, supposedly executed for insubordination on the order of Iurii Dol'gorukii in 1665 (*Reysen*, 195).
66 Kurskov, *Vedushchee napravlenie*, 39. Don Cossacks under Vas'ka Us had in fact begun plundering along the northern reaches of the Don in 1666.
67 Kostomarov, 'Bunt',' 432–4; Avrich, *Russian Rebels*, 72.
68 Savory, *Iran*, 64, 75, 79–81.
69 Kostomarov, 'Bunt',' 435–9, 441; Loviagin, ed., *Posol'stvo*, 150; Konovalov, 'Ludwig,' 78n4.
70 Kostomarov, 'Bunt',' 440; Avrich, *Russian Rebels*, 74. The Russian rulers remained anxious about Persian perceptions: in 1675 Artamon Matveev interrogated the Polish envoy Gurdziecki to find out whether the shah believed that the tsar had dispatched Razin to Persia (*Krest'ianskaia voina* vol. 3, 343–4).
71 *Reysen*, *'Extract,'* anon., 3–4.
72 *DAK* vol. 6, 16.
73 Kostomarov, 'Bunt',' 449; Loviagin, ed., *Posol'stvo*, 150; *DAK* vol. 6, 17.
74 Kostomarov, 'Bunt',' 445, 447. *Reysen* equates Cossacks with brigands, echoing Olearius (see, for example, *Reysen*, 184; Olearius, *Travels*, 305).
75 Kostomarov, 'Bunt',' 446–7; *Reysen*, 198–9. It allowed though for another evocative engraving (*Reysen*, engraving between 198 and 199).
76 *Reysen*, 197–8; Kostomarov, 'Bunt',' 449, 452–3, 468; Konovalov, 'Ludwig,' 79; Avrich, *Russian Rebels*, 76; Loviagin, ed., *Posol'stvo*, 151, 156–7.
77 Kostomarov, 'Bunt',' 453–4. Butler speaks of March, but he likely uses the Julian calendar (*Reysen*, 'Extract [Butler],' 5).
78 Avrich, *Russian Rebels*, 78–9.
79 Kostomarov, 'Bunt',' 453; Konovalov, 'Ludwig,' 80.
80 Kostomarov, 'Bunt',' 454.
81 Konovalov, 'Ludwig,' 80; *Reysen*, 'Extract [Butler],' 6; Avrich, *Russian Rebels*, 79–80.
82 Kostomarov, 'Bunt',' 454–5
83 Kostomarov, 'Bunt',' 455; *Reysen*, 206.
84 It is highly likely that Coyett relied in his account of this episode on Struys, as, like Struys, he calls L'vov '*Elbof*' (compare Loviagin, ed., *Posol'stvo*, 154, with *Reysen*, 206). Also see Konovalov, 'Ludwig,' 80–1.
85 Kostomarov, 'Bunt',' 455; *Reysen*, 'Extract [Butler],' 6–7.

86 Kostomarov, 'Bunt',' 455–6. Butler thought they numbered 2,600 (*Reysen*, 'Extract [Butler],' 7).
87 Kostomarov, 'Bunt',' 457; Konovalov, 'Ludwig,' 81; Avrich, *Russian Rebels*, 83.
88 Konovalov, 'Ludwig,' 83.
89 Konovalov, 'Ludwig,' 82.
90 *Reysen*, 207; *Reysen*, 'Extract [Butler],' 7.
91 Kostomarov, 'Bunt',' 458. L'vov was executed a year later by the Cossack chiefs in Astrakhan (Konovalov, 'Ludwig,' 78n3).
92 Kostomarov, 'Bunt',' 459; Konovalov, 'Ludwig,' 84; *Reysen*, 207–8; *Reysen*, 'Extract [Butler],' 7–8.
93 Loviagin, ed., *Posol'stvo*, 151–2; *Reysen*, 206–7; *Reysen*, 'Extract [Butler],' 7; Avrich, *Russian Rebels*, 84.
94 Kostomarov, 'Bunt',' 457–9.
95 *Reysen*, 'Extract [Butler],' 8.
96 *Reysen*, 208. See as well Faber's fear of being discovered as a 'foreigner' (Konovalov, 'Ludwig,' 88). Later revolts by *strel'tsy* and Cossacks in Astrakhan and elsewhere showed similar hatred of '*nemtsy*' (Hughes, *Russia*, 455).
97 Van Nierop, 'Catholics,'109–11.
98 *Reysen*, 208–9; Kostomarov, 'Bunt',' 459; *Reysen*, 'Extract [Butler],' 8–9; *GAA NA* 4304 (Notaris Nicolaes Hemminck), p. 247.
99 *Reysen*, 'Extract [Butler],' 9; *Reysen*, 208–9.
100 *Reysen*, 209; Faber witnessed Helt's death, according to a statement he deposed in the *Posol'skii prikaz* in May 1676 (*Krest'ianskaia voina* vol. 3, 355). In *Reysen*, Struys remembered how Helt managed to postpone the sloop's departure for a few hours, but is silent about the man's further fate. Confirmation that Dirk Pietersz fell at the defence of Astrakhan is found in his widow's petition of a few years later (*Krest'ianskaia voina* vol. 4, 135). In Butler's letter, the '*Schipper*' (Helt) is mentioned as staying with him, and Brandt is identified by name; the captain mentions two unnamed sailors, who must be Dirk Pietersz and Trappen (*Reysen*, 'Extract [Butler],' 8–9).
101 *Reysen*, 208.
102 His narrative omits some scenes depicted in an engraving in the book, showing the Cossack capture of Astrakhan, which was to be appriopriately placed within the text of Butler's letter (see *Reysen*, 'Bericht Aan den Boekbinder'), on page 17.
103 Man'kov, ed., *Zapiski* vol. 1, 126n33.
104 Orlenko, *Vykhodtsy*, prilozheniia, 305; *GAA NA* 4304 (Notary Nicolaes Hemminck), p. 227; *GAA NA* 4304 (Notary Nicolaes Hemminck), p. 247.
105 'Skazanie letopisi o grabe Astrakhani,' in *Materialy dlia istorii vozmushcheniia Sten'ki Razina* (Moskva, 1857), 241–61: 242. The use of this originally Western loanword is telling: *saldaty* was a word indicating non-traditional (originally Western) warriors.
106 Butler writes of 22 June (*Reysen*, 'Extract [Butler],' 12); Konovalov, 'Ludwig,' 83n3.
107 *Reysen*, 'Extract [Butler],' 13; Mankov, ed., *Zapiski* vol. 1, 78n20.
108 *Reysen*, 'Extract [Butler],' 13; Kostomarov, 'Bunt',' 463.
109 Kostomarov, 'Bunt',' 463–4.

110 Kostomarov, 'Bunt',' 464.
111 '[E]*en vader van veel godtloose kinderen,' Reysen*, 197.
112 *Reysen*, 'Extract [Butler],' 10–11; Kostomarov, 'Bunt',' 464; *Cherkesy* or Circassians was a generic name used for *Adyg* (of which the Kabardinians are a subgroup), hailing from the northern Caucasus region (see Khodarkovsky, *Russia's*, 15; Kappeler, *Russian Empire*, 179–80). Fabricius writes of a heroic and bloody defence by the few remaining Dutchmen (Konovalov, 'Ludwig,' 84–5).
113 *Reysen*, 'Extract [Butler],' 13, 20: Butler notes how many of the Persians' lives were spared, although he is not sure why.
114 Kostomarov, 'Bunt',' 465; Konovalov, 'Ludwig,' 85.
115 Kostomarov, 'Bunt',' 465; Konovalov, 'Ludwig,' 85; Loviagin, ed., *Posol'stvo*, 155–6.
116 Kostomarov, 'Bunt',' 466–7, 485–94.
117 *Reysen*, 'Extract [Butler],' 20–1.
118 Razin was not only supported by Russians and Cossacks, but also by various Finno-Ugrian peoples, Kalmyks, Chuvash, and Tatars who rose against Russian rule, making this more than internecine warfare among Orthodox Slavs (Kostomarov, 'Bunt',' 468, 471; Loviagin, ed., *Posol'stvo*, 157).
119 Kostomarov, 'Bunt',' 470.
120 Loviagin, ed., *Posol'stvo*, 157–8; Crummey, *Aristocrats*, 47, 138; Longworth, *Alexis*, 201–2.
121 Kostomarov, 'Bunt',' 470–1, 482–4, 494–5.
122 Kostomarov, 'Bunt',' 495.
123 *PRO* 91/3, 202 (letter to Mr Richard Daniels, Merchant, by Thomas Hebdon, from the Sloboda, of 6 June 1671); Kostomarov, 'Bunt',' 496–7.
124 Kostomarov, 'Bunt',' 500; Konovalov, 'Ludwig,' 91n4.
125 Poe, *Russian Elite*, vol. 1, 422.
126 Kostomarov, 'Bunt',' 501.
127 Kostomarov, 'Bunt',' 501–2; Konovalov, 'Ludwig,' 92–3. He was merciless, even having *foreign* officers beaten with *batogi* (a kind of stick; see Orlenko, *Vykhodtsy*, 183; Konovalov, 'Ludwig,' 93; Poe, *Russian Elite* vol. 1, 431).
128 *Hollandtsche Mercurius*, August 1670, 113; November 1670, 141; June 1671, 78; July 1671, 85–98.
129 *GAA NA* 4304 (Notary Nicolaes Hemminck), 227–227verso and *GAA NA* 4304 (Notary Nicolaes Hemminck), 247.
130 See further Chapter 14.
131 *Reysen*, 'Extract [Butler],' 30–1; *Reysen*, 209.
132 *Reysen*, 'Extract [Butler],' 8, 18–20, 29; Mankov, ed., *Zapiski*, 82n47; Konovalov, 'Ludwig,' 74–5. Faber too was enslaved by a 'Tatar,' but merely for a few weeks (ibid., 75, 90).
133 *Reysen*, 'Extract [Butler],' 15–19; Konovalov, 'Ludwig,' 86, 88n1; Mankov, ed., *Zapiski* vol. 1, 25–6. The story of this foiled escape was corroborated by *strel'tsy* during Iakov Odoevskii's investigation of the revolt (see Popov, 'Istoriia vozmushcheniia,' 62–3; *Krest'ianskaia voina* vol. 3, 237–8, 268; *DAK* vol. 4, 417–18).
134 *Reysen*, 'Extract [Butler],' 21–2; Konovalov, 'Ludwig,' 89, 89n2; Mankov, ed., *Zapiski* vol. 1, 31–2.

135 In the extract from his letter Butler confuses several times the spelling of Russian Terki (Terskii Gorod) with Dagestani Tarku (*Reysen*, 'Extract [Butler],' 27–9).
136 Konovalov, 'Ludwig Fabritius's Account,' 90.
137 *Reysen*, 'Extract [Butler,' 29; Mankov, ed., *Zapiski* vol. 1, 33–4, 81n39, 81n40. It is dubious whether 400 rubles, the equivalent of 2,000 guilders, was paid to his captors, for it is unclear why someone would pay such an enormous sum for Faber, or how he convinced his saviour that he could pay him back.
138 Konovalov, 'Ludwig Fabritius's Account,' 90–1; Richter, *Geschichte* vol. 3, 177–8. For Derbent's cultural importance, see M. Kemper, A. Sixsaidov, 'Einleitung,' in *Die Islamgelehrten Daghestans*, eds M. Kemper and A. Sixsaidov (Berlin, 2004), 8, and Kemper, Sixsaidov, eds, *Islamgelehrten*, 36–7n20.
139 *Reysen*, 'Extract [Butler],' 28–9; Butler called the population '*Kalmuk*,' the Dutch word used for Kalmyk, but they were, in fact, Kumyks (see Khodarkovsky, *Russia's Steppe Frontier*, 68–9).
140 They just missed encountering Struys, who had departed for Shemakha with his master, as we shall see in the next chapter.
141 *Reysen*, 'Extract [Butler],' 29–32. Butler's account regarding de Vries and Arentsz slightly differs from that of the VOC agents, see *NA VOC Daghregister* 1285, 289verso, 290; W. Coolhaas, ed., *Generale Missieven van Gouverneurs-Generaal en Raden aan Heren XVII der Verenigde Oostindische Compagnie* vol. 3 (Den Haag, 1968), 775–6; Konovalov, 'Ludwig,' 91; *Reysen*, 260.

9 A Dutch Slave in Asia

1 This fascination is reflected in Dutch painting: see H. Goetz, 'Persians and Persian Costumes in Dutch Painting of the Seventeenth Century,' *The Art Bulletin* 3 (1938) 280–90. On the ambiguous self/other binaries between Western Europe and West-Asian Empires, see Aune, 'Early Modern European,' 122.
2 See Dapper, *Asia*; Olearius, *Podrobnoe*. A publication by van Meurs comparable to Dapper's *Asia*, Montanus's account of the Dutch embassies to the Japanese court, had an auction catalogue price of more than 9 guilders in 1687, half of Struys's monthly takings as a sailmaker at sea (see Schama, *Embarrassment*, 619; A. Montanus, *Gedenkwaerdige gesantschappen der Oost-Indische Maetschappy in 't Vereenigde Nederland, aen de kaisaren van Japan* [Amsterdam: van Meurs, 1669]).
3 Said, *Orientalism*, 39–40.
4 See Said, *Orientalism*, 50–2, and Aune, 'Early Modern European,' 121.
5 Pagden, 'Europe and the World Around,' 15.
6 Said, *Orientalism*, 55, 58, 62. For the more traditional view of the world still prevalent before 1700, see Tillyard, *World Picture*, or Thomas, *Man*. For the traditional Christian view of Islam, see for example R. Fletcher, *The Cross and the Crescent* (London, 2003), 10, 17–19.
7 Matthee, *Politics of Trade*, 154–5.

8 See Hughes, *Russia*, 58–9.
9 Aune, 'Early Modern European,' 121.
10 Not merely was there a competition between English and Dutch merchants or Dutch and Russians in Iran, but also among Dutch merchants, as between the VOC interest and the Muscovy interest (for similar rivalries in the Ottoman Empire, see Aune, 'Early Modern European,' 126).
11 Chardin, *A New and Accurate Description* vol. 1, 41. See also Fletcher, *Cross*, 150–4.
12 See Hughes, *Russia*, 23.
13 The section is *Reysen*, 210–56.
14 Aune, 'Early Modern European,' 125–6.
15 *Reysen*, 210–11.
16 *Reysen*, 213. The second edition of Witsen's *Architectura* may have used Struys as a source to describe shipbuilding on the Caspian Sea (see Witsen, *Architectura* [1690], 268).
17 T. Barrett, *At the Edge of Empire* (Boulder, CO, 1999), 19. See also M. Khodarkovsky, 'Of Christianity, Enlightenment, and Colonialism: Russia in the North Caucasus,' *Journal of Modern History* 2 (1999) 394–430: 395.
18 *Reysen*, 213–14. As Butler recalled, the city had gone over to Razin's side and its *voevoda* Petr Semenovich Prozorovskii, a brother of the murdered governor of Astrakhan, was under house arrest (*Reysen*, 'Extract [Butler],' 28; Hellie, *Economy*, 289–90).
19 *Reysen*, 214–15; an island in this inlet was given Meindertsz's name as he had spotted it first, see maps of Caspian Sea, ibid., between 236 and 237.
20 *Reysen*, 215–18. *Reysen*'s description of the Adyg residing here is again almost literally copied from parts of Olearius's more detailed discussion (compare Olearius, *Podrobnoe*, 1002–4 with *Reysen*, 215–18). The Adyg(h) ethnos today includes Kabardinians and Adyg proper as well as Circassians (Cherkess); the Adyg reputation of friendliness and tolerance has survived until today (see S. Lyagusheva, 'Islam and the Traditional Moral Code of Adyghes,' *Iran and the Caucasus* 1 [2005] 29–35: 29n1, 34; see as well Kemper, *Herrschaft*, 116; Khodarkovsky, *Russia's*, 15). Most of *Reysen*'s coordinates of Struys's journey here are confused, as is reflected in the inaccurate albeit fairly detailed map as well (*Reysen*, 214–21; map between 236 and 237). Leo Bagrow has nevertheless noted that, even if it still showed a misshapen Caspian Sea, Struys's map was an improvement over earlier maps (L. Bagrow, 'Italians on the Caspian,' *Imago Mundi* 13 [1956] 2–10: 10).
21 After 1700, Circassian female beauty was to become legendary in Europe. *Reysen* has Struys only stay one night in this Circassian territory. It is unlikely that he was so heavily propositioned by the local women as *Reysen* claims. The inclusion of this part may have occurred to justify the engraving rendering a Kalmyk and a Circassian man (bare-breasted) woman, and child together (see *Reysen*, between 214 and 215).
22 Barrett, 'Lines,' 589–90; White, *Middle Ground*.
23 Kemper, Sixsaidov, 'Einleitung,' 8, 12.
24 See Pagden, 'Europe and the World Around,' 13–14. On the liminality of the borders between Islam and Christianity, see Faroqhi, *Ottoman Empire*, 2–3, and on the idea of borders in general ibid., 21.

25 Said, *Orientalism*, 1.
26 See for Muscovy's inclusion into Europe, L. Wolff, *Inventing Eastern Europe* (Stanford, CA, 1994), 11.
27 *Reysen*, 219. Pirates cruised the Caspian Sea throughout the seventeenth century (see Kotilaine, 'When the Twain,' 1054).
28 As Michael Kemper notes, historical Dagestan was located further south from the current Dagestani Republic of the Russian Federation; it covered a considerable part of what is now Azerbaijan (Kemper, *Herrschaft*, 22).
29 Khodarkovsky, *Russia's*, 15; Khodarkovsky, 'Of Christianity,' 400–1; Kemper, Sixsaidov, eds., *Islamgelehrten*, 45n43. Around 1400, the *shamakhila* emerged as rulers of the Dagestan coast (see ibid., 68, 68n89). The *shamkhal* in 1670–1 was either Çuban II or Mahmud (Kemper, *Herrschaft*, 122n31).
30 Barrett, *At the Edge*, 19–21; Faroqhi, *Ottoman Empire*, 49, 74; Khodarkovsky, *Russia's*, 36–7; Khodarkovsky, 'Of Christianity,' 409; Kemper, *Herrschaft*, 23, 115, 117–19; M. Polyevktov, 'Russian and Georgian Communications,' *Journal of Modern History* 3 (1930) 367–77: 369–70. Note how even the famous Turkish traveller and memoirist Evliya Çelebi (*c*.1611–*c*.1682) is unreliable about the *shamkhal*'s allegiance, in depicting a non-existent alliance between shamkhal and Crimean khan in the 1660s (Kemper, *Herrschaft*, 117–18n17).
31 Kemper, *Herrschaft*, 24–6. On the great diversity in terrain, see Barrett, 'Lines,' 582–3.
32 *Reysen*, 220–1; see as well Barrett, 'Lines,' 588.
33 *Reysen*, 221. *Reysen*'s map seems more accurate here, as it calls the people of the town of '*Boeinack*' '*Kamoksche Tartaren*' (*Reysen*, map between 236 and 237). In his petition to the tsar of March 1674, Klopper also called them Kumyks (see Orlenko, *Vykhodtsy*, 305). Among the Ottoman Turks, too, the inhabitants of the Caucasus enjoyed notoriety: 'Now, in Circassia and Abkhazia robbery is bravery; it is praised not blamed' (E. Çelebi, *The Intimate Life of an Ottoman Statesman*, ed. R. Dankoff [Albany, NY, 1991], 272). Derbent was a major centre for the slave trade in the area (Kostomarov, 'Bunt,' 435).
34 For Buynak, see Kemper, *Herrschaft*, 33.
35 Barendse, *Arabian Seas*, 130. Shah 'Abbas the Great had expanded his empire northward in his wars against the Ottoman Turks to include Derbent (see Polyevktov, 'Russian,' 372). 'While Holland was looked down upon as a republic ... , the Dutch knew how to compensate for their lack of official status with their wealth and naval power and thus gained more than a little respect in Persia' (R. Matthee, 'Persian Image of Europe,' in E. Yarshalter, ed., *Encyclopaedia Iranica* [available at http://www.iranica.com/newsite/, accessed on 12 January 2007]).
36 *Reysen*, 208–9, 222. Olearius notes that the Persians called some of the people in this region *Lesgi*, the name still used for one of the ethnic groups living in the region (Olearius, *Podrobnoe*, 975–6). Sunni Muslims, the Dagestani Kumyks and (Caucasian-language speaking) Lezgins lived in a coastal area stretching from Terskii gorod on the Terek in the north to Derbent in the south, some 280 kilometres.
37 H. Roodenburg, 'Social Control Viewed from Below,' in *Social Control in Europe*, eds H. Roodenburg and P. Spierenburg, vol. 1 (Columbus, OH, 2004), 145–58: 153.

38 *Reysen*, 223; *Reysen*, 'Extract [Butler],' 30. On the *utsmii*, the chief of the Dargins, see Khodarkovsky, *Russia's*, 15. Much of *Reysen*'s description of Dagestani customs is copied from Olearius (compare Olearius, *Podrobnoe*, 975–6, with *Reysen*, 220–3). Of course, *Reysen*'s narrative is interspersed with Struys's account of the adventures that befell him and his comrades.

39 See Spierenburg, *Judicial Violence*, 18; for the censorship of language, see Burke, *Toward*, 25–8. 'Body shame seems to have been more acute in Dutch society than elsewhere in Europe' (Burke, *Venice*, 89). Thus, somewhat later (in 1748–49), Cleland's *Fanny Hill* was written, according to its editor Peter Wagner, in a 'periphrastic style'(P. Wagner, 'Introduction,' in John Cleland, *Fanny Hill* (Harmondsworth, 1985) 7–30: 16, 25–9). Erotic or pornographic texts only began to be published widely, especially in French, in the last decades of the seventeenth century.

40 *Reysen*, 225.

41 *DAK* vol. 6, 11; *Reysen*, 'Extract [Butler],' 31.

42 *Reysen*, 'Extract [Butler],' 30. Writing that Maria was raped only on the second occasion, Struys remembered the sequence somewhat differently from Butler (*Reysen*, 223).

43 *Reysen*, 'Extract [Butler],' 30–1; *Reysen*, 223–4, 248.

44 Munster died in Isfahan in 1671, while Daniel Cornelisz and Pieter Bartelsz were presumed to be alive in 1676, when Albertsz had died; the fate of Poppes is unclear (Coolhaas, ed., *Generale Missieven* vol. 3, 775–6; *GAA NA* 4304 [Notary Nicolaes Hemminck], 227–227verso and 247).

45 For their liberation VOC Director de Haze already had spent 2891 *mahmudi* in vain (ten mahmudi was worth four guilders, thus more than 1100 guilders; Coolhaas, ed., *Generale Missieven*, vol. 3, 915 [31 January 1674]; *NA VOC Daghregister* 1291, 164 folio). François La Haye (de Haze) succeeded van der Dussen as Persian director in Gombroon-Bandar-e-'Abbas in 1672, and was succeeded in 1675 or 1676 by Bent, who lasted himself until 1680 or 1681 and was succeeded by Kasenbroot, who stayed until 1683 (Valentijn, *Oud en Nieuw Oost-Indië* vol. 5, 205). For Termundt's continued efforts in 1672, see *NA VOC* 10435 (*Missieven van de directeur en raad van Perzié*, Gamron to Batavia), 27 September 1672, and 7 October 1672.

46 See Khodarkovsky, 'Of Christianity,' 410; Olearius, *Podrobnoe*, 970.

47 *Reysen*, 223–4.

48 Khodarkovsky, 'Of Christianity,' 410; Kemper, *Herrschaft*, 24, 126–7. Dargins were also notorious for capturing people, primarily to ransom them (Kemper, *Herrschaft*, 48). The *utsmii* is called *usmi* in some transcriptions. In *Reysen*, Butler's letter differs from Struys's in some of its detail regarding the tribulations the Dutch sustained at the hands of the 'Tatars' (see *Reysen*, 'Extract [Butler],' 29–31).

49 *Reysen*, frontispiece, 224. Rudi Matthee notes how Struys observed here the first signs of the Dagestanis' customary use of bow and arrow giving way to firearms, indicating the belated arrival of the gunpowder revolution (R. Matthee, 'Unwalled Cities and Restless Nomads: Firearms and Artilery in Safavid Iran,' in *Safavid Persia*, ed. Melville, 389–416: 406)

50 This havoc had indeed been extensive, see Matthee, *The Politics*, 179.

51 Khodarkovsky, 'Of Christianity,' 404; Kemper, *Herrschaft*, 23.

52 *Reysen*, 224. For the location of the *utsmii*'s palace, see Kemper, *Herrschaft*, 43.
53 *Reysen*, 225–6.
54 Ibid., 225.
55 Ibid., 227. Perhaps Muhammad Khan (see Matthee, *The Politics*, 176).
56 *Reysen*, 228–32.
57 Ibid., 228.
58 Ibid., 229.
59 Ibid., 229–30; Struys, *Voyages*, 215.
60 *Reysen*, 230–2.
61 *Reysen*, 228. Voltaire referred to an unspecified French version in quarto, which may have translated the passage erroneously (footnote 70 states '*Voyage de Jean Struys,* in-4, p. 208'), see 'Ararat,' in Voltaire, *Dictionnaire Philosophique*, fifth edn. [1765] [available at http://www.voltaire-integral.com; accessed 6 August 2006]).
62 See Lloyd R. Bailey, 'Wood from 'Mount Ararat': Noah's Ark?,' *The Biblical Archeologist* 4 (1977) 137–46: 145, and compare to T. La Haye and J. Morris, *The Ark on Ararat* (Nashville, TN: 1976).
63 Bailey, 'Wood,' 138.
64 *Krest'ianskaia voina* vol. 3, 380. '[W]herever one went in the Arabian seas one was likely to encounter a European mercenary as surgeon, painter, or military advisor' (Barendse, *Arabian Seas*, 107).
65 'Self-medication' was common, see Dekker, *Humour*, 123–5. See also Schama, *Embarrassment*, 526.
66 Richter, *Geschichte* vol. 1, 12.
67 *Reysen*, 232–5.
68 See Aune, 'Early Modern European,' 125–6.
69 *Reysen*, 235.
70 Chardin specified that Iranian Muslim men turned to other women for sex once their partner was three or four months pregnant (Chardin, *New and Accurate* vol. 2, 12).
71 *Reysen*, 234. Indicative of Muslims' ability to discriminate among their captives, Ellen Friedman notes how on the Barbary Coast enslaved Christian shipwrights were highly valued (E.G. Friedman, 'Christian Captives at "Hard Labor" in Algiers, 16th–18th Centuries,' *International Journal of African Historical Studies* 4 [1980] 616–32: 623).
72 Maxwell, 'Afanasii,' 246, 256–7, 260, 265; Fletcher, *Cross*, 45–6.
73 *Reysen*, 289. On the Islamic prohibition of owning Muslim-born slaves, see R. Blackburn, 'The Old World Background to European Colonial Slavery,' *William and Mary Quarterly* 1 [1997] 65–102: 73–4.
74 For instance, see Scammell, 'European Exiles,' 643–4; Schmidt, *Innocence Abroad*, 279.
75 *Sharia* law forbade the enslavement of Muslims, but allowed the continued enslavement of converts (see J.F. Guilmartin, 'Ideology and Conflict: The Wars of the Ottoman Empire, 1453–1606,' *Journal of Interdisciplinary History* 4 [1988] 721–47: 730n13).
76 See Chapter 10.
77 Since the historical Jan Struys did not enjoy formal education, he evaded as a child the confessionalization project pursued by the Reformed Church,

Mennonites or Catholics, which may explain his somewhat eclectic and unpolished Christianity; if so, it happily matched the ghostwriter's portrayal of the Christian mindset of *Reysen*'s Struys, which well-nigh becomes ecumenical on occasion (see de Vries and van der Woude, *First*, 170).

78 As Prak notes, this humanist attitude was widespread in the Dutch Republic, although it may have eroded somewhat through the 'confessionalization' that gathered momentum in the course of the seventeenth century (Prak, *Dutch Republic*, 204–5, 210).

79 N.I. Matar, 'The Renegade in English Seventeenth-Century Imagination,' *Studies in English Literature, 1500–1900* 3 [1993] 489–505: 502.

80 See Gorski, *Disciplinary Revolution*, 20.

81 Perhaps they were slaves themselves, which might allow them such sexual relations; see Fletcher, *Cross*, 21, 115.

82 Prud'homme van Reine, 'Schittering,' 28–9.

83 See for example Prak, *Dutch Republic*, 53–5; Spierenburg, *Judicial Violence*, 113.

84 *Reysen*, 235–6.

85 *Reysen*, 235–6.

86 *Reysen*, 236–9. Map between *Reysen*, 236 and 237, and 'Bericht aan den Boek-binder.'

87 For the link between maps and power or ownership, see Cracraft, *Revolution*, 95.

88 Bagrow, 'Italians,' 10.

89 On Albertsz as scribe, see *RGADA* 50/42 [1668]), l.43. On the primitive contemporary Muscovite maps of the Caspian, see, for instance, J.F. Baddeley, *Russia, Mongolia, China*, 2 vols (New York, 1964; origin. 1919), vol. 1, cxxv–cxxvi, cxxxix, vol. 2, 215–17, and the various maps in both volumes. In the 1960s a copy of 'Struys's map' of the Caspian Sea was found in an Armenian Cathedral in Isfahan's Julfa neighbourhood (see R.A. Gardiner, 'John Struys "New Card of the Caspian Sea" 1668,' *Geographical Journal* 4 [1969] 631–2: 631).

90 For his (later) map of the Holy Land, see Dozy, 'Olfert,' 429.

91 *Reysen*, 236–7. While this is utter nonsense, the sea is far deeper in its southern part. Even if this part only counts for slightly less than one third of its surface-expanse, it contains two thirds of the Caspian's water (see de Planhol, 'Caspian Sea'). A century after *Reysen*'s publication, the idea that the water disappeared into subterranean channels draining into the Gulf was still taken seriously enough to be discussed by Buffon in his *Histoire naturelle* (for example, see George Louis le Clerc, Comte de Buffon, *Oeuvres complètes* vol. 1 [Paris, 1848], 197; George Louis le Clerc, Count of Buffon, *A Natural History, General and Particular* 9 vols [London: A. Strahan and T. Cadell, 1791], vol. 3, 329–30).

92 See Wallerstein, *Modern World System* vol. 2, 37–71.

93 See for instance Kamen, *Empire*, 430–1; Israel, *Dutch Primacy*, 320–1; de Vries and van der Woude, *First*, 465.

94 Just before he reached Gombroon in the spring of 1672, Struys crossed paths with one of his benefactors, VOC agent Kasenbroot, who was travelling from Isfahan to Surat, 'with his slaves' (*Reysen*, 363). See Cook, *Matters*, 186–9, 204–5.

95 Nevertheless, slaves 'were better treated under Islamic law than under Christian law' (Barendse, *Arabian Seas*, 116).
96 Aune, 'Early Modern European,' 121.
97 Davis, *Christian Slaves*, 53–4.
98 Davis, *Christian Slaves*, 6–7, 10, 15.
99 Khodarkovsky, *Russia's*, 21; Solov'ev, *Istoriia* vol. 3 (parts 11–15), 708; Hughes, *Russia*, 136; van Deursen, *Dorp*, 303–6. Whereas the Russians collected a tax, the Dutch held collections among their church communities.
100 Van Deursen, *Dorp*, 303–6; Davis, *Christian Slaves*, 20–1.
101 See Prud'homme van Reine, *Rechterhand*, 91–3.
102 Khodarkovsky, *Russia's*, 22.
103 See Barendse, *Arabian Seas*, 108–9.
104 *Reysen*, 242–3. He adds some more details later, see ibid., 289. Comparing Olearius, *Podrobnoe*, 965–9, and *Reysen*, 240–1, shows Struys's text to likely have been original here. For Derbent, see Kemper, *Herrschaft*, 46. Struys's observation about Barbary slaves was well founded; see van Deursen, *Dorp*, 303–6. I follow Willem Floor's transcription of Haji Biram's name here (Floor, 'Fact,' 59).
105 See for more P.C. Emmer, 'The History of the Dutch Slave Trade: A Bibliographical Survey,' *Journal of Economic History* 3 (1972), 728–47: 742, 745–6; see also J. Postma, 'The Dimension of the Dutch Slave Trade from Western Africa,' *Journal of African History* 2 (1972), 237–48, and other works by both authors.
106 *Reysen*, 289.
107 See Linebaugh, Rediker, *Many-Headed Hydra*, 160, and 'Natalie Zemon Davis,' in, *Visions of History*, eds H. Abelove et al. (New York, 1984) 97–122: 111.
108 *Reysen*, 243–5.
109 *Reysen*, 245.
110 See Dekker, *Humour*, 98–101. It resembles a tale in Cervantes's *Don Quijote*, a highly popular text in the seventeenth-century Republic, about Zoraida, a young woman who secretly maintains her Christian faith and escapes with a Christian slave to Spain (see Cervantes, *Don Quijote*, 272–95).
111 *Reysen*, 245–6.
112 *Reysen*, 246–7.
113 De Vries and Arentsz also gave Struys the tidings about the Brak family (*Reysen*, 248).
114 *Reysen*, 249–50.
115 According to *Reysen*, most of the Dutch captives at Boynak and in the *utsmii*'s settlement enjoyed such freedom, which had allowed de Vries and Arentsz to run away easily (*Reysen*, 247–8). Their colleagues even preferred this relative freedom over the uncertainty of a flight toward Derbent.
116 Fletcher, *Cross*, 20. The exception here were galley slaves (Davis, *Christian Slaves*, 75–7).
117 Fletcher, *Cross*, 20–1. On the Ottoman Empire, see for example Parker, 'Paying,' 278–83.

118 See *Reysen*, 247–8.
119 *Reysen*, 250–1, 254.
120 See Kemper, *Herrschaft*, 115. *Reysen* equates it with classical Media; it was situated where today Azerbaijan is.
121 For the previous safety of the Persian roads, see Barendse, *Arabian Seas*, 159, and Savory, *Iran*, 190; for the increasing dangers, see Floor, 'Fact,' 65.
122 *Reysen*, 251; Olearius, *Podrobnoe*, 497–9, 511.
123 *Reysen*, 251–2. Olearius's description may have been the source again, see Olearius, *Podrobnoe*, 962.
124 *Reysen*, 253; *Reysen*, between 252 and 253; Matthee, 'Unwalled Cities,' 402.
125 *Reysen*, 255.
126 On the appeal of the St Sebastian legend to Catholic sensibility in the Republic, see Westermann, *Worldly Art*, 96. Of course, baroque aesthetics, albeit originating in Catholic Europe, also appealed in Protestant regions, and Dutch Protestants continued to celebrate certain Catholic holidays such as St Martin's Day (November 11) and St Nicholas's Day (December 6).
127 *Reysen*, 254–5. Struys calls his faith '*Apostolisch Catholijk*', which stood for the Dutch Reformed Church: 'Catholic' means in this instance universal, and 'Apostolic' that all believers were equally Christ's disciples and were to behave as the apostles (see Gorski, *Disciplinary Revolution*, 21).
128 *Reysen*, 255.
129 Gurdziecki was his Polonized name, which was Gurdiia in Georgian; Russians called him Gurdsetskii.
130 *Reysen*, 'Extract (Butler),' 32; *Krest'ianskaia voina* vol. 3, 199–201, 332–3. See also R. Matthee, 'Gurdziecki, Bogdan,' in E. Yarshalter ed., *Encyclopaedia Iranica*, available at www.iranica.com/newsite (accessed on 12 January 2007). Gurdziecki had been sent by the Polish government to Isfahan in 1668, mainly to probe the possibilities of a Persian-Polish-Muscovite alliance against Turkey, a mission dispatched partially on Russian request; after arriving in late 1669 in Iran, he had set out on his return trip in June 1670, the month Astrakhan fell to Razin (Matthee, *Politics*, 195).
131 E.S. Zevakin, 'Persidskii vopros,' 144; Matthee, *Politics*, 195.
132 It seems that Gurdziecki's mission entailed another effort to reroute the silk trade northward as well (see Kotilaine, *Russia's*, 460).
133 *Reysen*, 255–6. It may confirm Faroqhi's suggestion that Muslim masters treated non-Muslim servants humanely (see Faroqhi, *Ottoman Empire*, 177).
134 This exchange rate is based on Butler's account, see *Reysen*, 'Extract (Butler),' 31.
135 See *Reysen*, 265.

10 Liberation

1 *Reysen*, 257.
2 *Reysen*, 'Extract [Butler],' 32; *Reysen*, 258; *Pan* means lord in English. Butler calls Paniegros 'Jan Gros,' also known as 'Martin Eudan', while he is called 'Ender' in Carmelite sources (*A Chronicle of the Carmelites in Persia and the Papal Mission of the XVIIth and XVIIIth Centuries*, 2 vols [London, 1939] vol. 1, 425).

3 *Reysen*, 259–60.
4 *Reysen*, 260, 271; see Chapter 8.
5 *Reysen*, 260–1; *Reysen*, 'Extract [Butler],' 31–2; Mankov, ed., *Zapiski* vol. 1, 37–9, 81n39.
6 *Reysen*, 261–2. *Reysen* unconvincingly explains Struys's improved circumstances by claiming that he had somehow won the trust of his Polish master (*Reysen*, 273).
7 *Reysen*, 261; *Reysen*, 'Extract [Butler],' 32.
8 *Reysen*, 261.
9 See Khodarkovsky, 'Of Christianity,' 401, 404.
10 Floor, Clawson, 'Safavid Iran's,' 347–8.
11 *Reysen*, 261.
12 *Reysen*, 261; *Reysen*, 'Extract [Butler],' 33.
13 *Reysen*, 'Extract [Butler],' 33; *Reysen*, 262. See as well *Krest'ianskaia voina* vol. 3, 199–201, 332.
14 *Reysen*, 270. See below.
15 *Reysen*, 275.
16 Coolhaas, ed., *Generale Missieven* vol. 3, 776; *Reysen*, 329.
17 *Reysen*, 275.
18 *Reysen*, 277; these letters would have travelled along one of the two silk routes to the Levant used at the time (see Matthee, *Politics*, xxi, map 5). If he did dispatch them, he must have dictated them, perhaps with the aid of others in his company. How he knew Molives is unclear, but he may have met during his travels in the Mediterranean.
19 *Reysen*, 285. Soon after he reached Isfahan, Termundt's companion 'Pieter Adriaensz Van Schevelingen' died, see *NA VOC Daghregister* 1285, 289 verso, folio 290 (a letter by Maetsuycker, Pit, Overtwater, Speelman, and Van Hoorn to the *Heeren XVII*, 31 January 1673).
20 *Reysen*, 'Extract [Butler],' 33–4.
21 Butler's letter as printed in the book is dated 6 March 1671 (*Reysen*, 'Extract [Butler],' 5). Possibly, some of the Catholic monks residing across Iran sent the epistle to Europe via the Levant, or, more likely, it reached Amsterdam via the VOC's good offices (see Barendse, *Arabian Seas*, 166–7). When it reached Amsterdam, Butler's story was past its newsworthiness and no longer held enough topicality to issue it as a pamphlet such as *Kort waerachtigh Verhaal*.
22 Coolhaas, ed., *Generale Missieven* vol. 3, 775–6n2. Butler became '*vaandrig*', or ensign, in the Company's service; perhaps he disembarked at Ceylon or southern India, for he already served in the VOC army on India's Coromandel Coast in March 1672 (*NA VOC Daghregister* 1288 [Ceylon] 411–12 and *NA VOC Daghregister* 1288 [Coromandel] 203).
23 *Reysen*, 262–70. Also see Olearius, *Podrobnoe*, 957–60, which shows that by this point in *Reysen*, Olearius's text is no longer the dominant source: Struys's personal recollections surface even in the chrorographic sections.
24 See Chapters 11 and 12.
25 *Reysen*, 286–7, and see engraving in ibid., between 286 and 287.
26 *Reysen*, 287–8.
27 He claims to have heard the woman's screams and seen her flayed body lying on the street (*Reysen*, 287).

28 Brancaforte, *Visions*, 102–7.

29 Brown, 'Reading Race,' 431–2; Dekker, *Humour*, 97–101.

30 *Reysen*, 233–5; Struys is also disparaging about two young Georgian women bought by Gurdziecki, who are said to have no qualms about their lost virginity (*Reysen*, 256).

31 Brown, 'Reading Race,' 432. See also Schama, *Embarrassment*, 400. See for the later connection between the 'Orient' and sexual temptation, Said, *Orientalism*, 167, 188, 190.

32 See H. van der Velden, 'Cambyses for Example: The Origins and Function of an *Exemplum Iustitiae* in Netherlandish Art of the Fifteenth, Sixteenth, and Seventeenth Centuries,' *Simiolus* 1 (1995) 5–39: 8–9, 18, 30; H. van der Velden, 'Cambyses reconsidered: Gerard David's *Exemplum Iustitiae* for Bruges Town Hall,' *Simiolus* 1 (1995) 40–62: 57n54, 58.

33 Olearius, *Podrobnoe*, 899; Brancaforte, *Visions*, 210n97; van der Velden, 'Cambyses Reconsidered,' 40n3. Floor, 'Fact,' 63: 'We know that people were flayed in Iran, and their skin stuffed with straw for the amusement and elucidation of the public.' For the Ottoman-Safavid flaying known to the Brownes, see R. Cawley, 'Sir Thomas Browne and His Reading' *PMLA* 2 (1933) 426–70: 441–2, 441–2n103, and 442n104.

34 For Dutch corporal punishments, see Spierenburg, *Judicial Violence*; see especially ibid., 202; Foucault, *Discipline*, 3–4.

35 Davies, *God's Playground* vol. 1, 134; Paul of Aleppo, *The Travels of Macarius, 1652–1660*, ed. L. Redding (NY, 1971), 75.

36 See for instance van Deursen, *Plain Lives*, 50, 53–5; Foucault, *Discipline*, 75, 77–8; Spierenburg, *Judicial Violence*, 11, 15–16, 76, 113–17, 173–7; J.H. Langbein, 'The Historical Origins of the Sanction of Imprisonment for Serious Crime,' *The Journal of Legal Studies* 1 (1976) 35–60: 48–51, 59–60.

37 *Reysen*, 289.

38 See Khodarkovsky, *Russia's*, 25.

39 See P.M. Holt, 'The Exalted Lineage of Ridwan Bey: Some Observations on a Seventeenth-Century Mamluk Genealogy,' *Bulletin of the School of Oriental and Asian Studies* 1–3 (1959) 221–30: 225.

40 *Reysen*, 256. The Georgian-born Gurdziecki, too, is mercenary and cruel, and lascivious Georgian girls had mesmerized Haji Biram Ali, according to Altijn.

41 *NA VOC Archief* 10435 (*Missieven van de directeur en raad van Perzië*, Gamron to Batavia), 27 September 1672 and 7 October 1672; *NA VOC Daghregister* 1279, 905verso (31 January 1672); Barendse, *Arabian Seas*, Appendix I, 502; Mankov, ed., *Zapiski* vol. 1, 39. *Zapiski* indicates that Faber collapsed the two stories of Struys's initial purchase by Gurdziecki (in the fall of 1670) and his compensation in the form of the horse almost a year later. Since the money was key, Termundt rather than Faber proved most pivotal in getting Struys and Meindertsz released, even if Faber probably acted as Termundt's proxy. *Reysen* suggests that the VOC's Persian Director Lucas van der Dussen was instrumental in this transaction (*Reysen*, 344; Valentijn, *Oud en Nieuw Oost-Indië* vol. 5, 205).

42 *Reysen*, 295.

43 *Reysen*, 261; *Reysen*, 'Extract [Butler],' 31–2. In July 1671, Termundt had also made sure to send Struys, Klopper, and Brandt word of his safe arrival at Ardabil, something which Butler had declined to do (*Reysen*, 290–1). Termundt's name is repeatedly mentioned as middleman in VOC correspondence; see for example *NA VOC Archief* 10435, 27 September 1672 and 7 October 1672; *NA VOC Daghregister* 1279, 905verso (31 January 1672); Coolhaas, ed., *Generale Missieven* vol. 3, 915; *NA VOC Daghregister* 1285, 290v (31 January 1673). His efforts on behalf of Willem Willemsz, Brak, Maria Brak, and their child (and possibly Popkes, Cornelisz, Bartelsz, and Albertsz) were in vain throughout 1671 and much of 1672.

44 *Reysen*, 295–7. Klopper, who was in poor health, preferred to remain with Gurdziecki; he reached Russia likely in 1672; in April 1674, he requested the tsar for the arrears due to him since 1670 (*Krest'ianskaia voina* vol. 3, 199–201, 332; Orlenko, *Vykhodtsy*, prilozheniia, 304–6).

45 Coolhaas, ed., *Generale Missieven* vol. 3, 775–6.

46 *NA VOC Daghregister* 1279, 905verso (31 January 1672).

47 *Reysen*, 297. Immediately prior to their caravan's departure from Shemakha, they met the just released Meindertsz, who found somewhat later his way back to Amsterdam via Isfahan, Bandar, and Batavia.

48 Struys's description of the journey is quite similar to that of Olearius (for example, compare Olearius, *Podrobnoe*, 560–5, with *Reysen*, 298–303). This similarity may be partially due to the fact that both indeed saw the same landscape. Different from Olearius, who had protection from the highest authorities, Struys travelled in a 'private' caravan which may have made the trip more risky; the roads in Persia had become less safe in the period between Struys's and Olearius's trip. Some parts are nevertheless directly copied from Olearius (compare *Reysen*, 303–4 with Olearius, *Podrobnoe*, 583–4). Ardabil lies at an altitude of 1500 metres above sea level (Savory, *Iran*, 1). On the caravanserais, see Savory, *Iran*, 190–1; *A Journey to Asia: Jean Chardin's Portrait of a Seventeenth-Century Empire*, ed. and trans. R.W. Ferrier (London, 1996), 29.

49 *Reysen*, 317–19. Most of Qazvin's description in *Reysen* is a truncated version of Olearius, *Podrobnoe*, 614. The description of Bairam (commemorating Ibrahim's sacrifice of Ishmail) is slightly more elaborate than Olearius's version, perhaps indicating some original additions from Struys's experience (see Olearius, *Podrobnoe*, 570–1, 829–31, and *Reysen*, 320–1).

50 *Reysen*, 305–6.

51 *Reysen*, 306–12; Olearius, *Podrobnoe*, 585–94. The description of the royal graves in *Reysen* matches that of Olearius, perhaps because Struys's observations hardly differed from it; Olfert Dapper had already described the shahs' graves and Ardabil in his *Asia* (Dapper, *Asia*, 44–7). On the Safavid dynasty, see Savory, *Iran*, 5–8.

52 *Reysen*, 313–15. For the slow pace and continuous trading of caravans, see Barendse, *Arabian Seas*, 156–7.

53 *Reysen*, 315–17. *Reysen* copies the description of the town and its environs from Olearius (see Olearius, *Podrobnoe*, 605–7).

54 *Reysen*, 319.

55 *Reysen*, 328.

56 Earlier, Butler had stayed with them (*Reysen*, 'Extract [Butler],' 34).

57 Persian silk was not a highly valued commodity for the VOC, according to Braudel (Braudel, *Civilization* vol. 3, 217). Nevertheless, the Company's sale of Chinese and Benghal silk might have suffered from the competition with Persian silk imported via Muscovy. During the 1620s, Dutch merchants exported Persian silk that had arrived in Russia via Astrakhan, thus competing with the VOC's attempts to export Persian silk via Bandar-e-'Abbas (see Kotilaine, *Russia's*, 74–5).

58 Confirming Schama's observation regarding Dutch charitable sentiment (see Schama, *Embarrassment*, 579).

59 *Reysen*, 330.

60 *Reysen*, 330–9; Olearius, *Podrobnoe*, 715–33.

61 *Reysen*, engraving between 332 and 333. Also known as the Naghsg-i Jahan Square, a UNESCO World Heritage site today.

62 Ibid., 330; Olearius, *Podrobnoe*, 716.

63 Savory, *Iran*, 176.

64 *NA VOC Daghregister* 1285, folio 290 (31 January 1673).

65 Adopting Pieter van Dam's outline of the reckoning of wages with crews (or their dependents) of lost and captured ships, Struys was entitled to sufficient compensation for his troubles during this return trip to Holland to settle his outstanding debts with the VOC (van Dam, *Beschryvinge* vol. 1, part 1, 708–10; Jones, *Anglo-Dutch Wars*, 34). Van Dam was then the VOC's legal counsel.

66 *NA VOC Daghregister* 1279, 905verso (31 January 1672).

67 The Dutch usage of Go(a)mbroon derived from 'Gamru', originally the caravanserai of the island of Hormuz, but was renamed *Bandar mobrak ye-abbasi*, the 'felicitous port of Abbas', in 1624 (see Barendse, *Arabian Seas*, 47). From it, a highway was developed, along which there were a series of caravanserais, through Lar, Shiraz, Kirman, leading to Isfahan, which was the road along which Struys seems to have travelled.

68 *Reysen*, 341–2; see X. de Planhol, 'Bandar-e-'Abbas,' in E. Yarshalter, ed., *Encyclopaedia Iranica* (available at http://www.iranica.com/newsite/, accessed on 12 January 2007).

69 *Reysen*, 343. Already before his departure, Brandt and Faber left for the north-east, endeavouring to reach Muscovy again; Astrakhan had been captured by tsarist forces in late November 1671; they managed to reach the port somewhere in April or May 1672 (see *Polnoe sobranie zakonov'* vol. 1, 868; Man'kov, *Zapiski* vol. 1, 6, 8).

70 *Reysen*, 344–6.

71 *Reysen*, 347–50, and ibid., between 348 and 349.

72 *Reysen*, 350.

73 *Reysen*, 351–2.

74 See F. Richard, 'L'Apport des Missionnaires européens à la conaissance de l'Iran en Europe et de l'Europe en Iran,' in *Études Safavides*, ed. Calmard (Paris, 1993) 251–66: 254, 264.

75 Floor has established that these people actually lived in Iran at the time (see Floor, 'Fact,' 64).

76 Richard, 'L'Apport,' 353. Since he was unable to write, he probably used a scribe. The text cannot hint at this, since it would undermine the pretence that Struys was *Reysen*'s writer. But it is puzzling why it was felt

necessary to tell the reader that Struys also wrote a letter to Haji Biram Ali in Turkish during his travels in southern Iran, unless a scribe draw up such an epistle as well (see *Reysen*, 357).

77 *Reysen*, 357–8.

78 *Reysen*'s direct reference to the penis (called 'manhood,' *mannelijkheyt*) is unusually explicit (only in the discussion of Iranian eunuchs is there another mention of the '*roede*' ['rod'] see *Reysen*, 288). The text otherwise uses euphemisms in discussing sexuality, following the periphrastic style typical for its day (see Dekker, *Humour*, 106, or John Cleland, *Fanny Hill* [Harmondsworth, 1985], 100, 151). Given the much more frank description of genitals in the jokes of the Dutch patrician van Overbeke's contemporary unpublished manuscript, we can assume a large discrepancy between spoken and printed language regarding sex. Seventeen-century Dutch writing about sexuality was not as circumspect as it became by the Victorian Age, as is obvious from the printed work of W.G. van Focquenbroch and the private poems of his friend, the Alkmaar Protestant minister Ulaeus (see Karel Bostoen, 'Mogelijk een vroom gelaat en een vroom gewaad, maar beslist on-vrome praat: de aankomend predikant Johannes Ulaeus in zijn Alkmaars 'Collegij',' *Fumus* 1 [2005] 22–55). Struys's more explicit or heartier manner of expression would have been censored by ghostwriter and publisher (for the hazy borders between Early Modern folk and elite cultures, see P. Burke, *Popular Culture in Early Modern Europe*, revised edn. [Aldershot, 1994], 23–9, 270–81; Burke, *Venice*, xx).

79 *Reysen*, 358.

80 See also Foucault, *Discipline*, 32; Spierenburg, *Judicial Violence*, 39, 113–14, and *passim*.

81 See Aphra Behn, *Oroonoko* in *The Norton Anthology of English Literature*, vol. C, eds. L. Lipking and J. Noggle (London, 2006; origin. 1688), 2183–226: 2226.

82 van Nierop, 'Catholics,' 109–11; Spierenburg, *Judicial Violence*, 39.

83 *Reysen*, 360.

84 *Reysen*, 361–3.

85 Matthee, *Politics*, 176–7; Savory, *Iran*, 241.

86 *Reysen*, 364. *Reysen* indicates 28 March 1672 as arrival, but this seems impossibly fast if he left Isfahan in early March.

87 *Reysen*, 365, and engraving between 365 and 366. The Dutch conducted more trade here than the French or English at the time. *Reysen* remarks that the Dutch were the 'strongest and most respected' (*sterkste en best-gesienste*) merchants of the port (ibid., 367). The port had about 1,500 houses at the time (see de Planhol, 'Bandar-e-'Abbas'). The *Nuysenburgh* (*Nuissenburgh*) was a *pinas* (yacht) which served the VOC between 1664 and 1678; it anchored in Batavia in March 1671 under the captaincy of Kornelis Temminck (see www.vocsite.nl/schepen/detail.html? id=11799). VOC documents confirm that Struys sailed on it to Batavia (*VOC* 10435 (*Missieven van de Directeur en Raad van Perzie* [7 October 1672], 26verso). Like the *Nuysenburgh*, the *Alphen*, likewise depicted on the engraving, was indeed a VOC ship at the time (built on the VOC's wharves in Amsterdam in 1662), captained by Jan Kop. Leaving Batavia on 4 February 1673, the *Alphen* was captured about 11 May 1673 in the battle with the English near

St Helena that *Reysen* depicts further down (see www.vocsite.nl/schepen/detail.html? id=10023; *Reysen*, 370; van Dam, *Beschryvinge* vol. 1, part 1, 481).

88 The VOC lodge was a quadrangle surrounded with a wall and with towers, as Decker's engraving shows; it still survived by 1900, see A.W. Stiffe, 'Ancient Trading Centres of the Persian Gulf: VI. Bandar Abbas,' *Geographical Journal* 2 (1900) 211–15: 211.

89 The text hardly exaggerated, for in 1583 Ralph Fitch had barely survived his sojourn at Ormuz, suffering from the 'flux' (likely dysentery, see M. Edwardes, *Ralph Fitch, Elizabethan in the Indies* (New York, 1973), 28). Chardin, too, described Bandar's climate in the most abject terms (Ferrier, ed., *A Journey*, 41). Most merchants arrived from the interior in October; and Bandar-e-'Abbas was almost abandoned during summer. Fevers in the area were common (Barendse, *Arabian Seas*, 46, 48; Breet, *Oost-Indische Voyagie*, 232).

90 *Reysen*, 367.

91 *Reysen*, 367–8.

92 *Reysen*, 367–8. This seems to mean the carpenter of the *Nuysenburgh*, not one of the *Oryol*'s carpenters (*Reysen*, 368).

93 *Reysen*, 368–9.

94 *Reysen*, 369. We can verify *Reysen*'s story with archival evidence: In a letter from the VOC agents in Persia to Maetsuycker of October 1672, the departure from Gambroon of Struys for Batavia was noted, when he had already reached Java (*NA VOC* 10435, 7 October 1672).

95 *Reysen*, 370.

96 See van Dam, *Beschryvinge* vol. 1, part 1, 557. This was likely a ship built on the Amsterdam wharves of the VOC in 1665 (see van Dam, *Beschryvinge* vol. 1, part 1, 481).

97 *Reysen*, 370; see Boxer, *Dutch Seaborne Empire*, Appendix II, 301. The ship was a prize captured from the French (and thus called sometimes the '*Franse*' *Europa*); it left Batavia on the same day that Struys suggests in his book, 4 February 1673; after stopping at the Cape, the *Europa* was indeed captured off St Helena in May 1673 (see www.vocsite.nl/schepen/detail.html?id=12021; Bruijn et al. eds, *Dutch-Asiatic Shipping*, 96).

98 *Reysen*, 370–1; Bruijn et al. eds, *Dutch-Asiatic Shipping*, 96.

99 *Reysen*, 371.

100 *Reysen*, 371. On this, Koker and Gotskens/Godskens, see Boxer, *Dutch Merchants*, 362–3, 383–4.

101 *Reysen*, 370–2; De Jonge, *Geschiedenis* vol. 2, 471–2. See for the trunk, E.M. Jacobs, 'De Verenigde Oost-Indische Compagnie: een veelkantig handelsbedrijf,' *Tijdschrift voor Geschiedenis* 4 (2002) 525–43: 528. The booty acquired by the English here was great (see Jones, *Anglo-Dutch Wars*, 212).

102 *Reysen*, 373.

103 *Reysen*, 374–5; see www.vocsite.nl/schepen/detail.html?id=10023, www.vocsite.nl/schepen/detail.html?id=11061, and van Dam, *Beschryvinge* vol. 1, part 1, 486. Struys calls the second ship *Ter Veer* (*Reysen*, 375).

104 *Reysen*, 376–7; Boxer, *Dutch Merchants*, 373.

105 Jones, *Anglo-Dutch Wars*, 211. Treatment of POWs in this age is an area that deserves more study (see also Faroqhi, *Ottoman Empire*, 119).

106 Despite the state of war between the Republic and England at the time, 'packets' (mailboats) continued to sail between Harwich and Hellevoetsluis (see Jones, *Anglo-Dutch Wars*, 197). The war between England and the United Provinces ended in February 1674 (ibid., 216).

11 *Reysen*'s Readers

1 The Estates of Holland issued the exclusive rights to the publication of van der Heiden's *Vervarelyke schipbreuk*, Schouten's *Oost Indische Voyagie* and *Reysen* to van Meurs and van Someren for fifteen years in September 1675 (see *Reysen*, where the licence follows the title page). Struys, however, had already departed for Muscovy in July 1675, when the manuscript was far from ready to print (see Chapters 12 and 14).

2 On the economic downturn, see Israel, *Dutch Primacy*, 292–304. Note the decline in book printers and sellers in the 1670s in de Vries and van der Woude, *First*, 318, Table 8.6. Mikhail Bakhtin noted rightfully how 'each literary genre within an epoch or trend is typified by its own special concepts of the addressee of the literary work, a special sense and understanding of its reader, listener, public or people'; *Reysen* serves as a good example of this point (M. Bakhtin, *Speech Genres and Other Late Essays* [Austin, TX, 1986], 98).

3 Paul Hazard made the classic case for the sea change of the European mindset in the late seventeenth century in his *The European Mind* (Cleveland, 1963). Adopting some of Foucault's ideas, Hayden White pinpoints a sea change occurring around this time in 'the human sciences', in which the predominant metaphorical trope gives way to a metonymic dominance (as may be seen in Witsen's thought or Buffon's work; see H. White, 'Foucault Decoded: Notes from Underground,' *History and Theory* 1 [1973] 23–54: 45–7; see below in this chapter and in the next two chapters for Witsen and Buffon). If we adopt White's ideas, it seems that *Reysen*'s appeal collapsed when the metonymic trope was replaced by that of the synecdoche.

4 There were numerous portolans and maps available in the Republic, see for instance R. Tavernier, *Russia and the Low Countries* (Groningen, 2006) 99–102.

5 See the earlier remarks in the Introduction and Chapter 1 about van Nijenrode's report on Thailand as well as Witsen, *Moscovische Reyse*, and Kurts, *Sochinenie Kil'burgera*, 73, 76, and H.E. Ellersieck, 'Russia under Aleksei Mikhailovich and Fedor Alekseevich 1645–1682: The Scandinavian Sources,' Unpubl. Ph.D. Diss. (Los Angeles, CA: UCLA, 1955).

6 On this, see Verhoeven, Verkruijsse, 'De vele gezichten,' 41. This reading public was potentially as large as 10% of the population, but in effect rather smaller (since 10 to 15% of the Dutch boys attended some form of secondary education, see De Vries and van der Woude, *First*, 170; see also Schama, *Embarrassment*, 121).

7 Henry Fielding, *Joseph Andrews and Shamela*, ed. Judith Hawley (London, 1999), 198; Brenner, 'Mythos,' 32–3; Luke 15: 11–32 *King James Bible*.

8 Its length and presumably its price will have prevented a 'broad public' from reading *Reysen*, even if sailors' stories were popular among all layers of

Dutch society (see 'Verantwoording,' in *Iournael ofte Gedenckwaedige beschrijvinghe*, eds Verhoeven and Verkruijsse, 19–33: 20).

9 'A more civilized pen than my own' is given credit in the preface, see 'Voorreden,' in *Reysen*, 2.

10 A. Goeteeris, *Journael der legatie ghedaen inde Jaren 1615 ende 1616* ('s-Gravenhage: Aert Meuris, 1619); J.P. Danckaert, *Beschrijvinghe van Moscovien ofte Ruslandt* (Amsterdam: Broer Jansz, 1615); J.P. Danckaert, *Beschrijvinghe van Moscovien ofte Ruslandt*, second edn. (Amsterdam: G.J. Saeghman, 1646); J.P. Danckaert, *Reyze, ofte voyagie, door Moscovien ofte Rus-Landt* (Dordrecht: n.p., 1652); Olearius, *Beschrijvingh* (Amsterdam: Benjamyn and Roest, 1651); Olearius, *Beschrijvingh* (Utrecht: L. Roeck, 1651); Olearius, *Persiaensche Reyse* (Amsterdam: Broer Jansz., 1651); Herberstein's work, the first lengthier description of Muscovy in Dutch, had been translated in 1605, an edition nowadays exceedingly rare (I did not find a copy in any of the research libraries I visited); it was a translation of S. von Herberstein, *Rerum moscoviticarum commentarii* (Vienna: n.p., 1549). For a list of most works in Western languages, see M. Poe, *Foreign Descriptions of Muscovy* (Columbus, OH, 1995).

11 H. Gerritsz, *Beschryvinghe vander Samoyeden landt* (Amsterdam: n.p., 1612). Until 1633 Gerritsz was the official cartographer of the VOC (C. Koeman, *Joan Blaeu and his Grand Atlas* [Amsterdam, 1970], 23).

12 See G.E. Orchard, 'Introduction,' in I. Massa, *A Short History of the Beginnings and Origins of These Present Wars in Moscow*, trans. and ed. G.E. Orchard (Toronto, 1982) ix–xxiv: xxiii; Elias Herckmans, *Een historischen verhael van de voornaamste beroerten des keyserrychs van Russia* (Amsterdam, 1625); *Kort waerachtigh Verhaal*. The Herckmans text was listed as existing in manuscript among the Imperial Library's collection in St Petersburg by Uhlenbeck in the nineteenth century (and dated Amsterdam, 1625); it seems never to have been published (see Uhlenbeck, *Verslag*, 3). For the dearth of original Dutch publications on Muscovy, see Naarden, 'Dutchmen,' 4. For an overview of pamphlets and books, see Tavernier, *Russia*, 365–73. On Witsen see also Locher, de Buck, 'Inleiding,' in Witsen, *Moscovische reyse*, xix–lxxv; Witsen, *Moscovische Reyse*; Rietbergen, *Europe*, 310–11; Gebhard, *Het leven*; M. Peters, 'Nepotisme, patronage en boekopdrachten bij Nicolaes Witsen (1641–1717), burgemeester van Amsterdam,' *Lias* 1 (1998) 83–134; Peters, 'From the Study'; A.N. Kirpichnikov, *Rossiia xvii veka v risunkakh i opisaniiakh gollandskogo puteshestvennika Nikolaasa Vitsena* (St Petersburg, 1995).

13 Danckaert, *Beschrijvinghe van Moscovien*, second edn., 41–2.

14 *Het ellendigh leven der Turken, Moscoviters en Chinesen, aende Christenheyt vertoont* ('s-Gravenhage: Iohannes Sonnevelt, 1664). While this is the epitome of a sort of early Orientalist caricature of these foreign cultures, it is evident from its foreword that the book hardly intended to present an honest description of those empires; it rather aimed at contrasting the 'True Freedom' of the stadtholderless government of the Republic with despotism. Saeghman reprinted his edition of Danckaert's *Beschrijvinghe* in 1660, 1663, and 1665.

15 See G. Hornius, *Orbis Politicus* (Leiden: F. Lopes de Haro and C. Driehijsen, 1667).

16 P. Valkenier, *'t Verwerd Europa* (Amsterdam: H. and D. Boom, 1675). Wittram, erroneously, thought that Valkenier's work's first edition was published in 1668 (see Wittram, *Czar* vol. 1, 36).

17 Olearius, *Beschrijvingh*; Olearius, *Persiaensche Reyse*. A popular foreign treatise on Persia was found in Pietro della Valle's work (see della Valle, *Viaggi*).
18 See Dapper, *Asia*.
19 Haberland, 'Einführung,' 27.
20 See also the next chapter for some some additional reasons for van Meurs and van Someren's interest in the project.
21 For this curiosity, see van den Boogaart, *Civil and Corrupt*, 3; Brenner, 'Mythos,' 17–18; Haberland, 'Einführung,' 25.
22 J. Huygen van Linschoten, *Discours of Voyages into Ye East and West Indies* (Amsterdam, 1974); Gerrit de Veer, *Waerachtige beschryvinghe van Willem Barents* (Den Haag, 1917; origin. Amsterdam, 1598–9); Bontekoe, *Memorable Description*.
23 As was exemplified by the enduring popularity of the folk hero Tijl Uilenspiegel's stories (see *Ulenspiegel* [Antwerp: van Hoochstraten, *c*.1530]; *Historie van Thijl Ulespiegel* [Amsterdam: Broer Jansz., *c*.1640]). See also Blussé, 'Op zoek,' 13.
24 Burke, *Venice*, xx; Burke, *Popular Culture*, 270–81. Ginzburg's hypothesis (following Bakhtin) of a circular relationship between the cultures of the dominant and subordinate classes seems apt in envisioning this phenomenon (see Ginzburg, *Cheese*, xii, xvi–ii). He suggests that a repression and effacement of popualr culture by the elite began toward 1600 (ibid., 126).
25 Thus van der Heijden's *Vervarelyke schip-breuk*; among the most famous were the books by Jan Huygen van Linschoten and W.Y. Bontekoe: See Linschoten, *Discours of Voyages*, and Bontekoe, *Memorable Description*.
26 Sailmakers are counted among the poorest layers of Holland's society by historians, as we saw in Chapter 3.
27 Dekker, 'Van Grand Tour,' 13, 15.
28 Even the unfashionable beard of the tied-up 'Struys' in the engraving suggests that this is not member of the Dutch elite. See the next chapter for more on this engraving. I agree with Linda Colley here who argues that 'many of the seventeenth- and eighteenth-century British captivity narratives which were published, as distinct from being preserved in manuscript, were by women or artisanal men [since it was easier for] these groups to testify to an experience of vulnerability in print than it was for officer-class males' (Linda Colley, 'Going Native, Telling Tales: Captivity, Collaborations and Empire,' *Past and Present* 168 [2000] 170–93: 176).
29 See Schama, *Embarrassment*, 121.
30 Bunyan, *Pilgrim's Progress*.
31 See R.W. Ferrier, ed., *A Journey to Asia* (London, 1996), xii.
32 Eisenstein, *Printing Press*, 99; Rietbergen, *Europe*, 412–13; Olearius, *Offt begehrte beschreibung*; White, 'Foucault Decoded,' 47.
33 P. Rietbergen, 'Varieties of Asia? European perspectives, *c*.1600–*c*.1800,' *Itinerario* 3–4 (2001) 69–89: 80–1; see also Stagl, *History*, 81.
34 See Blussé, 'Op zoek,' 16.
35 See the dedication of Dapper to Johan de Witt in his work on the Dutch embassies to China (Dapper, *Gedenkwaerdig Bedryf*, 'Opdracht,' n.p.); see also Lanni, 'Une cartographie ethnique.' See among others for this curiosity van den Boogaart, *Civil and Corrupt*, 3; Brenner, 'Mythos,' 17–18; Haberland, 'Einführung,' 25; Cook, *Matters*, 17–21.

36 The term was coined by Clifford Geertz (see C. Geertz, *The Interpretation of Cultures* [New York, 1973] 3–32). Van Meurs's model was probably Pliny's work (see Cook, *Matters*, 22).
37 For the economic 'crash', see Israel, *Dutch Primacy*, 292. Van Meurs's competitive spirit is evident from his attempts to buy the services of Coenraet Decker in the early 1670s, when Decker was under contract with the partnership of van Someren and van Waasbergh(e) (see M. Kleerekoper, W. van Stockum, *De boekhandel te Amsterdam*, 2 vols ['s-Gravenhage, 1914–1916], vol. 1, 422, and vol. 2, 1341–2). See also P. Rietbergen, 'Zover de aarde reikt. De werken van Johan Nieuhof (1618–1672) als illustratie van het probleem der cultuur- en mentaliteitsgeschiedenis tussen specialisatie en integratie,' *De zeventiende eeuw* 1 (1986) 17–40: 25.
38 See also Febvre, Martin, *Coming*, 89.
39 See Chapter 2. Van Meurs was keenly aware of the cost of engravings, as he had earned a living as an engraver previously (see *GAA NA* 1104 [Notary J. van de Ven], 23 April 1653, which identifies van Meurs as engraver [*plaatsnijder*]).
40 While he likely printed no more than about 800 copies of each of the first two editions of the description of Africa by Olfert Dapper, by 1678 he had at least 250 copies of the second (folio) edition of 1676 in stock showing continued halting sales of expensive books; see Dapper, *Naukeurige Beschrijvinge*, second edn.; see A. Jones, 'Olfert Dapper et sa description de l'Afrique,' in *Objets interdits* (Paris, 1989) 76–81: 76. See also Lanni, 'Une cartographie ethnique.'

12 *Reysen*'s Creation and its Creators

1 R. Barthes, *Image, Music, Text*, ed. and trans. S. Heath (London, 1977), 146.
2 The printer was almost certainly Christoffel Cunrades, who has been identified as the printer of *Reysen*'s 1678 German-language edition (see N. Malcolm, *Aspects of Hobbes* [Oxford, 2002], 362, 362n95)
3 For a brief overview, see van Delft and de Wolf, eds, *Bibliopolis*, 57–106. Van Meurs was an engraver before he became a publisher (see F.G. Waller, *Biographisch Woordenboek van Noord Nederlandsche Graveurs* [Amsterdam, 1974], 222).
4 The title on the first printed page was actually different.
5 On the role of such a 'paratext', see G. Genette, 'Introduction to the Paratext,' *New Literary History* 2 (1991) 261–72.
6 For a fine analysis of the frontispiece and some of the other engravings, see Brancaforte, *Visions of Persia*.
7 Brancaforte, *Visions*, 106.
8 Pierre Dan, *Historie van Barbaryen en des zelfs Zeerovers*, trans. G. van Broekhuizen (Amsterdam: Jan ten Hoorn, 1684).
9 See the frontispiece of T.J. van Bragt (Bracht), *Het bloedig toneel* (Amsterdam: Van der Dyster & Co, 1685); Andrew Wheatcroft, 'Orientalism: The Impact of the Image of the East on the West,' paper delivered at *SHARP Annual Conference*, The Hague (12 July 2006).
10 See Brancaforte, *Versions*, 97–103.

11 For this work and an analysis of this tradition, see van der Velden, 'Cambyses for Example,' 30.
12 For the full text of the title page, see the Bibliography.
13 See for the genre Chapter 13.
14 See Febvre, Martin, *Coming*, 102; see *Reysen*, 'Bericht Aan den Boek-Binder.'
15 *Reysen*, *'Extract,'* anon.; *Reysen*, 'Extract' [Butler]. It is not clear how these letters ended up in the possession of van Meurs and van Someren.
16 In an e-mail to the author, Marja Smolenaars noted how 'engravings often were re-used, even if they had not originally been produced for a certain text, ... The copper and engraving amounted to an expensive investment ... '(Marja Smolenaars, e-mail to author, 26 June 2006; translation mine). For the plates, see Dozy, 'Olfert Dapper,' 430, as well as Febvre, Martin, *Coming*, 102–3. P.A. Tiele, ed., *Bijdragen tot eene Nederlandsche Bibliographie uitgegeven door het Frederik Muller-fonds* vol. 1 (Amsterdam, 1884), 233, identifies C. Decker and J. Kip as engravers (for Kip, see the next note). Kip and Decker also worked for van Meurs on Dapper's 1677 work on the Near East; it is not unlikely, given van Meurs's apparent personal authorship of the map of Palestina in this work, that he himself engraved the map of the Caspian Sea in Struys's book (see Dozy, 'Olfert,' 429; Dapper, *Naukeurige Beschrijvinge*, second edn. [Amsterdam: van Meurs, 1676]). Decker was born in Amsterdam around 1650, and buried there in 1685 (see Waller, *Biographisch Woordenboek*, 76, 149).
17 See Dozy, 'Olfert,' 431, and *Reysen*, between 96 and 97. Johannes (Jan) Kip was born in Amsterdam in 1652/3, and died in Westminster, London in 1722 (see Waller, *Biographisch Woordenboek*, 175).
18 See *GAA DTB* 491, p.11.
19 See Linschoten, *Discours of Voyages*. See Schmidt, *Innocence Abroad*, 161–3; J.H. van Linschoten, *Beschryvinghe van de gantsche kuste van Guinea* (Amsterdam: Cornelis Claesz, 1596). Although based on the captain's ship log, Bontekoe's work was edited by a scribe under the auspices of the publisher (Verhoeven, Verkruijsse, 'De vele gezichten,' 45–7).
20 *Reysen*, 206.
21 *Reysen*, 'Voorreden aan de Lesers,' 2.
22 For the Dutch fondness for 'intrepid nautical heroes', see Schama, *Embarrassment*, 28–34.
23 Ghostwriters became a common phenomenon in eighteenth-century France, perhaps emulating these earlier Dutch practices (see Eisenstein, *Printing Press*, 146).
24 For other candidates who ultimately have to be discounted as ghostwriters, see K. Boterbloem, 'The Genesis of Jan Struys's "Perillous Voyages" and the Business of the Book Trade in the Dutch Republic,' *Papers of the Bibliographical Society of America* 1 (2008) 5–28. Dapper, together with Arnoldus Montanus, had already worked for van Meurs as a compiler of an account of a Dutch Embassy to China; see E.J. van Kley, 'Asian Religions in Seventeenth-Century Dutch Literature,' *Itinerario* 3–4 (2001) 54–68: 58; Johannes Nieuhof, *Het gezantschap der Neerlandtsche Oost-Indische Compagnie* (Amsterdam: van Meurs, 1665). Among works published by Dapper before 1672 were *Gedenkwaerdig bedryf*; *Asia*; *Naukeurige Beschrijvinge* (first edn.).
25 Dapper, *Historische beschryving*. On Dapper's medical training, see Dozy, 'Olfert,' 420.

26 For the relation between printers-publishers and authors-scholars, see Eisenstein, *Printing Press*, 18, 23.

27 Dapper, *Asia*, 3–3verso; in a somewhat analogous case, Nicolaas Witsen (who survived the Orangist purges of 1672, see below) excised references to Cornelis de Witt in the second edition of his *Scheepsbouw*, of 1690, from which he also stripped the anglophobic parts (see W. Nijhoff, 'De anglophobie van Nic. Witsen en verschillende redactiën van zijn Scheepsbouw, 1671,' *Het Boek* 14 [1925] 88–96: 88–91; Witsen, *Scheeps-Bouw*; Witsen, *Architectura navalis*). Dapper's description of China and the Dutch Embassy sent to the Qing emperor in the 1660s had been dedicated to Johan de Witt in 1670 (Dapper, *Gedenkwaerdig Bedryf*, 'Opdracht').

28 See Prud'homme van Reine, *Schittering*, 301–8.

29 See W. Troost, *Stadhouder-Koning Willem III* (Hilversum, 2001), 95; Prud'homme Van Reine, *Schittering*, 301–8).

30 *Reysen*, 371.

31 Dapper, *Naukeurige Beschrijvinge der Afrikaensche gewesten*, second edn.; Dapper, *Naukeurige beschryving van gantsch Syrie*. That the book on Africa was reprinted virtually unchanged in 1676 lends further credence to the idea that Dapper had time and opportunity to dedicate to other writing projects (despite his title, he does not seem to have practised as a doctor).

32 Lach and van Kley call Dapper 'indefatigable', churning out 'long, encyclopedic and discursive' works (see Lach and Van Kley, *Asia* vol. 3, 493). Montanus, the other author who published similar works with van Meurs, was not under suspicion of political opposition and continued to publish under his own name, and with other publishers as well (see A. Montanus, *Kerkelyke historie van Nederland* [Amsterdam: Cornelis and Jan van Zwol, 1675]). Montanus had been an Orangist in opposition to the de Witts in the 1660s, see A. Montanus, *'t Leven en bedryf der prinsen van Oranje* (Amsterdam: A. van den Heuvel, S. Imbrecht, 1664). On him see Reinier H. Hesselink, 'Memorable Embassies: The Secret History of Arnoldus Montanus' *Gedenkwaerdige Gesantschappen*,' *Quaerendo*, 1–2 (2002) 99–123. For some of his work for van Meurs, see Montanus, *Gedenkwaerdige gesantschappen*; A. Montanus, *De Nieuwe en onbekende weereld* (Amsterdam: J. van Meurs, 1671).

33 See E. Haitsma Mulier, 'De eerste Hollandse stadsbeschrijvingen uit de zeventiende eeuw,' *De zeventiende eeuw* 2 (1993) 97–116: 103; Jones, 'Olfert Dapper,' 73–4; Brugmans, *Geschiedenis* vol. 3, 158–63, 166–7.

34 Witsen was good friends with Gillis Valckenier, the most outstanding Amsterdam regent who in 1672 chose the side of Orange (see Roorda, *Partij*, 181, 250; in fact, Roorda calls him Valckenier's [former] *'creatuur,'* ibid., 250). Witsen proved his loyalty and value to the prince by his role in defending the country against the French in 1672 and beyond, while he earned further credibility in William III's eyes by his protest against attempts to strike a compromise with France (A.J. van der Aa, *Biographisch Woordenboek der Nederlanden* [Haarlem, 1852–78] vol. 7, 111, and vol. 1, 283). In 1676, Witsen was the Holland Estates' army representative at several battles, and oversaw the army's provisions and pay (see van Nimwegen, *Deser landen crijchsvolck*, 294, 310). On his numerous other responsibilities, see Peters, 'From the Study,' 1.

35 On the politics of book dedications, especially involving Witsen, see Peters, 'Nepotisme,' 83, 103, 105. See *Reysen*, 'Opdracht.'

36 A. Jones, 'Decompiling Dapper: A Preliminary Search for Evidence,' *History in Africa* 17 (1990) 171–209: 171.
37 See Dozy, 'Olfert,' 415, 419–20. On the various argots, see Burke, *Toward*, 26–7. Although the text contains instances of a sort of folksy style and popular sayings hinting at folk wisdom, *Reysen* utterly lacks sailors' salty language: '[Dutch patricians] expressed a heightened sensitivity towards violence and verbal fieriness' (Spierenburg, *Judicial Violence*, 18).
38 The other two were Van der Heiden, *Vervarelyke schip-breuk*, and Schouten, *Oost Indische Voyagie*.
39 See for South Africa, Breet, *Oost-Indische Voyagie*, 35–6, 465–73; for the Indonesian Archipelago, ibid., 39–109, 427–58; for Aracan (Burma), ibid., 117–63; for Sri Lanka (Ceylon), ibid., 173–87, 280–99; on south India, ibid., 190–221, 233–80; on the Bengal region, ibid., 319–33, 360–426; on Malaysia, ibid., 343–60. Both Schouten and Struys write about Taiwan (Formosa), although Schouten's description is a second-hand account of the island's loss in 1661 to Zheng Chenggong, also known as Coxinga (Breet, *Oost-Indische Voyagie*, 164–71). Schouten describes the Japanese in far greater detail than Struys (Breet, *Oost-Indische Voyagie*, 335–43). He provides a brief overview of Persia (ibid., 227–31). Van der Heiden's tale was more an adventure story about a shipwrecked crew.
40 Dapper was more generous (albeit far from meticulous) at crediting his sources in books appearing under his own name. Dapper's *Asia*'s thus cited Olearius (whose text had been available in Dutch translation, of course), the Swede Peter Petrejus's account (mainly focussed on Muscovy), and Pietro della Valle, but used other sources that Dapper did not acknowledge (for example, Dapper, *Asia*, 60, 66, 90, 102–4, 132, show how he uses Olearius, *Offt begehrte beschreibung*, P. Petrejus, *Regni Muschovitici sciographia* [Stockholm: I. Meurer, 1614–15], and della Valle, *Viaggi*; Petrejus's book was available in German as P. Petrejus, *Historien und bericht von dem grossfürstenthum Muschkow* [Tipis Bavaricis: Lipsiae, 1620]). Similar to *Reysen*, Dapper's *Asia* discusses the graves of the Persian shahs and some of the towns (for instance Ardabil, see Dapper, *Asia*, 44–7; see Chapter 10). Adam Jones points out in his analysis of Dapper's book on Africa how '[h]e used at least a hundred published sources and several unpublished ones; moreover, instead of lifting whole passages from one book, he often based a single paragraph on two or three different sources' (Jones, 'Decompiling,' 171).
41 Compare for example the description of Dagestani habits in both texts, see Olearius, *Podrobnoe*, 975–6, with *Reysen*, 220–3; see further Chapters 9 and 10.
42 See Chapter 7.
43 See note 40 above.
44 See della Valle, *Viaggi*, as well as Chapters 1 and 2, note 37 of this chapter, and Dozy, 'Olfert,' 428.
45 See Chapter 1.
46 Jones, 'Decompiling,' 172. Witsen's private library contained more than 2,000 volumes at the time of his death in 1717; although it will have been smaller in 1675, it was likely one of the collections mined by Dapper (see Peters, 'From the Study,' 13, Table 3).
47 As *Reysen*'s frontmatter shows (*Reysen*, [2]). They received exclusive rights to the book's reprinting for 15 years in the Province of Holland. As van Meurs

specialized in lavish and costly publications with many engravings, he seems to have repeatedly requested copyright protection, which was rather exceptional (see Dozy, 'Olfert,' 427). The 1678 German version of *Reysen* was copyright-protected for the Empire for a term of six years, as were the two parallel texts, see *Johann J. Straußens Reisen* (Amsterdam: von Meurs und von Sommern, 1678), frontmatter. In 1678, 96 members of the Amsterdam publishers' guild signed a petition to halt the printing of pirated versions, a first step toward more meaningful copyright protection, see I.H. van Eeghen, *De Amsterdamse Boekhandel, 1680–1725*, vol. 5, no.1 (Amsterdam, 1978), 31–2.

48 The consistency of Western texts' image of Muscovy as first set out by Baron Sigismund Herberstein (and perhaps Giles Fletcher in English) is remarkable, whether written in Latin, German, French, Swedish, English or Dutch (Herberstein, *Rerum moscoviticarum*; G. Fletcher, *Of the Russe Commonwealth*, eds R. Pipes and J. Fine, [Cambridge, MA, 1966]).

49 For a thoughtful discussion of the rather different concepts of imitation and plagiarism in Antiquity and the Early Modern period, see M. Randall, 'Appropriate(d) Discourse: Plagiarism and Decolonization,' *New Literary History*, 3 (1991), 525–41: 527–8. See also P. Kewes, ed., *Plagiarism in Early Modern England* (Basingstoke, 2003).

50 De Vries and van der Woude, *First*, 170–1.

51 *Herodoot van Halikarnassus negen boeken der historien*, Uit het Grieks vertaelt door Dr O.D. (Amsterdam: Hieronymus Sweerts, 1665). It had also been dedicated to Nicolaas Witsen (see Dozy, 'Olfert,' 423).

52 *Reysen*, 83, 111–12, 349–50. The purported Latin document given to Struys by Alessandro on Ararat may have been a Dapperite concoction as well (see *Reysen*, 230–2).

53 Katherine George, 'The Civilized West Looks at Primitive Africa: 1400–1800, a Study in Ethnocentrism,' *Isis* 1 (1958) 62–72: 65.

54 See Chapter 13.

55 *Reysen*, engraved frontispiece, 1.

56 See Chapter 2.

57 The variation in script is indicative of the transition from Gothic to Latin print, although Gothic lingered for a while since it was the type taught to children at primary school well into the eighteenth century (Dekker, *Humour*, 34; see also Febvre, Martin, *Coming*, 83).

58 See Chapter 14.

59 Hair's remarks seem nevertheless apposite here: 'All this suggests that Dapper's editing was, by modern standards, sometimes casual and slapdash; that is, it was much as one would expect from an editor of that age producing books of this kind' (see P.E.H. Hair, 'Barbot, Dapper, Davity: A Critique of Sources on Sierra Leone and Cape Mount,' *History in Africa* 1 [1974] 25–54: 35–6).

60 The licence by the Estates of Holland and West-Vrieslandt was awarded on 23 September 1675; the Klenk Embassy had left Amsterdam in July 1675 (see *Reysen*, [2]; Loviagin, ed., *Posol'stvo*, 13).

61 See Chapter 14.

62 In a personal style, hinting at Struys's direct involvement in composing the passage, *Reysen* refers to the death of Tsar Aleksei Mikhailovich in February 1676 (*Reysen*, 166).

63 Coyett's book, judging from the preface by its publisher Jan Claesz ten Hoorn, appeared anonymously late in 1677 (it is dated 20 November 1677, see Loviagin, ed., *Posol'stvo*, 4). Ten Hoorn was to become a major publisher in Amsterdam (see Dekker, *Humour*, 52). Contemporary Dutch curiosity about Muscovy was heightened, as can be seen from the pamphlet *Relaes van't gepasseerde voor ende op de Inkomste ende Receptie van den Heere van Klenck* ('s Gravenhage: Jacobus Scheltens, 1676).
64 See Chapter 13. Bert van Selm noted that books of this kind were printed in editions of about 1,000 copies (Bert van Selm, ' "... te bekomen voor een Civielen prijs". De Nederlandse boekprijs in de zeventiende eeuw als onbekende grootheid,' *De zeventiende eeuw* 1 [1990] 98–116: 103; see also Febvre, Martin, *Coming*, 219). See the Bibliography.
65 Smith, the first London publisher, had the plates at his disposal.
66 But such rewards, even for actual authors, remained meagre (see Febvre, Martin, *Coming*, 162–3).
67 Dapper's texts are still used as a source for the historical study of some regions; see, for example, *Objets interdits*. In France, a *Fondation Dapper* exists. In the Netherlands, his name is associated with a street in Amsterdam immortalized in a famous poem by J.C. Bloem ('Domweg gelukkig in the Dapperstraat,' see J.C. Bloem, *Gedichten* [Amsterdam, 1979]) and with a collection of vitriolic essays by W.F. Hermans (*Het evangelie van O. Dapper Dapper* [Amsterdam, 1973]).
68 Recently, Struys's account of his travels in Russia was twice republished in Russian (see Kotoshikhin et al., *Moskoviia*, and J.J. Struys, D. Butler, *Tri puteshestviia*, ed. A.I. Tsepkov [Riazan' 2006]).

13 Genre and the Test of Time

1 Perhaps the travel account as such can be seen as an ancestor of the epistolary novel (see R. Zuber, M. Cuénin, *Littérature française* vol. 4 [Paris, 1984], 127).
2 See Adams, *Travel Literature*, 176–7. For example, Pietro della Valle's travel account had been published in the form of letters (see della Valle, *Viaggi*).
3 The model for it were engravings by Rubens; see Brancaforte, *Visions*, 97–107 and A. Hauser, *Social History of Art* vol. 2 (London, 1951), 161–4. On Rubens see further, for example, Hauser, *Social History* vol. 2, 205–7.
4 Hazard, *La crise*; Israel, *Radical Enlightenment*. The debate between Ancients and Moderns that raged among intellectuals was another sign of this transition; for instance, see R. Pomeau, J. Erhard, *Littérature Française* vol. 5 (Paris, 1984) 88–9; Jozien Driessen argues that the Ancients had already lost the battle in the Netherlands before 1700 (see Driessen-van het Reve, *Kunstkamera*, 42). For similar changes in publishing, see Eisenstein, *Printing Press*, xv.
5 Similarly, Schmidt discerns the 1670s as a 'turning point' in the Republic's history (Schmidt, *Innocence Abroad*, 315).
6 Schmidt, *Innocence Abroad*, 315.
7 See Danckaert, *Beschrijvinghe van Moscovien*, 41–2. Nevertheless, homophobia simmered, erupting around 1730 (Schama, *Embarrassment*, 601–7).

8 Olearius's book especially seems a forerunner of this subgenre (see Olearius, *Offt begehrte beschreibung*). See J. Cook, *A Voyage Toward the South Pole*, 2 vols (London: Strahan and Cadell, 1777); L. de Bougainville, *A Voyage Around the World*, trans. J.R. Foster (London: Nourse and Davies, 1772).

9 Maczak, *Travel*, 142. Following the ideas of Marijke Barend, Verhoeven and Verkruijsse note a development of the genre in the Republic from travel journals (popular *c.*1590–1660) to travel descriptions (*c.*1660–*c.*1710) to imaginary travel narratives (after *c.*1700; see Verhoeven, Verkruijsse, 'De vele gezichten,' 39).

10 See Introduction and Chapter 11. Rietbergen remarks how Thomasius went as far as to argue in 1693 that 'travel had [...] become unnecessary now that books could be obtained so easily' (Rietbergen, *Europe*, 293).

11 See Introduction and Chapter 12. Stéphane Mund details the common elements in travel descriptions of Russia, also noting the habit of most texts to move between narrative and description (see Mund, *Orbis Russiarum*, 109–48). See also Verhoeven, Verkruijsse, 'De vele gezichten,' 39.

12 Poe, '*A People*,' 7; Herberstein, *Rerum moscoviticarum*. See also Haberland, 'Einführung,' 27; Th. Locher, 'Het beeld van Rusland in de zestiende-eeuwse Europese beschrijvingen,' *Tijdschrift voor Geschiedenis* 70 (1957) 289–308: 294; A. Morozov, 'Parusnyi Master Ian Streis i ego puteshestvie,' in *Tri puteshestviia*, ed. Morozov, 23–38: 32.

13 Brenner, 'Mythos,' 26–7.

14 Stagl, *History*, 82.

15 E. Harth, 'An Official "Nouvelle",' in M.M. de Lafayette, *The Princess of Clèves*, ed. and trans. J.D. Lyons (London, 1994) 230–40: 232.

16 Olearius, *Offt begehrte beschreibung*; Olearius, *Vermehrte Neue Beschreibung*.

17 'They formed the collective luggage of the Republic of Letters' (Mund, *Orbis Russiarum*, 237–8; translation my own).

18 As a comparison shows between the naming of rivers and seas by Herodotus and Olearius (Olearius, *Travels*, 110; Herodotus, *The Histories*, trans. Robin Waterfield [Oxford, 1998]).

19 *Reysen* mentions Troy on four pages (see *Reysen*, 83, 85, 111, 112). It should be remembered that there was at the time only a traditional consensus where Troy had stood, as its archeological rediscovery had to await Heinrich Schliemann's enterprise in the nineteenth century.

20 J. Paul Hunter, 'The "Occasion" of *Robinson Crusoe*,' in Daniel Defoe, *Robinson Crusoe* (New York, 1975) 365–78: 373–4; for a discussion of the genre, see also Adams, *Travel Literature*, 165, 208–9, and Aune, 'Early Modern European,' 120.

21 See Maczak, *Travel*, 142; Stagl, *History*, 50–60, 81–2; Hunter, 'The "Occasion",' 376, 376n9.

22 Daniel Defoe, *Robinson Crusoe*, ed. Michael Shinagel (New York, 1975); Swift, *Gulliver's Travels*.

23 Cornelis de Bruyn, *Reizen over Moskovie* (Amsterdam, 1714).

24 On the early eighteenth-century rise of British periodicals and reviews and the journalists behind them, see R. Porter, *Enlightenment* (London, 2000), 79–80, 160–1.

25 Cervantes, *Don Quijote*. See 'Picaresque Novel,' in W. Harmon and C. Holman, *A Handbook to Literature*, eight edn. (Upper Saddle River, NJ, 1999), 389–90. Whether or not consciously, it is likely that *Der abenteuerliche Simplicissimus*

by Grimmelshausen exerted an influence on *Reysen*'s writer, too (see Adams, *Travel Literature*, 201; H. von Grimmelshausen, *Der abenteuerliche Simplicissimus* [Halle, 1889; origin., 1669]). On the popularity of the picaresque novel in the Republic, see J. Vles, *Le roman picaresque hollandais* ('s Gravenhage, 1926), *passim*; Dekker, *Humour*, 22. G.A. Bredero, not the least of the Dutch writers of the age, wrote picaresque plays, and even the unctuous moralist and Grand Pensionary Jacob Cats (1577–1660) wrote a play based on Miguel de Cervantes's *La Gitanilla* (*The Little Gypsy*), see J. Cats, *Het Spaens Heydinnetje* (Zwolle, 1963; origin. 1637). Most of the works were either translations of Spanish and French originals, or derivative of them.

26 Vles, 'Roman picaresque,' 55, 61n1, 75n1, 95n1, 109, 118n1, 127. Vles notes how *De Twee Vermaarde Fortuyns Kinderen* (Delft: n.p., 1681), may have been based on real-life characters who fought as army officers in the Dutch war of 1672–78 (Vles, 'Roman picaresque,' 132–4, 132n6).

27 For this predilection, see Schama, *Embarrassment*, 28–34, 248–9; Verhoeven, Verkruijsse, 'De vele gezichten,' 39. Famous examples are G. de Veer, *Waerachtige beschryvinghe van Willem Barents* (Den Haag, 1917; origin. Amsterdam, 1598–9); E. Herckmans, *Der zee-vaert lof* (Amsterdam: Wachter, 1634); Bontekoe, *Memorable Description*; Linschoten, *Discours of Voyages*.

28 As Schmidt points out, meanwhile, Linschoten's *Itinerario*, perhaps the first Dutch account of its kind in the Republic, was not quite original either: it was modelled after the Dutch version of the fictive (although widely believed to be genuine) travels of John Mandeville (Schmidt, *Innocence Abroad*, 9). While the Mandeville stories originated in the Late Middle Ages, only a few years before Linschoten's work a Dutch version had been published in Antwerp; see *Die wonderlijcke reyse van Jan Mandevijl* (Antwerpen: Wed. Guilleam van Parijs, 1592).

29 Schama, *Embarrassment*, 28–34; see Chapters 4, 10 and 11.

30 Homer, *Odyssey*, trans. S.O. Andrew (London, 1953); 'Sindbad the Seaman and Sindbad the Landsman,' in *The Arabian Nights*, ed. and trans. R.F. Burton (New York, 2001) 331–89. Luke 15: 11–32; Jonah; Job, *King James Bible*, as well as the Pentateuch.

31 An early Dutch example is Constantijn Huygens (1596–1687), secretary of two princes of Orange, who kept a private diary (see Dekker, *Childhood*, 23). The classic example of the first modern diary is that of Samuel Pepys (1633–1703), a contemporary of Struys; see S. Pepys, *The Diary of Samuel Pepys*, 10 vols, eds R. Latham and W. Matthews (Berkeley, CA, 1970–1983).

32 Dekker, *Childhood*, 17; see Saint Augustine, *The Confessions of Saint Augustine* (London, 1961); Bunyan, *Pilgrim's Progress*.

33 Dekker, *Childhood*, 17–18; see as well Adams, *Travel Literature*, 165, 169–70.

34 Dekker, *Childhood*, 18; Linschoten, *Discours of Voyages*; Bontekoe, *Memorable Description*.

35 J.J. Rousseau, *Les Confessions* 2 vols (Paris, 1973).

36 *Reysen*'s narrator describes the impulsive manner in which the young sailmaker embarks on his first two voyages in an avuncular tone (deliberately?) hinting at Struys's hard-won wisdom and middle age in 1675–6.

37 See Bakhtin, *Rabelais*, 145–6, 160, 164–72, 191. Hearty comedies of G.A. Bredero deploy such language rather more faithfully (see G.A. Bredero, *De kluchten* [Amsterdam, 1926]).

38 Such as the ugliness of the Kalmyks ('their mugs are very wide'), or Tatar 'men's eyes [that] are shrunk like old wives' (*Reysen*, 185, 192).
39 See *Reysen*, where the licence follows the title page.
40 Or books that already had sold well earlier in a different edition (see van Delft and de Wolf, eds, *Bibliopolis*, 80).
41 This is evident, too, from the publication by the publishers themselves of a German and French translation: *Johann J. Straußens Reisen*; *Les Voyages de Iean Struys* (Amsterdam: Veuve de Jacob van Meurs, 1681). It may be noted as well that the book was still valued a century later, in 1788, when the edition published by van Esveldt of the 1740s cost four guilders, which was the same for a 1691 print of Olearius's work in Dutch; Voltaire's *Candide* (the Geneva 1759 edition) was priced at a mere one guilder (see R. Arrenberg, *Naamregister van de bekendste en meest in gebruik zijnde Nederduitsche boeken* [Rotterdam: G.A. Arrenberg, 1788], 385, 496, 552; Voltaire, *Candide* [Geneva: n.p., 1759]).
42 Two copies may be the same original Dutch version of 1676 (see www.nlr.ru/rlin/Rossican.php [accessed 20 February 2007]).
43 de Bruyn, *Reizen over Moskovie.*
44 The 1740s may have been a watershed; after this decade the popularity of other seemingly perpetual bestsellers faded as well (see Schama, *Embarrassment*, 83, 599).
45 Willem Floor has scrutinized the book's Persian section's translations into German, French, and English, and considers the German version most faithful, while he deems the English translation slightly more accurate than the French one. In comparing other parts of *Reysen* with its translations, this verdict seems indeed accurate (see Floor, 'Fact,' 66–8).
46 Braudel, *Civilization*, vol. 3, 197; see also Burke, *Toward*, 15. For the general shift from Latin to French as the language of the Republic of Letters, see as well Eisenstein, *Printing Press*, 138, and Febvre, Martin, *Coming*, 233. There were still a few Latin descriptions of Muscovy written around this time. The Imperial envoy to Moscow Lyseck-Lizeck (who visited Moscow at the same time as the Klenk embassy) wrote one (see A. Lyseck, *Relatio eorum, quae circa Sacrae Caesareae Majestatis as Magnum Moscorum Czarum* [Salzburg: n.p., 1676]). *Reysen* seems to have never been translated into Latin, Spanish, or Italian.
47 *Joh. Jansz. Straußens reysen*; *Les Voyages de Iean Struys.*
48 See J.J. Struys, *Unglückliche Schiffs-Leute, oder mirkwirdige Reise zwenzing Holländeren, welche auss Befehl des christlichen Reussichen Keisers, in der Moscau, ein grosses Schiff gebauet, die Kaspische See damit zubefaaren*, ed. Jakob Redinger (Zürich: Heinrich Müller, 1679). See also Poe, *Foreign Descriptions*, 173–4; Adelung, *Kritisch-literärische Übersicht* vol. 2, 345–6; and Morozov, 'Parusnyi Master,' 24–7.
49 Leipzig: K.F. Köhler, 1797 (translated from a French version), and Gotha and Erfurt: Heunigsche Buchhandlung, 1832.
50 A. Anemone, 'Monsters of Peter the Great: The Culture of the St. Petersburg Kunstkamera in the Eighteenth Century,' *Slavic and East European Journal* 4 (2000) 583–602: 586–8. See as well Cook, *Matters*, 23, 28–30, 141–2.
51 (Anonymous), Review of 'Les Voyages de Jean Struys en Moscovie, Tartarie, Perse, & plusieurs autres Païs étrangers, avec des remarques sur la qualité, la

Religion & c. de tous ces Pays, & la Relation d'un naufrage effroyable, par M. Glanius. A Amsterdam. Et se trouvent à Paris chez Antoine Cellier, 1681,' *Journal des Sçavans* (Lundi 21 Juillet 1681) 150–2. This first 'academic journal' (it was connected to the *Académie Française*) had been published since 1665 (see Febvre, Martin, *Coming*, 235–6).

52 Review of 'Les Voyages,' 151–2. See Chapter 8.

53 Such as J. Thévenot, *Voyage de Levant* (Paris: n.p., 1665); Bernier, *Histoire*; Chardin, *Le couronnement*; Tavernier, *Les Six Voyages*.

54 G.J. Ames, 'Colbert's Indian Ocean Strategy,' *French Historical Studies* 3 (1990) 536–59: 536–9, 549, 555, 559; Zuber, Cuénin, *Littérature française*, 127–8.

55 A.A.M. Awad, 'The Gulf in the Seventeenth Century,' *Bulletin of the British Society for Middle Eastern Studies* 2 (1985) 123–34: 128–9.

56 See (Anonymous), 'From the Journal des Scavans, set forth July 24. 1681., "Les Voyages de Jean Struys en Muscovie,"' *Weekly Memorials for the Ingenious* (January 16, 1682) 5–7.

57 *The perillous and most unhappy voyages of John Struys* (London: Samuel Smith, 1683); *The voiages and travels of John Struys* (London: Abel Swalle, 1684).

58 Anderson, *Britain's Discovery*, 51.

59 Behn, *Oroonoko*; Defoe, *Robinson Crusoe*; Swift, *Gulliver's Travels*. On the growing reading public in England and Defoe's popularity, see Porter, *Enlightenment*, 73. For knowledge about Muscovy, English seventeenth-century readers could choose from works by Willoughby-Chancellor, Jenkinson, Fletcher, Horsey, John Milton, and Collins, to mention but a few of the more famous English authors (see Poe, *Foreign Descriptions*, 195–202).

60 P. Avril, *Reize door verscheidene Staten van Europa en Asia*, trans. H. van Quellenburgh (Utrecht: Anthony Schouten, 1694), 33, 45 (original: P. Avril, *Voyage en divers etats d'Europe et d'Asie* [Paris: Jean Boudot], 1693). See *Reysen*, 188. It is possible that the miles used in *Reysen* were not the 'German' miles used elsewhere in the book; a bewildering array of measurements had the epithet of 'mile' in the Early Modern Age.

61 On the Bollandists, see Knowles, 'Presidential Address.'

62 See [P. Rabus], 'Reize van Avril,' *De Boekzaal van Europe* (July–August 1693) 106–15: 109–10; [P. Rabus], 'Berigt wegens Jan Jansz Struis,' *De Boekzaal van Europe* (May-June 1694) 562–5.

63 'Or otherwise the person who embellished that famous journey under his name' ('*of anders de gene, die op zijn naam die bekende reize verziert heeft*,' Rabus, 'Reize,' 109).

64 See 'Berigt.'

65 *Reysen*, 227.

66 Witsen, *Noord en Oost Tartarye*.

67 My translation of '*Men weet door ervarenheit, dat veele Reizigers, onmagtig om zelfs de penne te voeren* [think of *the beschaafde Penne*], *en 't geen hen is ontmoet in geschrift te stellen, slechts hier en daar iets van optekenende, zonder schikking, vervolg, of t' zaamenhang, wanneer zy t' huis koomen den eenen of anderen Schryfklerk zoeken, om hunne reizen uit veele andere, die omtrent dien zelven streek hebben gehouden, op te maaken; en dus 't geene van de vorige Reizigers met opmerkinge gezien, en lange voorhenen beschreven is, weer voor wat nieus de waereld op te disschen, en breet uit te weiden, dat men zelf zoo veele wonderen ontdekt, en vreemde dingen gezien heeft, als 'er in de Reizen, op hunnen*

naam uitgegeven, geboekt staan. ... Wat beklimmen andere niet al te hooge toppen van gebergten, en wat ontmoeten ze verder al wonderheden op dien steilen tocht; ... ' (E. Ysbrants Ides, *Drie-Jaarige Reize naar China* [Amsterdam: Pieter de Coup, 1710; origin. 1704], 'Voorberegt,' 1–2); the reprint maintains the foreword by the publisher of the first edition, Halma (see Peters, 'Nepotisme,' 100).

68 It is possibly reminiscent of the Dutch *lingua franca* that prevailed along the Baltic shores in those days.

69 Peters, 'Nepotisme,' 100, 100n50; Keuning, 'Nicolaas Witsen,' 107. Ides had been Russian ambassador to China (see Wittram, *Czar* vol. 2, 609n1). The 1747 auction of Witsen's book collection (which perhaps by then was already less complete than at the time of his death thirty years earlier) indicates that he had owned two German and one Dutch editions of the work of Adam Olearius, one of Coyett's account, Struys's first print in Dutch, Avril's book in Dutch, and six copies of his own book on 'Tartary' in its 1692 edition, and five copies of the same book in its 1705 edition (see *Bibliotheca sive catalogus librorum* [Amsterdam: Salomon & Petrus Schouten, 1747], 17, 18, 40).

70 *Reysen*, engraving between 348 and 349. '*A propos du Dessein de Persepolis par Sebastien Serlio, je ne puis m'empêcher d'observer l'impudence avec laquelle on publie des Planches de Dessein le plus grossierrement inventées, comme de vraies représentations. Il y en a plusieurs comme cela dans une Relation de Perse, qui porte le nom de Voyage de Jean Struys. Celle, entre les autres, qu'il appelle le Tombeau Royal de Persepolis, n'a pas un trait de Persepolis; & ce qui est tout à fait extravagant, le Dessein n'a pas un Trait de la Description pour laquelle il est fait; cependant le titre du livre porte que les Planches ont été dessinées par l'Auteur*' (*Voyages de Monsieur le Chevalier Chardin, en Perse, et autres Lieux de l'Orient* vol. 3 [Amsterdam: Jean Louis de Lorme, 1711], 107). See also P.G. Adams, *Travellers and Travel Liars* (Berkeley, CA, 1962), 226, 235.

71 See Keuning, 'Nicolaas Witsen,' 105, 107, and map on 106; *Reysen*, map between 236 and 237. See P. Pekarskii, *Nauka i literatura v Rossii pri Petre Velikomu*, 2 vols (Sint-Petersburg, 1862) vol. 1, 6–7; Witsen, *Noord en Oost Tartarye*. Marion Peters suggests that Struys may have supplied Witsen with maps for his work on Eurasia, but Witsen's possible use of any maps Struys made, or brought back, or of Struys's cartographical notes, is dubious as Struys was at best half-literate, and Witsen even declined to use 'Struys's' (superior) map of the Caspian Sea (see Peters, 'Nepotisme,' 108–10).

72 See Witsen, *Noord en Oost Tartarye*; Witsen, *Architectura navalis* (1690), 198; Peters, 'Nepotisme,' 110). Witsen did write in *Noord en Oost Tartarye* that, at the time of the conquest of Astrakhan by Razin, 'some Hollander sailors, who earlier had been in his Tsarist majesty's service, fled along the Dagestan coast sailing the Caspian Sea, who almost all there were sold as slaves', but then carelessly adds that '[the] prince of Boinak raped, in her husband's presence and that of many others, a Netherlands' Woman, whom he kept as a concubine', and he further wrote in the margin that '[t]his woman has afterwards become so deranged, that she did not want to be released: I knew her in the year 1666 [*sic!*: Witsen was there in 1665] in Moscow' (Witsen, *Noord en Oost Tartarye*, 560; my translation). Witsen appears to have relied much more on the erudite Engelbert Kaempfer (see Gerhard Bonn, *Engelbert Kaempfer*

[1651–1716] [Frankfurt, 2003], 85; Karl Meier-Lemgo, *Engelbert Kämpfer* [Stuttgart, 1937], 155), and on Olearius, to whose work he refers repeatedly (Witsen, *Noord en Oost Tartarye*, 562, 713–24).

73 See *Bibliotheca sive Catalogus Librorum*, 40.

74 De Bruyn, *Reizen over Moskovie*. De Bruyn refers to Herberstein, Olearius, Carlisle (that is, Guy Miege) and Allison, but ignores Struys even when criticizing the poor renditions supplied in the engravings by other travellers such as della Valle and da Silva Figueroa of Persepolis (de Bruyn, *Reizen over Moskovie*, 'Aen de Lezer,' 1).

75 P. Gay, *The Enlightenment* vol. 2 [New York, 1977], 152. See Georges Louis Leclerc, comte de Buffon, *Natural History*, trans. W. Smellie, second edn. vol. 3 (London: Strahan and Cadell, 1785), 66, 80, 88, 89, 98, 119, 120, 145, 162, 310. Its first edition began to appear in 1749 (see Gay, *Enlightenment* vol. 2, 152). Buffon's project brought Witsen's desires for an all-encompassing system to fruition (see Rietbergen, 'Varieties,' 80–1; see also Eisenstein, *Printing Press*, 99). See Chapter 7 as well.

76 Montesquieu, *Lettres persanes* (Paris, 1973; origin., 1721).

77 See Arrenberg, *Naamregister*, 496. For the oblivion, see (Ackerdijck W.C.), 'Verhandeling over den Nederlandschen reiziger Jan, [sic] Jansen Struijs,' *De Fakkel* (Rotterdam, 1826), 92. As an altogether not very reliable source (although Struys was deemed an intrepid man), Struys was still mentioned in the *Biographie universelle, ancienne et moderne* vol. 44 (Paris: L.G. Michaud, 1826), 80–1, and he also received brief mention in A.J. van der Aa, ed., *Biographisch Woordenboek* vol. 6, 329.

78 Ackerdijck writes that Struys lacks any 'early' civilization ['*beschaving*'], see Ackerdijck, 'Verhandeling,' 92.

79 W. van Saerdam, *Jan Jansen Struys. De omzwervingen van een Zaansch zeilmaker* (Den Haag, 1929); D.J. Douwes, ed., *De schriklijke reis van Jan Jansz. Struys* (Zaandijk, 1974). Both books were written by people hailing from the same region of which Struys was a native, the Zaanstreek, a part of the Noorderkwartier.

80 N. Novikov, ed., *Drevnaia rossiiskaia bibliofika* vol. 1 (January–June 1773), 18–23: 18; see also N. Novikov, ed., *Drevnaia rossiiskaia bibliofika* vol. 2 (1788), 463–71.

81 Iurchenko, 'O puteshestviia,' 265–6. The tsar also ordered a translation of Olearius's account.

82 Iurchenko, 'O puteshestviia,' 267, 267n15; Pekarskii, *Nauka i literatura* vol. 1, 224. In 1824 in the journal *Severnyi arkhiv*, A. Kornilovich' stated that in the private library of Count F.A. Tolstoi a handwritten document of 1701 was preserved with notes for the translation into Russian of the travels and the two letters contained in *Reysen* (A. Kornilovich', 'Ian Iansen' Streis' (okonch.),' *Severnyi arkhiv* 7 [1824] 26–40: 40).

83 Iurchenko, 'O puteshestviia,' 269; Kornilovich', 'Ian Iansen' Streis' (okonch.),' 40.

84 Popov, 'Istoriia vozmushcheniia,' 61.

85 See also Adelung, *Kritisch-literärische Übersicht*, 345–6.

86 Popov, 'Istoriia vozmushcheniia,' 61–2.

87 See Scheidegger, *Perverses Abendland*, 27. Solov'ev was more interested in the naval part, while Kliuchevskii used it as a source for Razin's revolt (see

S.M. Solov'ev, *Istoriia Rossii s drevneishikh vremen*, book 6, vols 11–12 [Moscow, 1961; origin. St Petersburg: *c.*1851]; V.O. Kliuchevskii, *Skazaniia inostrantsev o Moskvoskom gosudarstve*, second edn. [Petrograd, 1918]).

88 Kliuchevskii, *Skazaniia*. Kliuchevskii's modest advocacy of the continued use of Western travel accounts as legitimate sources for Russian history has been adopted by many Western-based historians who have edited collections of primary sources on Muscovy (see, for example, Anthony Cross, ed., *Russia under Western Eyes* [London, 1971], 121, which provides a reproduction of one of Struys's engravings).

89 'Puteshestvye po Rossii gollandtsa Streisa,' *Russkii arkhiv* 1 (1880) 17–108. Iurchenko also provided an introduction, see 'Predislovie k Russkomu perevode: Pervoe puteshestvie po Rossii gollandtsa Streisa,' *Russkii arkhiv'* 1 (1880) 5–16.

90 'Kopiia s' pis'ma, pisannogo neizvestnym' litsam' na karable "Orle"... 24 Sentiabria (...) 1669 g.,' *Russkii arkhiv'* 1 (1880), 109–11; 'Kopiia s' pis'ma Davida Butlera, ... , s' spisanyem' vziatiia Astrakhani,' *Russkii arkhiv'* 1 (1880), 111–28.

91 A. Morozov, ed., *Tri puteshestviia Ia.Ia. Streis* (Moskva, 1935).

92 See for example Kotoshikhin et al., *Moskoviia*.

93 See the Introduction.

94 Colley, 'Going Native,' 191.

95 As proposed by van Deursen, *Dorp*, 311, and J. Huizinga, 'De wetenschap der geschiedenis,' in *Verzamelde werken* vol. 7 (Haarlem, 1948–53) 104–72: 145.

14 'Any Soil is the Fatherland for a Courageous Man'

1 *'Omne solum forti viro patria'* Latin inscription on the *'Alte Münze'* house in Friedrichstadt, completed for the Remonstrant leader and religious refugee Adolf van Wael in 1626. It is now the domicile of the local museum. Its meaning seems appropriate for Struys's life and his choice to end his days in Friedrichstadt.

2 At best, Struys might have looked forward to the home for 'aged Sea-men' at Enkhuizen that impressed Sir William Temple during a visit in the 1660s; Temple rather overrated this home's comfort, however, and there were usually limited spaces in such institutions available; see Temple, *Observations*, 88.

3 She may have engendered a separation, or excluded him from the rights to part of her house, because of his prolonged absence (see for the occurrence of such actions by women Schama, *Embarrassment*, 405).

4 For the size of such payments, see van Delft and de Wolf, eds, *Bibliopolis*, 72–3.

5 These were also lean economic times (see Israel, *Dutch Republic*, 881).

6 In the words of Lodewijk Faber (Konovalov, 'Ludwig,' 91); the city surrendered to Miloslavskii on 27 November 1671 (Julian Calendar; see *Polnoe sobranie zakonov'* vol. 1, 868).

7 Konovalov, 'Ludwig,' 91–2. The benign treatment of Faber disappeared once Iakov N. Odoevskii succeeded Miloslavskii as *voevoda* of Astrakhan. Odoevskii was suspicious of Faber's activities in 1670. Next to Faber, Brandt, who

accompanied Faber, probably received some money as well, while Klopper and Dirk Pietersz's widow had petitioned the tsar for their arrears before 1675. Communications between Moscow and Amsterdam were regular: by this time, news and letters from Muscovy by way of the landroute (Wilno-Königsberg–Berlin–Hamburg–Amsterdam), which was slightly faster than the searoute via Riga, took approximately four weeks to reach the Republic (see Wittram, *Czar* vol. 1, 45). And news, of course, also reached Holland through ships sailing on Arkhangel'sk.

8 *Reysen*, 'Opdracht.'

9 *Reysen*, 173. Muscovite archival records published in the nineteenth century indicate payments to tsarist servitors involved in the *Oryol* project (van Bockhoven, Schak, Butler, van Sweeden's widow) in April and May 1669 in Moscow (*DAK* vol. 5, 276–80). That little was paid to the crew after Nizhnii Novgorod (see Chapters 7 and 8) may be inferred from a query by the *Kazanskii Prikaz* to the *Posol'skii Prikaz* of 26 May 1670 as to whether the ship had reached Astrakhan after its departure from Dedinovo the previous year. This query hints at a complete breakdown of communications between Moscow and the port even before Razin's revolt had spread along the Volga (*DAK* vol. 5, 283). The roaming Cossacks north of Astrakhan prevented Moscow from sending any funds toward the Volga mouth; and even if it was up to *voevoda* Prozorovskii to pay the men, he is unlikely to have had sufficient funds when much of Astrakhan's economic activity was interrupted by Cossack unrest.

10 See *NA SG* 1550–1796, 282, pp. 3–3v. The Russian records identify him merely as *pushkar'*: cannonneer (*RGADA* 50/1/9, 1675–6, l.184). In a note of 19 June 1679 demanding payment of wages to the wives of Amsterdam sailors who served the Danish King Christian V, the Dutch Estates-General underlined their poverty by suggesting that *even* constable's wives had been reduced to begging (*Rigsarkivet* [from here: *RA*] *RA Denmark* 301 *TKUA Nederlandene Brevveksling danske Kongehus* 1648–97, 70–5).

11 Loviagin, 'Vvedenie,' clvii, although ibid., clx, notes 7 January 1676; Loviagin, ed., *Posol'stvo*, 15. The Russians urged the Imperial Embassy to depart Moscow quickly to make room for the Dutch party of van Klenk; see 'Skazanie Adolfa Lizeka,' 367.

12 See G.V. Forsten', 'Datskie diplomaty pri Moskovskom dvore vo vtoroi polovine xvii veka,' *Zhurnal' ministerstva narodnago prosveshcheniia* (Sentiabr' 1904) 110–81; Loviagin, ed., *Posol'stvo*, 174.

13 Loviagin, 'Vvedenie,' ix–xi, cxlix–cli. Apart from Coyett's work, a very detailed report by van Klenk to the Estates General (*Verbael*) is available in the Dutch national archives: *NA SG* 8586 (*Verbael*). By November 1676, the dice had been rolled: The first Dutch resident in Moscow, van Keller (a former member of the van Klenk embassy who had stayed behind), wrote to the Estates-General that the Turkish-Tatar threat was given priority at the tsar's court, see *NA SG 7364* (16 November 1676). The Russians were at war with the Ottoman Empire from 1676 to 1681, but they had already engaged in hostilities since the early 1670s (see Oparina, *Inozemtsy*, 327).

14 See Uroff, 'Kotoshikhin,' 87, 94, 125–31. In the Russian view, a republic was suspect and weak by nature. The Russians were not the only ones who had difficulty placing the Dutch among the European states, see W. Roosen,

'Early Modern Diplomatic Ceremonial: A Systems Approach,' *Journal of Modern History* 3 (1980) 452–76: 461.

15 Twice he met Aleksei and twice Fyodor III; see Loviagin, 'Vvedenie,' cxxxvii–i, cxliii–iv.

16 Kotilaine, *Russia's*, 462–3. Klenk's main interlocutors in Moscow were M.Iu. Dolgorukov (d. 1682), A.S. Matveev, *dumnyi d'iak* G.K. Bogdanov (exiled in 1682, d. after 1686), *d'iak* V.I. Bobinin (d. after 1686) and *d'iak* E.I. Ukraintsev (d. 1708) (*NA SG* 8586 [*Verbael*], 64–5; see for these individuals, Birgegard, ed., *Sparwenfeld's*, 293n394, 295n409; Crummey, *Aristocrats*, 55, 89, 197). A.A. Vinius served as translator. Dolgorukov and Matveev were the boyars supervising foreign affairs (mainly via the *Posol'skii Prikaz*), while the other three were secretaries of the foreign office.

17 *RGADA* 50/1 (1675–6), ll.468–468ob.; *RGADA* 50/1 (1675–6), ll.642–3. See also Loviagin, 'Vvedenie,' xxvi-vii, n11, and Morozov, 'Parusnyi Master,' 24.

18 It seems indeed that van Klenk thus also presented not just Struys's request, but also the case of Meindertz, the wives, widows, and orphans before the tsar's negotiators. In 1676, relatives and guardians visited a notary in Amsterdam to ensure that a written affadavit existed that Struys was their executor (see *GAA NA* 4304 [Notary Nicolaes Hemminck], 227–227verso and *GAA NA* 4304 [Notary Nicolaes Hemminck], 247, and Chapter 8). It is not clear why this group visited the notary only on 7 May 1676 (almost a year after the van Klenk embassy took Struys to Moscow for a second time); Struys was identifed in the notary's document as currently *living* in Moscow, which was not the case, but may indicate that he was suspected of collecting all the money for himself. Struys's brief as the groups' representative suggests that they had formed a sort of 'bos', a 'mutual benefit fund which made payments to contributors on sickness, accident or to survivors on death of the member' (Unger, *Dutch Shipbuilding*, 21).

19 See for this habit Uroff, 'Kotoshikhin,' 134; sable was 'the centerpiece of the Russian fur trade', see Hellie, *Economy*, 62. This gift, however, would have amounted to no more than one or two sable pelts for someone of his modest position among the retinue; the pelts were valued at three rubles a piece in Russia at this time (Struys's gift would have been worth 15 guilders, therefore); of course, the sable fetched a much higher price in Holland (see Kotilaine, 'When the Twain,' 1061, Table 11.1). Altogether, then, the hides may have yielded Struys at best a couple of hundred guilders.

20 Loviagin, 'Vvedenie,' clvii.

21 My translation of '*Hier is een hollander genaemt Jan Janse Struijs, welke met de heer Klenck voor hofmeester in Moscovie geweest is, deselve maeckt voor den Coninck van denemarcken een model van een schip dat niet soude cunnen in de gront geschooten werden, hij heeft voores. model waerdig gemaeckt, ende den Coninck getoont, die het wel gevalt, en oock oordeelt practicable te zijn; heeft oock mijn versocht dat ick het wilde comen befestigen* [...], *hetwelcke ick gedaen hebben, met mijn nemende een ervarende Hollandsche meester scheepstimmerman, die het werck prijst, ende meent dat het sal cunnen aengaen waeromme den Coninck geresolveert heeft een groot schip naer dat model te laten bouwen om te zien of het int groot soo wel sal practicable sijn als int kleijn; oock soo gaet hij voor den intentie te hebben van te cunnen maecken dat de Splinters, die den cogel*[s] *veroorsaken geen schade aen het scheepswerk sullen cunne doen*' (*NA SG* 7260

[*liasen Denemarken; resident Jacob le Maire to the Estates General*], 22 October 1678). Jacob le Maire represented Dutch interests in Copenhagen from 1657–59 and 1660–79 (see O. Schutte, ed., *Repertorium der Nederlandse vertegenwoordigers* [Den Haag, 1976], 241–4, 286).

22 See also Prud'homme van Reine, *Schittering*, 359, 365, 370, 372.

23 *RA Denmark* 510 (*Admiraletet-Søetaten*) 1673–1848: *Generalkopibog* 238 (1678), where in the table of contents for 1678 (Register) under the letter J Struys appears. The terms of his contract are found in ibid., 258; the date of its signing was 23 May 1678. He was to serve on the warship Prince George under the command of Admiral Jens Rodstehn. Admirals and vice-Admirals Bielcke, Rodstehn, and Niels or Jens Juel signed for the Danish side, confirming his recruitment (there are two people who initial, neither of whom appears to be Struys). Prud'homme van Reine writes of four naval captains, one captain of a fireship and one lieutenant who join Tromp in leaving Danish service in May 1678, who were replaced by others recruited in the Republic, one of whom apparently was Struys (Prud'homme van Reine, *Schittering*, 374).

24 The Dutch were well versed in producing ship models (see Unger, *Dutch Shipbuilding*, 41).

25 *Reysen*, 293.

26 The discussion in *Reysen* may have resembled the tale Struys told King Christian seven years later; I was unable to find any other records in Copenhagen or elsewhere of Struys's audience with the king about which Le Maire reports in 1678.

27 See Witsen, *Achitectura navalis* (1690), 198–9.

28 Jones, *Anglo-Dutch Wars*, 53.

29 Witsen, *Architectura navalis*, 198–9.

30 Prud'homme van Reine notes how the average sailor in 1676 earned 14 guilders, about 5½ rijksdaalders (see Prud'homme van Reine, *Schittering*, 359). Another source (from October 1678) notes Struys receiving a wage that has more than doubled, possibly because by then he was building his ship for the king (see *RA Denmark* 509 *Søkrigskancelliet (Søetaten)* 1676–7 *Kongelige ekspeditioner verdrørende Søetaten* 50 *Søkrigskancelliet* 50. *Kgl. ekspeditioner* 1678, no. 112 [1678]).

31 *RA Denmark* 509 *Søkrigskancelliet (Søetaten)* 1676–7 *Kongelige ekspeditioner verdrørende Søetaten* 50 *Søkrigskancelliet* 50. *Kgl. ekspeditioner* 1678, no. 112 (1678) and no. 84 (1679).

32 See *RA Denmark* 509 *Søkrigskancelliet Uddrag af Generalkommissariatets kgl. befalingsbøger* (1679–1729), no. 1019 (*Historiske efterretninger om søetaten sagligt ordene oplysninger*).

33 See for example Solov'ev, *Istoriia*, vol. 3 (parts 11–15), 1063–5.

34 On Brandt, see Hughes, *Russia*, 81. On Brandt's regular occupation, see van der Aa, *Biographisch Woordenboek* vol. 1, 366; Honig, *Geschiedenis* vol. 2, 98.

35 See *Kniga ustav' morskoi o vsem' chto kasaetsia dobromu upravleniiu, v' bytnosti flota na mor'* (St Petersburg: Sankt'piterburgskaia Tipografiia, 1720; in Russian and Dutch). Peter's recollection became part of the Naval Statute of 1720 (see F. Veselago, 'Botik' dedushka Russkago flota,' in *Svedeniia o botike dedushke russkago flota za 200 let' s' 1688 po 1888 god'* [St Petersburg, 1888] 19–43: 19). See as well Wittram, *Czar* vol. 2, 27.

36 *Kniga ustav' morskoi*, 28–9. Butler (*Boetler'*), the *Oryol*, and the yacht (as well as a *'galjoet,'* which may refer to the vessel built at Astrakhan in 1670) are all named in the treatise. It also recalled how both *'Ivan' Termunt"* and the 'ship's carpenter and constable *Karshten Brant"* had returned to Muscovy from Persia after their flight from Astrakhan (ibid., 29, 34–41. *Svedeniia o botike*, 6–9).

37 Hans de Hofman, *Den Danske Atlas Eller Konge-Riget Dannemark* (Kiøbenhavn, 1781) vol. 7, part 2, 707. Nowadays Friedrichstadt is located in Kreis Nordfriesland in the Federal Republic of Germany. Frederikstadt/Friedrichstadt was in Schlesvig on the border with Holstein and almost encircled by Gottorp territories, and thus factually located outside of Dithmarschen, even if in 1694 Rabus wrote how Struys lived in this region (see the maps on 107 in K. Jespersen, O. Feldbaek, *Revanche og Neutralitet: 1648–1814* [København, 2002], 107; Rabus, 'Berigt,' 562–5).

38 At that time, the duchy was also considered to be a fief of the King of Denmark (see E. Ekman, 'The Danish Royal Law of 1665,' *Journal of Modern History* 2 [1957] 102–7: 105). For the complications regarding sovereignty over (parts of) Schleswig and Holstein, see P. Torntoft, 'William III and Denmark-Norway, 1697–1702,' *The English Historical Review*, 318 (January 1966), 1–25.

39 De Hofman, *Den Danske Eller* vol. 7, part 2, 707.

40 See Olearius, *Travels*, 33 and 33n1. Oral permission was granted in 1619, while the written licence to establish the town was issued in 1620 (J.P. Trap, *Statistisk-topographisk Beskrivelse of Hertugdømmet Slesvig* vol. 2 [Kjøbenhavn, 1864], 619). It seemed Hugo de Groot (Grotius) planned to move there after his escape from captivity in the Republic (ibid., 700).

41 See map between 701 and 702 in Trap, *Statistisk-topographisk Beskrivelse*. In 1756 the town had 452 families, 20 Remonstrant, 38 Mennonite, 14 Catholic, 24 Jewish, and 356 Lutheran (ibid. 702). Once he had affirmed his rule over the town, King Frederick IV (r. 1699–1730) of Denmark ordered its city council to consist of 4 Lutherans and 2 Remonstrants and 2 Mennonites (previously it had been all Remonstrant), two of whom were mayors (ibid., 702–3).

42 Faber was the Swedish envoy who persuaded the shah to increase the shipping of silk via the Caspian Sea in the 1680s (see Birgegard, ed., *J.G. Sparwenfeld's Diary*; Kotilaine, *Russia's*, 332–3).

43 See Chapter 6. A Jan Jansen was buried in the seventeenth-century Lutheran Church in Friedrichstadt, but whether this is Jan Struys or someone else is unclear (see 'Der lutherische Kirchhof,' *Mitteilungsblatt der Gesellschaft für Friedrichstädter Stadsgeschichte* 10 [1976–77] 1–14: 13).

44 Rabus, 'Berigt,' 562. Another sign of his status and wealth may be that he was buried in a grave inside the Lutheran Church in Friedrichstadt, if the Jan Jansen interred there is indeed Struys.

Conclusion

1 See Israel, *Dutch Primacy*, 1, 3–4; Adams, *Familial*, 46–7.

2 Although Roma were targeted (Schama, *Embarrassment*, 595–6).

3 Schama, *Embarrassment*, 596.

4 It utterly lacks reference to '[o]mens, portents, and oracles,' which Schama found to be the norm, signs that were even taken seriously by Pieter Rabus (see Schama, *Embarrassment*, 147).
5 See Bekker, *De betoverde wereld*.
6 A project that was only successfully executed in the nineteenth century.
7 See van Deursen, *Dorp*, 298–300.
8 It should be noted, too, that *Reysen* has few references to superstition, which Schama argues to be widespread among sailors (Schama, *Embarrassment*, 148).
9 Schama, *Embarrassment*, 121.

Bibliography

The trilogy's full titles

Struys, J.J. *Drie aanmerkelijke en seer rampspoedige Reysen, door Italien, Griekenlandt, Lijflandt, Moscovien, Tartarijen, Meden, Persien, Oost-Indien, Japan, en verscheyden andere Gewesten. Waar in vertoont werden, behalven een nauwkeurige/en tot omstandige beschrijvinge der gemelde Landen/en 't geen tot haar nature gehoort/seer wonderlijke/en waarachtige toevallen den Auteur overgekomen door Schipbreuken/ Plonderingen/Slavernije onder de Turken, en Persianen, sware Hongers-noot/Pijniging/en andere ongemakken. Aangevangen anno 1647. en voor de Derde, of laatste Reys t'Huys gekomen 1673. begrijpende soo in alles den tijdt van 26 jaren. Nevens twee Brieven, particulierlijk verhandelende het overgaan van Astracan, en 't geene aldaar omtrent is voorgevallen; En daar in ook een verhaal der elenden/en sware ongemakken/uytgestaan by D. Butler, door hem selfs geschreven uyt Ispahan.* Met verscheydene curieuse koopere Platen, door den Auteur selfs na het leven geteekent, verçiert (t' Amsterdam, By Jacob van Meurs, op de Keysers-Graft, en Johannes van Someren, in de Kalverstraat, 1676. Met Privilegie).

Heiden, Frans Jansz van der. *Vervarelyke schip-breuk van't Oost-Indisch jacht ter Schelling, onder het landt van Bengale* (Amsterdam: J. Meuss [*sic*] and J. van Someren, 1675).

Schouten, Wouter. *Oost Indische Voyagie, Vervattende veel voorname voorvallen en ongemeene vreemde geschiedenissen, bloedige Zee- en Landt-gevechten tegen de Portugeesen en Makassaren: Belegering, Bestorming, en Verovering van veel voorname Steden en Kasteelen* (Amsterdam: J. van Meurs and J. van Someren, 1676).

Reysen's first translations

Joh. Jansz. Strauszens sehr schwere, wiederwürtige, und denckwürdige reysen, durch Italien, Griechenland, Lifland, Moscau, Tartarey, Meden, Persien, Türckey, Ost-Indien, Japan und unterschiedliche andere Länder. Trans. A.M. [Andreas Müller] (Amsterdam: J. von Meurs and J. von Sommern, 1678).

Struys, J.J. *Unglückliche Schiffs-Leute, oder mirkwirdige Reise zwenzing Holländeren, welche auss Befehl des christlichen Reussichen Keisers, in der Moscau, ein grosses Schiff gebauet, die Kaspische See damit zubefaaren*. Ed. Jakob Redinger (Zürich: Heinrich Müller, 1679). The first pirated edition.

Les Voyages de Iean Struys en Moscovie. en Tartarie. aux Indes, et en d'autres pays étrangers. Trans. Monsieur Glanius (Amsterdam: Veuve de Jacob van Meurs, 1681).

The perillous and most unhappy voyages of John Struys; through Italy, Greece, Lifeland, Moscovia, Tartary, Media, Persia, East-India, Japan, and other places in Europe, America and Asia. ... To which are added 2 narratives sent from Capt. D. Butler, relating to the taking in of Astrachan by the Cosacs (London: Samuel Smith, 1683).

The voiages and travels of John Struys through Italy, Greece, Muscovy, Tartary, Media, Persia, East-India, Japan, and other countries in Europe, Africa, and Asia (London: Abel Swalle, 1684).

NB: Below only publishers of texts before 1800 (as far as known) are indicated to help identifying such rare texts.

Aa, A.J. van der. *Biographisch Woordenboek der Nederlanden* (Haarlem, 1852–78).
(Ackerdijck W.C.). 'Verhandeling over den Nederlandschen reiziger Jan Jansen Struijs.' *De Fakkel* (Rotterdam, 1826).
Adams, Julia. *The Familial State: Ruling Families and Merchant Capitalism in Early Modern Europe* (Ithaca, NY, 2005).
Adams, Percy G. *Travelers and Travel Liars* (Berkeley, 1962).
——. *Travel Literature and the Evolution of the Novel* (Lexington, KY, 1983).
Adelung, F. *Kritisch-literärische Übersicht der Reisenden in Rußland bis 1700, deren Berichte bekannt sind.* 2 vols (St Petersburg, 1846).
Amburger, Erik. *Die Familie Marselis* (Giessen, 1957).
Ames, G.J. 'Colbert's Indian Ocean Strategy.' *French Historical Studies* 3 (1990) 536–59.
Anderson, B. *Imagined Communities*. Rev. edn. (London, 2006).
Anderson, Matthew S. *Britain's Discovery of Russia, 1553–1815* (London, 1958).
Anderson, R.C. *Naval Wars in the Levant, 1559–1853* (Princeton, NJ, 1952).
Andrade, T. 'Political Spectacle and Colonial Rule: The *Landdag* on Dutch Taiwan, 1629–1648.' *Itinerario* 3 (1997) 57–93.
——. 'The Rise and Fall of Dutch Taiwan, 1624–1662.' *Journal of World History* 4 (2006) 429–50.
Anemone, A. 'The Monsters of Peter the Great: The Culture of the St. Petersburg Kunstkamera in the Eighteenth Century.' *Slavic and East European Journal* 4 (2000) 583–602.
Anisimov, E. *The Reforms of Peter the Great* (Armonk, NY, 1993).
Appleby, J.H. 'The Royal Society and the Tartar Lamb.' *Notes and Records of the Royal Society of London* 1 (1997) 23–34.
Arel, M.S. 'The Archangelsk Trade.' In *Modernizing Muscovy*. Eds. J. Kotilaine, M. Poe (London, 2004) 175–201.
Arkheograficheskaia kommissia. *Akty istoricheskie*. 5 vols (St Petersburg, 1841–2).
——. *Dopolneniia k aktam istoricheskim*. 12 vols (St Petersburg, 1846–77).
Arrenberg, R. *Naamregister van de bekendste en meest in gebruik zijnde Nederduitsche boeken* (Rotterdam: G. Arrenberg, 1788).
Augustine, Saint. *The Confessions of Saint Augustine* (London, 1961).
Aulnoy (Marie-Catherine) Madame d'. *Relation du voyage d'Espagne* (La Haye: H. van Bulderen, 1691).
Aune, M.G. 'Early Modern European Travel Writing after Orientalism.' *Journal for Early Modern Cultural Studies* 2 (2005) 120–138.
Avrich, P. *Russian Rebels: 1600–1800* (New York, 1972).
Avril, P. *Voyage en divers états d'Europe et d'Asie* (Paris: Jean Boudot, 1693).
——. *Reize door verscheidene Staten van Europa en Asia, ..., gedaan sedert den Jaare 1685. tot 1692*. Trans. H. van Quellenburgh (Utrecht: A. Schouten, 1694).
Avvakum. *Archpriest Avvakum: The Life Written by Himself*. Trans. and ed. K.N. Brostrom (Ann Arbor, MI, 1979).
Awad, A.M. 'The Gulf in the Seventeenth Century.' *Bulletin of the British Society of Middle Eastern Studies* 2 (1985) 123–34.

Baddeley, J.F. *Russia, Mongolia, China*. 2 vols (New York, 1964; first edn., 1919).
Bagrow, Leo. 'Semyon Remezov: A Siberian Cartographer.' *Imago Mundi* 11 (1954) 111–25.
——. 'Italians on the Caspian.' *Imago Mundi* 13 (1956) 2–10.
Bailey, Lloyd R. 'Wood from 'Mount Ararat': Noah's Ark?' *The Biblical Archeologist* 4 (1977) 137–46.
Bakhtin, M.M. *Rabelais and His World* (Cambridge, MA, 1968).
——. *Speech Genres and Orther Late Essays* (Austin, TX, 1986).
Bantysh-Kamenskii, N.N. *Obzor vneshnikh snoshenii Rossii s derzhavami inostrannymi*. 4 vols (Moscow, 1894–1902). (Vol. 1: *Avstriia, Angliia, Vengriia, Gollandiia, Daniia, Ispaniia* [Moscow, 1894]).
Barbour, V. *Capitalism in Amsterdam in the Seventeenth Century* (Baltimore, 1950).
Barendse, R. *The Arabian Seas: 1640–1700* (Armonk, NY, 2002).
Baron, S.H. 'Who Were the Gosti?' *California Slavic Studies* 7 (1973) 1–40.
——. 'Osip Nepea and the Opening of Anglo-Russian Commercial Relations.' *Oxford Slavonic Papers*. New Series 11 (1978) 42–63.
——. 'Shipbuilding and Seafaring in Sixteenth-Century Russia.' In *Essays in Honor of A.A. Zimin*. Ed. D.C. Waugh. (Columbus, OH, 1983) 102–29.
——. *Explorations in Muscovite History* (Hampshire, 1991).
——. 'A.L. Ordin-Nashchokin and the Orel Affair.' In S.H. Baron, *Explorations in Muscovite History*, 1–22.
Barran, Th. Review of *Une infinie brutalité: L'image de la Russie dans la France des XVIe et XVIIe siècles* by M. Mervaud and J.-C. Roberti. *Slavic and East European Journal* 1 (1994) 205–7.
Barrett, Th.M. 'Lines of Uncertainty: The Frontiers of the North Caucasus.' *Slavic Review* 3 (1995) 578–601.
——. *At the Edge of Empire: The Terek Cossacks and the North Caucasus Frontier, 1700–1860* (Boulder, CO, 1999).
Barsov, Pavel. 'Predislovie perevodchika.' In A. Olearius, *Podrobnoe opisanie puteshestviia golshtinskago posol'stva v Moskoviiu i Persiiu*. Trans. P. Barsov'. (Moskva, 1870) i-xiv.
Barthes, R. *Image, Music, Text*. Ed. and trans. S. Heath (London, 1977).
Basilevich, K.V. 'Novotorgoyyi ustav, 1667 g.' *Izvestiia Akademii Nauk SSSR. Otdelenie obshchestvennykh nauk* 2 (1932) 589–622.
Bayle, P. *Projet et fragment d'un dictionnaire critique* (Rotterdam: R. Leers, 1692).
Begichev', K.N. *Kavkazskie puteshestvenniki proshlykh stoletii. Iogan' Ioganson Striuis' (1670 g.)* (Tiflis, 1900).
Behn, Aphra. *Oroonoko, or the Royal Slave*. In *The Norton Anthology of English Literature*. Vol. C: *The Restoration and the Eighteenth Century*. Eds. L. Lipking and J. Noggle (London, 2006) 2183–226. (Original: *Oroonoko, or The Royal Slave, a true History* [London: Canning, 1688]).
Behr, Johann von der. *Reise nach Java, Vorder-Indien, Persien und Ceylon, 1641–50*. Ed. S.P. L'Honoré Naber (Den Haag, 1930; original: Jena: U. Spaltholtzen, 1668).
Bekker, B. *De betoverde weereld, zijnde een grondig ondersoek van't gemeen gevoelen aangaande de geesten* (Amsterdam: van den Dalen, 1691–3).
Belov, M.I. 'Rossiia i Gollandiia v poslednei chetverti XVII v.' In *Mezhdunarodnye sviazi Rossii v XVII-XVIII vv*. Ed. L. Beskrovnyi (Moskva, 1966) 58–83.
Benjamin, W. *Illuminations* (New York, 1969).
Bernier, F. *Histoire de la dernière révolution des états du Grand Mogol* (Paris: N.p., 1670).

Bibliopolis: Geschiedenis van het gedrukte boek in Nederland. Eds. M. van Delft, C. de Wolf (Zwolle, 2003). English edition available at www.bibliopolis.nl/handboek/zoekscherm_handboek_en.html (Accessed 20 November 2007).

Bibliotheca sive Catalogus Librorum ... Nicolaus Witsen (Amsterdam: Salomon & Petrus Schouten, 1747).

Bijdragen tot eene Nederlandsche Bibliographie uitgegeven door het Frederik Muller-fonds. Ed. P.A. Tiele. Vol. 1. *Land- en volkenkunde* (Amsterdam, 1884).

Biographie universelle, ancienne et moderne vol. 44 (Paris: Michaud, 1826) 80–1.

Blackburn, R. 'The Old World Background to European Colonial Slavery.' *William and Mary Quarterly* 1 (1997) 65–102.

Blaeu, Joan. *Atlas Maior van 1665* (Köln, 2005).

Blaeu's the Grand Atlas of the 17th-Century World. Ed. J. Goss (New York, 1991).

Bloem, J.C. *Gedichten* (Amsterdam, 1979).

Blussé, L. 'Op zoek naar een verdwenen manuscript.' In *Johan Nieuhofs Beelden v an een Chinareis 1655–1657*. Eds. L. Blussé, R. Falkenburg (Middelburg, 1987) 9–20.

——. 'No Boats to China: The Dutch East-India Company and the Changing Pattern in the China Sea Trade, 1635–1690.' *Modern Asia Studies* 1 (1996) 51–76.

Bogoiavlenskii, S.K. 'Moskovskaia Nemetskaia sloboda.' *Izvestiia AN SSSR. Seriia istorii i filosofii* 3 (1947) 220–32.

Bogoslovskii, M.M. *Petr I. Materialy dlia biografii* vols 1 and 2 (Moskva, 1941–2).

Bonn, G. *Engelbert Kaempfer (1651–1716): Der Reisende und sein Einfluß auf die europäische Bewußtseinsbildung über Asien* (Frankfurt, 2003).

Bontekoe, W. *Memorable Description of the East Indian Voyage, 1618–25*. Ed. P. Geyl (London, 1929; Dutch origin.: Hoorn: J.J. Deutel, 1646).

Boogaart, E. van den. *Civil and Corrupt Asia: Image and Text in the Itinerario and the Icones of Jan Huygen van Linschoten* (Chicago, 2003).

Booy, E.P. de. *De weldaet der scholen. Het plattelandsonderwijs in de provincie Utrecht van 1580 tot het begin der 19e eeuw* (Utrecht, 1977).

Bostoen, K. 'Mogelijk een vroom gelaat en een vroom gewaad, maar beslist onvrome praat: de aankomend predikant Johannes Ulaeus in zijn Alkmaars "Collegij".' *Fumus* 1 (2005) 22–55.

Boterbloem, K. 'The Genesis of Jan Struys's "Perillous Voyages" and the Business of the Book Trade in the Dutch Republic.' *Papers of the Bibliographical Society of America* 1 (2008) 5–28.

Bougainville, L. de. *A Voyage Around the World*. Trans. J.R. Foster (London: Nourse and Davies, 1772).

Bourdieu, P. *Outline of a Theory of Practice* (Cambridge, 1977).

Boxer, C.R. *The Dutch Seaborne Empire* (New York, 1965).

——. *Dutch Merchants and Mariners in Asia, 1602–1795* (London, 1988).

Boyarskaia kniga 1658 goda. Ed. N.M. Rogozhin (Moskva, 2004).

Braddick, M.J. 'Government, War, Trade, and Settlement.' In *The Origins of Empire*. Ed. N. Canny (Oxford, 1998) 286–308.

Bragt (Bracht), T.J. van. *Het bloedig toneel, of Martelaers Spiegel der Doops-Gezinde of weerloose Christenen* (Amsterdam: H. Sweerts, J. ten Hoorn et al., 1685).

Brancaforte, Elio. *Visions of Persia: Mapping the Travels of Adam Olearius* (Cambridge, MA, 2003).

Brands, M.C. Review of *Op het Breukvlak van Twee Eeuwen* by Jan Romein. *History and Theory* 1 (1970) 116–21.

(Brandt, Gerard). *The Life of Michiel de Ruyter, Admiral of Holland* (London: Dorman Newman, 1677).

Braudel, F. *Civilization and Capitalism: 15th to 18th Centuries*. 3 vols (New York, 1981–2).

Bredero, G.A. *De kluchten*. Ed. A.van Rijnbach (Amsterdam, 1926).

Breet, M. *De 'Oost-Indische Voyagie' van Wouter Schouten* (Zutphen, 2003).

Brenner, Peter J. 'Der Mythos des Reisens. Idee und Wirklichkeit der europäischen Reisekultur in der Frühen Neuzeit.' In *Neue Impulse der Reiseforschung*. Ed. Michael Maurer (Berlin, 1999) 13–61.

Brockey, L.M. *Journey to the East* (Cambridge, MA, 2007).

Bronnen tot de geschiedenis van het bedrijfsleven en het gildewezen van Amsterdam. Ed. J.G. van Dillen. Vol. 3 (Den Haag, 1974).

Brown, Laura. 'Reading Race and Gender: Jonathan Swift.' *Eighteenth-Century Studies* 4 (1990) 425–43.

Brown, P.B. 'Bureaucratic Administration in Seventeenth-Century Russia.' In *Modernizing Muscovy*. Eds. J. Kotilaine and M. Poe (London, 2004) 57–78.

Brugmans, H. *Geschiedenis van Amsterdam*. Second edn. 6 vols (Utrecht, 1972–3).

Bruin, Cornelis de. *Reizen over Moskovie: door Persie en Indie* (Amsterdam: R. and G. Wetstein et al., 1714; origin.: Amsterdam: W. and D. Goeree, 1711).

Bruyn, C. Le. *Travels into Muscovy, Persia and Part of the East-Indies, Containing An Accurate Description of whatever is most remarkable in those Countries. ... To which is added, An Account of the Journey of Mr. Isbrants, Ambassador from Muscovy, through Russia and Tartary, to China; together with remarks on the Travels of Sir John Chardin, and Mr. Kempfer, and a Letter written to the Author on that Subject*. 2 vols (London: A. Bettesworth et al., 1737).

Buffon, George Louis le Clerc, Count of. *A Natural History, General and Particular*. Trans. W. Smellie. second edn. 9 vols (London: Strahan and Cadell, 1785).

——. *A Natural History, General and Particular*. Trans. W. Smellie. 9 vols (London: Strahan and Cadell, 1791).

——. *Oeuvres complètes*. Vol. 1 (Paris, 1848).

Bunyan, John. *The pilgrim's progress from this world, to that which is to come* (London: N. Ponder, 1678).

Burke, P. *Popular Culture in Early Modern Europe*. Revised edn. (Aldershot, 1994).

——. *Venice and Amsterdam: A Study of Seventeenth-Century Elites*. Second edn. (Cambridge, 1994).

——. *Towards a Social History of Early Modern Dutch* (Amsterdam, 2005).

Buruma, I. *Murder in Amsterdam* (New York, 2006).

Bushkovitch, P. *The Merchants of Moscow, 1580–1650* (Cambridge, 1980).

——. *Peter the Great: The Struggle for Power, 1671–1725* (Cambridge, 2001).

Calmard, J. 'Shi'i Rituals and Power II. The Consolidation of Safavid Shi'ism: Folklore and Popular Religion.' In *Safavid Persia: The History and Politics of an Islamic Society*. Ed. Charles Melville (London, 1996) 139–90.

Campbell, W. *Formosa under the Dutch, Described From Contemporary Records* (London, 1903).

Caron, F., J. Schouten. *A True Description of the Kingdoms of Japan and Siam*. Ed. C.R. Boxer. New York, 1971 (first edn.: London: S. Broun, J. L'Ecluse, 1663).

Cats, J. *Het Spaens Heydinnetje* (Zwolle, 1963; originally premiered in 1637).

Cawley, R. 'Sir Thomas Browne and His Reading.' *PMLA* 2 (1933) 426–70.

Çelebi, E. *The Intimate Life of an Ottoman Statesman: Melek Ahmed Pasha (1588–1662)*. Ed. R. Dankoff (Albany, NY, 1991).

Cervantes, M. de. *Don Quijote*. Trans. B. Raffel (New York, 1999).

Chardin, J. *Le couronnement de Soleïmaan Troisième roy de Perse* (Paris: Claude Barbin, 1671).

Chardin, John. *A new and accurate description of Persia, and other eastern nations. ... Illustrated with a great number of cuts of habits, towns, beasts, birds, ruins, prospects at sea, &c.* 2 vols (London: A. Bettesworth et al., 1724).

Christensen, S.T. 'Introduction.' *Scandinavian Journal of History* 3–4 (2003) 151–64.

A Chronicle of the Carmelites in Persia and the Papal Mission of the XVIIth and XVIIIth Centuries. 2 vols (London, 1939).

Clay, C.G.A. *Economic Expansion and Social Change: England, 1500–1700*. 2 vols (Cambridge, 1984).

Cleland, John. *Fanny Hill or Memoirs of a Woman of Pleasure*. Ed. P. Wagner. Harmondsworth, 1985 (First edition: London, G. Fenton, 1749).

Colley, Linda. 'Going Native, Telling Tales: Captivity, Collaborations and Empire.' *Past and Present* 168 (2000) 170–93.

Collins, S. *The Present State of Russia*. Ed. Marshall Poe (London: Dorman Newman, 1671).

Cook, Harold J. *Matters of Exchange: Commerce, Medicine and Science in the Dutch Golden Age* (New Haven, 2007).

Cook, James. *A Voyage Toward the South Pole, and Round the World*. 2 vols (London: Strahan and Cadell, 1777).

Court, Pieter de la. *The True Interest and Political Maxims of the Republic of Holland*. New York, 1972 (first Dutch edition: Amsterdam, J.C. van der Gracht, 1662; first English edition: London: J. Nourse, 1746).

(Coyett, Balthasar). *Historisch Verhael of Beschryving van de Voyagie gedaan onder de Suite van den Heere Koenraad van Klenck* (Amsterdam: Jan Claesz. ten Hoorn, 1677).

Cracraft, J. *The Revolution of Peter the Great* (Cambridge, MA, 2003).

Crummey, R.O. *Aristocrats and Servitors: The Boyar Elite in Russia, 1613–1689* (Princeton, NJ, 1983).

Dalrymple, W. 'Homer in India: The Oral Epics of Rajasthan.' *The New Yorker* (20 November 2006) 48–55.

Dam, Pieter van. *Beschryvinge van de Oostindische Compagnie*. Ed. F.W. Stapel. 7 vols (Den Haag, 1927–1954).

Dan, Pierre. *Historie van Barbaryen en des zelfs Zeerovers, behelzende een beschrijving van de koninkrijken en steden Algiers, Tunis, Salé en Tripoli*. Trans. G. van Broekhuizen (Amsterdam: Jan ten Hoorn, 1684).

Danckaert, J.P. *Beschrijvinghe van Moscovien ofte Ruslandt* (Amsterdam: Broer Jansz, 1615).

——. *Beschrijvinghe van Moscovien ofte Ruslandt* (Amsterdam: G. Saeghman, 1646).

——. *Reyze, ofte voyagie, door Moscovien ofte Rus-Landt* (Dordrecht: V. Caeymacx, 1652).

Dapper, O. *Historische beschryving der stadt Amsterdam* (Amsterdam: van Meurs, 1663).

——. *Herodoot van Halikarnassus negen boeken der historien, gezegt de Musen... Beneffens een beschrijving van Homeers leven, door den zelven Herodoot*. Uit het Grieks vertaelt door Dr. O.D. (Amsterdam: H. Sweerts, 1665).

——. *Naukeurige Beschrijvinge der Afrikaensche gewesten* (Amsterdam: van Meurs, 1668).
——. *Gedenkwaerdig Bedryf der Nederlandsche Oostindische Maatschappye* (Amsterdam: van Meurs, 1670).
——. *Asia of Naukeurige Beschryving van het Rijk des Grooten Mogols en een groot gedeelte van Indiën behelsende de landschappen van Kandahar, Kabul... beneffens een volkome Beschryving van geheel Perzie, Georgie, Mingrelie en andere Gebuurgewesten* (Amsterdam: van Meurs, 1672).
——. *Naukeurige Beschrijvinge der Afrikaensche gewesten.* Second edn. (Amsterdam: van Meurs, 1676).
——. *Naukeurige beschryving van gantsch Syrie en Palestyn of Heilige Lant.* (Amsterdam: van Meurs, 1677).
——. *Naukeurige beschryving der eilanden, in de Archipel der Middelantsche Zee, en ontrent dezelve, gelegen: waer onder de voornaemste Cyprus, Rhodus, Kandien, Samos, Scio, Negroponte, Lemnos, Paros, Delos* (Amsterdam: Wolfgangh, Waesbergen, et al., 1688).
——. *Naukeurige beschryving van Morea, eertijts Peloponnesus, en de eilanden, gelegen onder de kusten van Morea, en binnen en buiten de Golf van Venetiën: waer onder de voornaemste Korfu, Cefalonia, Sant Maura, Zanten* (Amsterdam: Wolfgangh, Waesbergen, et al., 1688).
Darnton, R. *The Literary Underground of the Old Regime* (Cambridge, MA, 1982).
Davies, D.W. *A Primer of Dutch Seventeenth Century Overseas Trade* (The Hague, 1961).
Davies, Norman. *God's Playground: A History of Poland.* Rev. edn. Vol. 1 (New York, 2005).
Davis, R.C. *Christian Slaves, Muslim Masters* (Basingstoke, 2004).
Defoe, Daniel. *Robinson Crusoe: An Authoritative Text; Background and Sources; Criticism.* Ed. Michael Shinagel (New York, 1975).
Dekker, R. 'Van Grand Tour tot treur- en sukkelreis: Nederlandse reisverslagen van de 16e tot begin 19e eeuw.' *Opossum* 13–14 (1994) 8–24.
——. *Childhood, Memory and Autobiography in Holland from the Golden Age to Romanticism* (New York, 2000).
——. *Humour in Dutch Culture of the Golden Age* (Basingstoke, 2001).
Demkin, A.V. 'Vvedenie.' In *Zapadnoevropeiskie kuptsy i ikh tovary v Rossii xvii veka.* Eds A.V. Demkin, A.A. Preobrazhenskii (Moskva, 1992) 3–11.
——. *Zapadnoevropeiskoe kupechestvo v Rossii v xvii v.* 2 vols (Moskva, 2004).
Deursen, A. Th. van. *Plain Lives in a Golden Age: Popular Culture, Religion and Society in Seventeenth-Century Holland* (Cambridge, 1991).
——. *Een dorp in de polder: Graft in de zeventiende eeuw.* Sixth edn. (Amsterdam, 2003).
Dittrich, Z.R. 'Illusies, misverstanden, wanklanken.' In *Rusland in Nederlandse ogen.* Eds J. Driessen et al. (Amsterdam, 1986) 33–50.
Dixon, S. *The Modernisation of Russia 1676–1825* (Cambridge, 1999).
Dozy, C.M. 'Olfert Dapper.' *Tijdschrift voor het Aardrijkskundig Genootschap* 3 (1887) 414–35.
Drevnaia rossiiskaia bibliofika. Ed. N. Novikov. Vol. 1 (January-June, 1773) 18–23.
——. Vol. 2. Part 3 (1788) 463–71.
Driessen-van het Reve, J.J. *De Kunstkamera van Peter de Grote* (Hilversum, 2006).

Dutch-Asiatic Shipping in the 17th and 18th Centuries. Eds J.R. Bruijn, F. Gaastra, and I. Schoffer, eds., 3 vols (Den Haag, 1979–1987).

Edwardes, M. *Ralph Fitch, Elizabethan in the Indies* (New York, 1973).

Eeghen, I.H. van. *De Amsterdamse boekhandel 1680–1725*. 5 vols (Amsterdam, 1960).

Eekhoff, W. 'Friezen in Rusland vóór en onder Keizer Peter den Groote.' *Nieuwe Friesche Volks-Almanak* (1859) 29–39.

Eekman, Th. 'Muscovy's International Relations in the Late Seventeenth Century: Johan van Keller's Observations.' *California Slavic Studies* 14 (1992) 44–67.

Eisenstadt, S.N. *Modernization: Protest and Change* (Englewood Cliffs, NJ, 1966).

Eisenstein, Elizabeth. *The Printing Press as an Agent of Change*. 2 vols in one. (Cambridge, 2005).

Ekman, E. 'The Danish Royal Law of 1665.' *Journal of Modern History* 2 (1957) 102–7.

Elias, J.E. *De vroedschap van Amsterdam, 1578–1795* (Amsterdam, 1963).

Elias, N. *The Civilizing Process*. Revised edn. (Malden, MA, 2000).

Het ellendigh leven der Turken, Moscoviters en Chinesen, aende Christenheyt vertoont. 's-Gravenhage: J. Tongerloo, 1664 (first edn.: 's-Gravenhage, I. Sonnevelt, 1663).

Ellersieck, H.E. 'Russia under Aleksei Mikhailovich and Fedor Alekseevich 1645–1682: The Scandinavian Sources.' Unpublished Ph.D. dissertation (Los Angeles, CA: UCLA, 1955).

Emerson, John. 'Sir John Chardin.' *Encyclopaedia Iranica*. Ed. Ehsan Yarshater. Available at http://www.iranica.com/newsite/. (Accessed 12 January 2007).

Emmer, P.C. 'The History of the Dutch Slave Trade: A Bibliographical Survey.' *Journal of Economic History* 3 (1972) 728–47.

Failed Transitions to Modern Industrial Society: Renaissance Italy and Seventeenth Century Holland. Eds F. Krantz and P.M. Hohenberg (Montreal, 1975).

Faroqhi, S. *The Ottoman Empire and the World Around It* (London, 2004).

Febvre, L. and H.J. Martin. *The Coming of the Book* (London, 1976).

'Feestelijke optocht in Wormer, ter ere van de vrede van Munster.' *Digitale Atlas Geschiedenis*. Available at http://www.digitaleatlasgeschiedenis.nl. (Accessed 18 February 2007).

Fielding, Henry. *Joseph Andrews and Shamela*. Ed. Judith Hawley (London, 1999).

Flacourt, Étienne de. *Histoire de la grande isle Madagascar*. Ed. C. Allibert. Paris, 1995 (first edn.: Troyes et Paris: N. Oudot et al., 1661).

Fletcher, G. *Of the Russe Commonwealth*. Eds R. Pipes and J. Fine (Cambridge, MA, 1966).

Fletcher, R. *The Cross and the Crescent* (London, 2003).

Floor, W. 'Fact or Fiction: The Most Perilous Journeys of Jan Jansz. Struys.' In *Études Safavides*. Ed. Jean Calmard (Paris, 1993) 57–68.

——. 'The Dutch and the Persian Silk Trade.' In *Safavid Persia: The History and Politics of an Islamic Society*. Ed. C. Melville (London, 1996) 323–68.

Floor, W., and P. Clawson. 'Safavid Iran's Search for Silver and Gold.' *International Journal of Middle East Studies* 3 (2000) 345–68.

Foner, E. 'American Freedom in a Global Age.' *American Historical Review* 1 (2001) 1–16.

Forsten', G.V. 'Datskie diplomaty pri Moskovskom dvore vo vtoroi polovine xvii veka.' *Zhurnal' ministerstva narodnago prosveshcheniia* (Sentiabr' 1904) 110–81.

Foucault, M. *Discipline and Punish: The Birth of the Prison* (New York, 1995).
Friedman, E.G. 'Christian Captives at 'Hard Labor' in Algiers, 16th-18th Centuries.' *International Journal of African Historical Studies* 4 (1980) 616–32.
(Anonymous) 'From the Journal des Scavans, set forth July 24. 1681. "Les Voyages de Jean Struys en Muscovie," ' *Weekly Memorials for the Ingenious* (January 16, 1682) 5–7.
Gaisinovich, A. 'Vvedenie.' In *Tri puteshestviia Ia.Ia. Streis*. Ed. A. Morozov (Moskva, 1935) 7–22.
Gardiner, R.A. 'John Struys "New Card of the Caspian Sea" 1668.' *Geographical Journal* 4 (1969) 631–2.
Gay, Peter. *The Enlightenment: An Interpretation*. 2 vols (New York, 1977).
Gebhard, J.F. *Het leven van Mr Nicolaas Cornelisz. Witsen (1641–1717)*. 2 vols (Utrecht, 1881).
Geertz, C. *The Interpretation of Cultures* (New York, 1973).
Gelder, A.M. van. *Amsterdamsche Straatnamen* (Amsterdam, 1915).
Generale Missiven van gouverneurs-generaal en raden aan Heren XVII der Verenigde Oostindische Compagnie. Vol. 3: *1655–1674*. Ed. W. Ph. Coolhaas ('s-Gravenhage, 1968).
Genette, G. 'Introduction to the Paratext.' *New Literary History* 2 (1991) 261–72.
George, Katherine. 'The Civilized West Looks at Primitive Africa: 1400–1800; a Study in Ethnocentrism.' *Isis* 1 (1958) 62–72.
Gerritsz, H. *Beschryvinghe vander Samoyeden landt* (Amsterdam: Hessel Gerritsz, 1612).
Gheyn, Jacob de. *The exercise of armes for caliures, muskettes, and pikes* (The Hague: n.p., 1607).
——. *Wapenhandelinghe van roers, musquetten ende spiessen* ('s-Gravenhage: n.p., 1607).
Ginzburg, C. *The Cheese and the Worms* (Baltimore, MD, 1980).
Glete, J. *War and the State in Early Modern Europe* (New York, 2002).
Goeteeris, Anthonis. *Journael der legatie ghedaen inde Jaren 1615 ende 1616* ('s-Gravenhage: Aert Meurs, 1619).
Goetz, H. 'Persians and Persian Costumes in Dutch Painting of the Seventeenth Century.' *The Art Bulletin* 3 (1938) 280–90.
Gol'dberg, A.L., A.G. Man'kov, S.Ia. Marlinskii. 'Izvestiia o vosstanii S. Razina v zapadnoevropeiskikh periodicheskikh izdaniiakh i khronikakh XVII v.' In *Inostrannye izvestiia o vosstanii Stepana Razina*. Ed. A.G. Man'kov (Leningrad, 1975) 80–91.
Goor, J. van. *De Nederlandse kolonien. Geschiedenis van de Nederlandse Expansie* (Den Haag, 1997).
Gordon, P. *Dnevnik 1659–1667*. Trans. and ed. D.G. Fedosov (Moskva, 2003).
Gorski, P.S. *The Disciplinary Revolution: Calvinism and the Rise of the State in Early Modern Europe* (Chicago, 2003).
Goubert, P. *The Course of French History* (London, 1991).
Goud uit graan: Nederland en het Oostzeegebied, 1600–1850. Ed. R. Daalder (Zwolle, 1998).
Grimmelshausen, H.J.C. von. *Der abenteuerliche Simplicissimus*. Halle, 1889 (Reprint of original: Mompelgart [Nürnberg]: J. Filion [W.E. Felssecker], 1669).
Grove, Richard H. *Green Imperialism* (Cambridge, 1995).

Guilmartin, J.F. 'Ideology and Conflict: The Wars of the Ottoman Empire, 1453–1606.' *Journal of Interdisciplinary History* 4 (1988) 721–47.

Gurliand', I.Ia. *Ivan' Gebdon' kommissarius' i rezident'* (Iaroslavl', 1903).

Gurney, J.D. 'Pietro Della Valle: The Limits of Perception.' *Bulletin of the School of Oriental and African Studies* 1 (1986) 103–16.

Haan, H. den. *Moedernegotie en grote vaart* (Amsterdam, 1977).

Haberland, D. 'Einführung des Herausgebers.' In A. Olearius, *Moskowitische und Persische Reise: die holsteinische Gesandtschaft beim Schah; 1633–1639*. Ed. Haberland: 13–46.

Habsmeier, M. 'Reisebeschreibungen als mentalitätsgeschichtliche Quellen.' In *Reisebeschreibungen als Quellen europäischer Kulturgeschichte*. Eds. A. Maczak and H.J. Teuteberg. (Wolfenbüttel, 1982) 1–32.

Hair, P.E.H. 'Barbot, Dapper, Davity: A Critique of Sources on Sierra Leone and Cape Mount.' *History in Africa* 1 (1974) 25–54.

Haitsma Mulier, E. 'De eerste Hollandse stadsbeschrijvingen uit de zeventiende eeuw.' *De zeventiende eeuw* 2 (1993) 97–116.

Hamilton, A. 'Introduction.' In *Friends and Rivals in the East*. Eds. A. Hamilton et al. (Leiden, 2000) 1–9.

Harmon, W., and C.H. Holman. *A Handbook to Literature*. Eighth edn. (Upper Saddle River, NJ, 1999).

Hart, S. *Geschrift en Getal* (Dordrecht, 1976).

Hart, M.C., 't. *The Making of a Bourgeois State* (Manchester, 1993).

Harth, E. 'An Official "Nouvelle".' In M.M. de Lafayette, *The Princess of Clèves*. Ed. and trans. J.D. Lyons (London, 1994) 230–40.

Hastings, A. *The Construction of Nationhood* (Cambridge, 1997).

Hauser, A. *The Social History of Art*. Vol. 2 (London, 1951).

Hazard, P. *La crise de conscience européenne: 1680–1715* (Paris, 1935).

Hellie, R. *Enserfment and Military Change in Muscovy* (Chicago, 1971).

——. *The Economy and Material Culture of Russia, 1600–1725* (Chicago, 1999).

——. 'The Expanding Role of the State.' In *Modernizing Muscovy*. Eds. Kotilaine and Poe: 29–55.

Herberstein, S. von. *Rerum moscoviticarum commentarii*. (Vienna: Johann Singriener, 1549; Dutch version, 1605).

Herckmans, Elias. *Een historischen verhael van de voornaamste beroerten des keyserrychs van Russia, onstaen door den Demetrium Ivanowyts* (Amsterdam, 1625 [ms.]).

——. *Der zee-vaert lof, handelende vande gedenckwaerdighste zee-vaerden met de daeraenklevende op en onderganghen der voornaemste heerschappijen der gantscher wereld* (Amsterdam: Wachter, 1634).

Herd, Graeme P. 'General Patrick Gordon of Auchleuchries – a Scot in Seventeenth Century Russian Service.' Unpublished PhD dissertation (Aberdeen: University of Aberdeen, 1994).

Hermans, W.F. *Het evangelie van O. Dapper Dapper* (Amsterdam, 1973).

Herodotus. *The Histories*. Trans. Robin Waterfield (Oxford, 1998).

Herport, A. *Reise nach Java, Formosa, Vorder-Indien und Ceylon, 1659–1668*. Ed. S.P. L'Honoré Naber (Den Haag, 1930; first edn.: Bern: G. Sonnleitner, 1669).

Hesselink, Reinier H. 'Memorable Embassies: The Secret History of Arnoldus Montanus' *Gedenkwaerdige Gesantschappen*.' *Quaerendo* 1–2 (2002) 99–123.

Hinton, R.W.K. *The Eastland Trade and the Common Weal in the Seventeenth Century* (Cambridge, 1959).

Historie van Thijl Ulespiegel (Amsterdam: Broer Jansz., *c.*1640).
Hobson, J.A. *The Evolution of Modern Capitalism* (London, 1894).
Hofman, Hans de. *Den Danske Atlas Eller Konge-Riget Dannemark*. Vol. 7. Part 2 (Kiøbenhavn: Godiche, 1781).
De Hollandtsche Mercurius (Haarlem: P. Casteleyn, 1668–76).
Holt, P.M. 'The Exalted Lineage of Ridwan Bey: Some Observations on a Seventeenth-Century Mamluk Genealogy.' *Bulletin of the School of Oriental and Asian Studies* 1–3 (1959) 221–30.
Homer. *Odyssey*. Trans. S.O. Andrew (London, 1953).
Honig, J. Jzs. Jr. *Geschiedenis der Zaanlanden*. 2 vols (Haarlem, 1849).
L'Honoré Naber, S.P. 'Vorwort.' In von der Behr, *Reise nach Java*: ix-xii.
Hooft, P.C. *Nederlandsche historien, sedert de ooverdraght der heerschappye van Kaizar Karel den Vyfden, op kooning Philips zynen zoon* (Amsterdam: Elzevier, 1642).
Hornius, G. *Orbis Politicus* (Leiden: F. Lopes de Haro, C. Driehuijsen, 1667).
Hosking, G. *Russia and the Russians* (London, 2001).
——. *Rulers and Victims: The Russians in the Soviet Union* (London, 2006).
Hughes, L.A.J. *Russia in the Age of Peter the Great* (London, 1998).
——. *Peter the Great: A Biography* (New Haven, CT, 2002).
Huizinga, J. *Homo Ludens* (Haarlem, 1938).
——. 'De wetenschap der geschiedenis.' In J. Huizinga, *Verzamelde werken*. Vol. 7 (Haarlem, 1948–53) 104–72.
Hullu, J. de. 'De matrozen en soldaten op de schepen der Oost-Indische Compagnie.' *Bijdragen tot de Taal-, Land- en Volkenkunde van Nederlandsch-Indie* 69 (1914) 318–65.
Hunter, J. Paul. 'The "Occasion" of *Robinson Crusoe*.' In Daniel Defoe. *Robinson Crusoe*. Ed. Michael Shinagel (New York, 1975) 365–78.
Ides, E. Ysbrants. *Drie-Jaarige Reize naar China*. 2nd. edn. (Amsterdam: Pieter de Coup, 1710; first edn.: Amsterdam: F. Halma, 1704).
Iournael ofte Gedenckwaedige beschrijvinghe van de Oost-Indische Reyse van Willem Ysbrantsz. Bontekoe van Hoorn. Eds G. Verhoeven, P. Verkruijsse (Zutphen, 1996).
Die Islamgelehrten Daghestans und ihre arabischen Werke. Eds M. Kemper and A. Sixsaidov (Berlin, 2004).
Israel, J. *Dutch Primacy in the World Trade* (Oxford, 1989).
——. *The Dutch Republic: Its Rise, Greatness, and Fall, 1477–1806* (Oxford, 1998).
——. *Radical Enlightenment: Philosophy and the Making of Modernity, 1650–1750* (Oxford, 2001).
Iurchenko, P. 'O puteshchestviia po Rossii Golandtsa Striusa.' *Russkii arkhiv'* 2 (1879) 265–9.
——. 'Predislovie k Russkomu perevode: Pervoe puteshestvie po Rossii gollandtsa Streisa.' *Russkii arkhiv'* 1 (1880) 5–16.
Jacobs, Els M. 'De Verenigde Oost-Indische Compagnie: een veelkantig handelsbedrijf.' *Tijdschrift voor Geschiedenis* 4 (2002) 525–43.
'Jan Struys, the Perillous and most Unhappy Voyages of John Struys ...' Ed. M.S. *Journal of the Siam Society* 94 (2006) 177–209.
Jansma, T.S. 'Olivier Brunel te Dordrecht.' *Tijdschrift voor Geschiedenis* 59 (1946) 337–62.
Jespersen, K., O. Feldbaek. *Revanche og Neutralitet: 1648–1814* (København, 2002).

Jones, Adam. 'Olfert Dapper et sa description de l'Afrique.' In *Objets interdits* (Paris, 1989) 72–81.

——. 'Decompiling Dapper: A Preliminary Search for Evidence.' *History in Africa* 17 (1990) 171–209.

Jones, J.R. *The Anglo-Dutch Wars of the Seventeenth Century* (London, 1996).

Jonge, J.C. de. *Nederland en Venetië* ('s-Gravenhage, 1852).

——. *Geschiedenis van het Nederlandsche Zeewezen*. 5 vols. Third edn. (Zwolle, 1869).

Jonker, J., K. Sluyterman. *At Home in the World Markets* (The Hague, 2000).

A Journey to Asia: Jean Chardin's Portrait of a Seventeenth-Century Empire. Ed. and trans. R.W. Ferrier (London, 1996).

Kaempfer, E. *Am Hofe des persischen Großkönigs, 1684–5* (Leipzig, 1940).

Kaiser, D. ' "Whose Wife Will She Be at the Resurrection?" Marriage and Remarriage in Early Modern Russia.' *Slavic Review* 2 (2003) 302–23.

Kamen, Henry. *Empire: How Spain Became a World Power, 1492–1763* (New York, 2004).

Kaplan, B.J. ' "Dutch" Religious Tolerance: Celebration and Revision.' In *Calvinism and Religious Toleration in the Dutch Golden Age*. Eds. R. Po-chia Hsia and H. Van Nierop (Cambridge, 2002) 8–26.

Kappeler, A. *The Russian Empire* (Harlow, 2001).

Kaptein, Herman. *Het Schermereiland. Een zeevarend plattelandsgebied 950–1800* (Bergen, 1988).

Kemper, M. *Herrschaft, Recht und Islam in Daghestan* (Wiesbaden, 2005).

Kemper, M., A. Sixsaidov. 'Einleitung.' In *Die Islamgelehrten Daghestans*. Eds Kemper and Sixsaidov: 3–24.

Keuning, J. 'Isaac Massa, 1586–1643.' *Imago Mundi* 10 (1953) 64–79.

——. 'Nicolaas Witsen as a Cartographer.' *Imago Mundi* 11 (1954) 95–110.

Khakkebord, L. 'Arkheologicheskie i kartograficheskie aspekty rannikh russko-gollandskikh torgovikh sviazei na Russkom Severe.' In *Niderlandy i Severnaia Rossiia*. Eds Iu.N. Besplatnikh et al. (St Petersburg, 2003) 7–13.

Khodarkovsky, M. 'Of Christianity, Enlightenment, and Colonialism: Russia in the North Caucasus, 1550–1800.' *Journal of Modern History* 2 (1999) 394–430.

——. *Russia's Steppe Frontier* (Bloomington, IN, 2002).

Kirpichnikov, A.N. *Rossiia xvii veka v risunkakh i opisaniiakh gollandskogo puteshestvennika Nikolaasa Vitsena* (St Petersburg, 1995).

Klebusek, M. 'The Business of News.' *Scandinavian Journal of History* 3–4 (2003) 205–13.

Kleerekoper, M.M., W.P. van Stockum. *De boekhandel te Amsterdam voornamelijk in de 17de eeuw*. 2 vols ('s-Gravenhage, 1914–16).

Klein, P.W. 'Nederland de eerste moderne economie? Een kritiek.' *Tijdschrift voor Geschiedenis* 109 (1996) 514–20.

Kley, E.J. Van. 'Asian Religions in Seventeenth-Century Dutch Literature.' *Itinerario* 3–4 (2001) 54–68.

Kliuchevskii, V.O. *Skazaniia inostrantsev o Moskvoskom gosudarstve*. Second edn. (Petrograd, 1918).

——. *Istoricheskie portrety*. Ed. V.A. Aleksandrov (Moskva, 1990).

Klyuchevsky, V.O. *Peter the Great*. Trans. L. Archibald (London, 1958).

Knevel, P. 'Armed Citizens: The Representation of the Civic Militias in the Seventeenth Century.' In *The Public and Private in Dutch Culture of the Golden Age*. Eds A.K. Wheelock, A. Seeff (Newark, 2000) 85–99.

Kniga ustav' morskoi o vsem' chto kasaetsia dobromu upravleniiu, v' bytnosti flota na mor' (St Petersburg, 1720).

Knowles, M.D. 'Presidential Address: Great Historical Enterprises I. The Bollandists.' *Transactions of the Royal Historical Society*. Fifth series 8 (1958) 147–66.

Koeman, C. *Joan Blaeu and His Grand Atlas* (Amsterdam, 1970).

Kohn, H. *The Idea of Nationalism* (New York, 1967).

Kollmann, N.S. *Kinship and Politics: The Making of the Muscovite Political System, 1345–1547* (Stanford, CA, 1987).

Koningsbrugge, H. van. 'Der Niedergang des niederländischen Handels mit Rußland und dem Baltikum in den Jahren 1710–1721.' In *Around Peter the Great*. Eds C. Horstmeier et al. (Groningen, 1997) 87–91.

Konovalov, S. 'Ludwig Fabritius's Account of the Razin Rebellion.' *Oxford Slavonic Papers* 6 (1955) 72–101.

'Kopiia s' pis'ma, pisannogo neizvestnym' litsam' na karable,Orle' ... 24 Sentiabria (...) 1669 g.' Trans. P.O. Iurchenko. *Russkii arkhiv'* 1 (1880) 109–11.

'Kopiia s' pis'ma Davida Butlera, ... , s' spisanyem' vziatiia Astrakhani.' Trans. P.O. Iurchenko. *Russkii arkhiv'* 1 (1880) 111–28.

Kornilovich, A. 'Gollandets Ian Iansen Streis (otryvok iz 'Opyty istorii', Puteshestvie po Rossii').' *Severnyi arkhiv* 7. Part 10 (1824) 26–40.

Kort waerachtigh Verhaal, van de bloedige Rebellye in Moscovien, aen gerecht, door den Grooten Verrader en Bedrieger Stenko Rasin (Haarlem: Pieter Casteleyn, 1671).

Kostomarov, N.I. 'Bunt' Sten'ki Razina.' In N.I. Kostomarov, *Sobranie sochinenii*. Vol. 1 (Sankt-Peterburg, 1903) 405–505.

Kotilaine, J. 'When the Twain Did Meet: Foreign Merchants and Russia's Economic Expansion in the Seventeenth Century.' Unpublished Ph.D. Dissertation (Cambridge, MA: Harvard University, 2000).

——. *Russia's Foreign Trade and Economic Expansion in the Seventeenth Century* (Leiden, 2005).

——. 'Mercantilism in Pre-Petrine Russia.' In *Modernizing Muscovy*. Eds Kotiliane and Poe: 143–73.

Kotilaine, J., M. Poe. 'Modernization in the Early Modern Context: The Case of Muscovy.' In *Modernizing Muscovy*. Eds. Kotilaine and Poe: 1–8.

Kotoshikhin, G.K. P. Gordon, J. Streis, Tsar' Aleksei Mikhailovich. *Moskoviia i Evropa* (Moskva, 2000).

Kovrigina, V.A. 'Inozemnye kuptsy-predprinimateli Moskvy Petrovskogo vremeni.' In *Torgovlia i predprinimatel'stvo v feodal'noi Rossiii*. Eds L.A. Timoshina, I.A. Tikhoniuk (Moskva, 1994) 190–213.

——. *Nemetskaia sloboda Moskvy i ee zhiteli v kontse xvii-pervoi chetverti xviii vv.* (Moskva, 1998).

Kozlovskii, I.P. *Andrei Vinius', sotrudnik Petra Velikogo (1641–1717 g.)* (Sankt-Peterburg, 1911).

Kraan, A. van der. 'On Company Business: The Rijckloff van Goens Mission to Siam, 1650.' *Itinerario* 2 (1998) 42–84.

Kreslins, Janis. 'Linguistic Landscapes in the Baltic.' *Scandinavian Journal of History* 3–4 (2003) 165–74.

Krest'ianskaia voina pod predvoditel'stvom Stepana Razina. 4 vols (Moscow, 1954–76).

Kukanova, Nina G. *Ocherki po istorii russko-iranskikh torgovykhotnoshenii v XVII-pervoi polovine XIX veka* (Saransk, 1977).

Kulisher, I.M. *Ocherk istorii russkoi torgovli* (Peterburg, 1923).

Kurskov, Iu.V. *Vedushchee napravlenie obshchestvennoi mysli i proekty gosudarstvennykh preobrazovanii Rossii 40–60-kh godov XVII veka* (Chita, 1973).

Kurts, B.G. *Sochinenie Kil'burgera o 'russkoi torgovle v tsarstvovanie Alekseia Mikhailovicha.'* (Kiev, 1915).

La Haye, T., and J. Morris. *The Ark on Ararat* (Nashville, TN, 1976).

Labadie, J. de. *erheffingen des geestes tot Godt, op de drie voornaemste tijden van den Christelyken dagh* (Amsterdam: J. Van Someren, 1667).

Lach, D.F., and E. Van Kley. *Asia in the Making of Europe*. Vol. 3: *A Century of Advance* (Chicago, 1993).

Lahana, Martha Luby. 'Novaia Nemetskaia Sloboda: Seventeenth Century Moscow's Foreign Suburb.' Unpublished Ph.D. Dissertation (University of North Carolina at Chapel Hill, 1983).

Landsman, N.C. 'The Middle Colonies: New Opportunities for Settlement, 1660–1700.' In *The Origins of Empire*. Ed. Canny: 351–74.

Langbein, J.H. 'The Historical Origins of the Sanction of Imprisonment for Serious Crime.' *Journal of Legal Studies* 1 (1976): 35–60.

Lanni, D. 'Une cartographie éthnique.' *Africultures: Le site et la revue de référence des cultures africaines* (2004; available at http://www.africultures.com//popup_article.asp? no=4025&print=1; accessed 25 February 2007).

LeDonne, John P. *The Grand Strategy of the Russian Empire, 1650–1831* (Oxford, 2004).

Lenin, V.I. *Imperialism: The Highest Stage of Capitalism*. In V.I. Lenin, *Collected Works*. Vol. 22 (Moscow, 1964) 185–304.

Liechtenhan, Francine-Dominique. *Les trois christianismes et la Russie* (Paris, 2002).

Linde, A. van der. *Isaac Massa van Haarlem: een historische studie* (Amsterdam, 1864).

Linebaugh, P., and M. Rediker. *The Many-Headed Hydra* (Boston, 2000).

Linschoten, J. Huygen van. *Beschryvinghe van de gantsche kuste van Guinea* (Amsterdam: Cornelis Claesz, 1596).

——. *Discours of Voyages Into Ye East & West Indies*. Amsterdam, 1974 (reprint of London: Wolfe, 1598).

Liubimenko, I. 'Trud inozemtsev v Moskovskom gosudarstve.' *Arkhiv Istorii Truda v Rossii* 6–7 (1923) 52–74.

Locher, Th. 'Het beeld van Rusland in de zestiende-eeuwse Europese beschrijvingen.' *Tijdschrift voor Geschiedenis* 70 (1957) 289–308.

Locher, Th.J.G., P. de Buck. 'Inleiding.' In N.C. Witsen, *Moscovische reyse 1664–5. Journaal en aentekeningen* (Den Haag, 1966–7) xix–lxxv.

Longworth, P. *Alexis: Tsar of All the Russias* (NY, 1984).

——. 'The Role of Westerners in Russia's Penetration of Asia, 17th–18th Century.' In *The Place of Russia in Eurasia*. Ed. Gyula Szvak (Budapest, 2001) 207–13.

Loviagin, A.M. 'Vvedenie.' In *Posol'stvo Kunraada fan' Klenka*. Tr. and ed. A.M. Loviagin: iv-clxxvi.

'Der lutherische Kirchhof.' *Mitteilungsblatt der Gesellschaft für Friedrichstädter Stadsgeschichte* 10 (1976–77) 1–14.

Lyagusheva, S. 'Islam and the Traditional Moral Code of Adyghes.' *Iran and the Caucasus* 1 (2005) 29–35.

Lynn, J. *The Wars of Louis XIV, 1667–1714* (London, 1999).

Lyseck, Adolph. *Relatio eorum, quae circa Sacrae Caesarae Majestatis ad Magnum Moscorum Czarum Ablegatos* (Salzburg: J.B. Mayr, 1676).

Maczak, Antoni. *Travel in Early Modern Europe* (Cambridge, 1995).

Le Magasin de l'Univers. Eds C. Berkvens-Stevelinck et al. (Leiden, 1992).

Malcolm, N. *Aspects of Hobbes* (Oxford, 2002).

Mans, Raphaël du. *Estat de la Perse en 1660*. Ed. Ch. Scheffer (Paris, 1960).

Marx, K. *The Capital*. Ed. F. Engels. 3 vols (New York, 1967).

Massa, I. *A Short History of the Beginnings and Origins of These Present Wars in Moscow*. Ed. G. Edward Orchard (Toronto, 1982).

Matar, N.I. 'The Renegade in English Seventeenth-Century Imagination.' *Studies in English Literature, 1500–1900* 3 (1993) 489–505.

Matthee, R. *The Politics of Trade in Safavid Iran: Silk for Silver, 1600–1730* (Cambridge, 1999).

——. 'Gurdziecki, Bogdan.' In *Encyclopaedia Iranica*. Ed. E. Yarshalter. (Available at http://www.iranica.com/newsite/. Accessed on 12 January 2007).

——. 'Persian Image of Europe.' *Encyclopaedia Iranica*. Ed. E. Yarshalter. (Available at http://www.iranica.com/newsite/. Accessed on 12 January 2007).

——. 'Unwalled Cities and Restless Nomads: Firearms and Artillery in Safavid Iran.' In *Safavid Persia*. Ed. Melville: 389–416.

Maxwell, M.J. 'Afanasii Nikitin: An Orthodox Russian's Spiritual Voyage in the *Dar al-Islam*, 1468–75.' *Journal of World History* 3 (2006) 243–66.

McCants, A.E. *Civic Charity in a Golden Age* (Urbana, IL, 1997).

McNeill, William H. *Europe's Steppe Frontier, 1500–1800* (Chicago, 1964).

——. *Venice: The Hinge of Europe, 1081–1797* (Chicago, 1974).

Meier-Lemgo, Karl. *Engelbert Kämpfer, der erste deutsche Forschungsreisende, 1651–1716* (Stuttgart, 1937).

Merklein, J.J. *Reise nach Java, Vorder- und Hinter-Indien, China und Japan, 1644–53*. Ed. S.P. L'Honoré Naber. Den Haag: Nijhoff, 1930 (Reprint of second edition: Nürnberg: M. und J. Endters, 1672).

Mervaud, M., J.-C. Roberti. *Une infinie brutalité: L'image de la Russie dans la France des XVIe et XVIIe siècles* (Paris, 1991).

Met Man en Macht: De Militaire Geschiedenis van Nederland 1500–2000. Eds J.R. Bruijn, C.B. Wels (Baarn, 2003).

Milton, G. *Nathaniel's Nutmeg* (London, 1999).

A Miracle Mirrored: The Dutch Republic in European Perspective. Eds K. Davids and J. Lucassen (Cambridge, 1995).

Mironov, B.N. *The Social History of Imperial Russia, 1700–1917*. Ed. B. Eklof. 2 vols (Boulder, CO, 2000).

Mistry, Rohinton. *A Fine Balance* (New York, 1997).

Modernizing Muscovy: Reform and Social Change in Seventeenth-Century Russia. Eds J. Kotilaine and M. Poe (London, 2004).

De Moedernegotie. Radio Documentary in eight parts. Eds K. Amsberg, P. de Buck (Hilversum, 1999).

Montanus, A. *'t Leven en bedryf der prinsen van Oranje* (Amsterdam: A. van den Heuvel, S. Imbrecht, 1664).

——. *Gedenkwaerdige gesantschappen der Oost-Indische Maetschappy in 't Vereenigde Nederland, aen de kaisaren van Japan* (Amsterdam: van Meurs, 1669).

——. *De Nieuwe en onbekende weereld: of Beschryving van America en 't zuid-land* (Amsterdam: van Meurs, 1671).

——. *Kerkelyke historie van Nederland* (Amsterdam: Cornelis and Jan van Zwol, 1675).

Montesquieu. *Lettres persanes*. Paris, 1973 (First edition, 1721).
Moor, J.A. de. 'Experience and Experiment: Some Reflections upon the Military Developments in 16th and 17th Century Western Europe.' In *Exercise of Arms: Warfare in the Netherlands, 1568–1648*. Ed. M. van der Hoeven (Leiden, 1997) 17–32.
Morozov, A. 'Parusnyi Master Ian Streis i ego puteshestvie.' In *Tri puteshestviia*. Ed. Morozov: 23–38.
Muliukin, A.S. *Priezd inostrantsev v Moskovskoe gosudarstvo* (St Petersburg, 1909).
Mund, Stéphane. *Orbis Russiarum* (Genève, 2003).
Naarden, B. 'Dutchmen and the European Image of Russia before 1917.' In *Russians and Dutchmen*. Eds J. Braat et al. (Groningen, 1993) 1–19.
'Natalie Zemon Davis.' In *Visions of History*. Eds H. Abelove et al. (New York, 1984) 97–122.
Niderlandy i Severnaia Rossiia. Eds Iu.N. Besplatnikh, Ia. Veluvenkamp, et al. (St Petersburg, 2003).
Nierop, H. van. 'Catholics and the Law in Holland.' In *Calvinism and Religious Toleration in the Dutch Golden Age*. Eds R. Po-chia Hsia and H. Van Nierop (Cambridge, 2002) 102–11.
Nieuhof, Johannes. *Het gezantschap der Neerlandtsche Oost-Indische Compagnie, aan den grooten Tartarischen cham* (Amsterdam: van Meurs, 1665).
Nimwegen, O. van. *Deser landen crijchsvolck* (Amsterdam, 2006).
Nolte, H. 'The Netherlands and Russia in the 17th Century.' *Review of the Fernand Braudel Center* 2 (1986) 230–44.
Nusteling, H.P.H. *Welvaart en werkgelegenheid in Amsterdam, 1540–1860* (Amsterdam, 1985).
Nijenrode, Cornelis van. 'Remonstrantie ende verthooninghe der gelegenheijt des conickrijck van Siam.' *Kronijk van het Historisch Genootschap gevestigd te Utrecht* 10 (1854) 176–91.
——. 'Vertoogh van de Gelegentheijd des Koninrijk van Siam.' *Kroniek van het Historisch Genootschap te Utrecht* 27 (1872) 279–318.
Nijhoff, W. 'De anglophobie van Nic. Witsen en verschillende redactiën van zijn Scheepsbouw, 1671.' *Het Boek* 14 (1925) 88–96.
Olearius, Adam. *Offt begehrte beschreibung der neuen orientalischen Reise* (Schleswig: Np., 1647).
——. *Beschrijvingh van de nieuwe Parciaensche Ofte Orientaelsche reyse*. Trans. Dirck Van Wageninge (Utrecht: L. Roeck, 1651).
——. *Beschrijvingh vande nieuwe Parciaensche, ofte orientaelsche reyse*. Trans. Dirck van Wageninge (Amsterdam: Benjamyn and Roest, 1651).
——. *Persiaensche Reyse uyt Holsteyn, door Lijflandt, Moscovien, Tartarien in Persien*. Second edn. (Amsterdam: Jan Jansz, 1651).
——. *Vermehrte Neue Beschreibung der Moskowitischen und Persischen reise* (Schleswig: J. Holwein, 1656).
——. *The voyages & travels of the ambassadors sent by Frederick duke of Holstein*. Trans. John Davies (London: Dring and Starkey, 1662).
——. *The Voyages and Travells of the Ambassadors*. Trans. John Davies. Second edn. (London: Starkey and Bassett, 1669).
——. *Podrobnoe opisanie puteshestviia golshtinskago posol'stva v Moskoviiu i Persiiu*. Tr. P. Barsov (Moskva, 1870).

——. *The Travels of Olearius in Seventeenth-Century Russia*. Trans. and ed. S.H. Baron (Stanford, CA, 1967).
——. *Moskowitische und Persische Reise: die holsteinische Gesandtschaft beim Schah; 1633–1639*. Ed. Detlef Haberland (Stuttgart, 1986).
Oparina, T.A. *Inozemtsy v Rossii XVI–XVII vv.* (Moscow, 2007).
Opstall, M.E. van. 'From Alkmaar to Ayudhya and Back.' *In All of One Company* (Utrecht, 1986) 108–20.
Orchard, G.E. 'Introduction.' In I. Massa, *A Short History*. Ed. Orchard: ix–xxiv.
The Origins of Empire. Ed. N. Canny (Oxford, 1998).
Orlenko, S.P. *Vykhodtsy iz Zapadnoi Evropy v Rossii XVII veka* (Moskva, 2004).
Ormrod, D. *The Rise of Commercial Empires: England and the Netherlands in the Age of Mercantilism, 1650–1770* (Cambridge, 2003).
Osipov, I. *Obzor svedenii ob izdaniiakh i perevodakh v xvii stoletii sochineniia Semiuelia Kollinsa 'The Present State of Russia.'* (Komi, Russian Federation, 2007).
Ottenheym, K.A. 'Dutch Contributions to the Classicist Tradition in Northern Europe.' *Scandinavian Journal of History* 3–4 (2003) 227–42.
Pagden, A. 'Europe and the World Around.' In *Early Modern Europe*. Ed. E. Cameron (Oxford, 2001) 1–28.
——. 'Europe: Conceptualizing a Continent.' In *The Idea of Europe*. Ed. A. Pagden. (Washington, DC, 2002) 33–54.
Parker, Ch.H. 'Paying for the Privilege.' *Journal of World History* 3 (2006) 267–96.
Parker, G. *The Military Revolution*. Second edn. (Cambridge, 2000).
——. *Success Is Never Final* (New York, 2002).
Paul of Aleppo. *The Travels of Macarius*, 1652–1660. Ed. L. Redding (New York, 1971).
Pekarskii, P. *Nauka i literatura v Rossii pri Petre Velikomu*. 2 vols (Sint-Petersburg, 1862).
Pepys, S. *The Diary of Samuel Pepys*. Ed. H.B. Wheatley. 2 vols (New York, 1942).
——. *The Diary of Samuel Pepys*. Eds R. Latham and W. Matthews. 10 vols (Berkeley, CA, 1970–1983).
Pérez-Mallaína, P. *Spain's Men of the Sea* (Baltimore, MD, 1998).
Peters, Marion. 'From the Study of Nicolaes Witsen (1641–1717): His Life with Books and Manuscripts.' *Lias* 1 (1994) 1–45.
——. 'Nepotisme, patronage en boekopdrachten bij Nicolaes Witsen (1641–1717), burgemeester van Amsterdam.' *Lias* 1 (1998) 83–134.
Petrejus, Peter. *Regni Muschowitici Sciographia* (Stockholm: I. Meurer, 1614–15).
——. *Historien und bericht von dem grossfürstenthumb Muschkow* (Tipis Bavaricis: Lipsiae, 1620).
Phillips, E.J. *The Founding of Russia's Navy: Peter the Great and the Azov Fleet, 1688–1714* (Westport, CT, 1995).
Phipps, Geraldine M. 'Britons in Seventeenth-Century Russia: A Study in the Origins of Modernization.' Unpublished Ph.D. Dissertation (Philadelphia: University of Pennsylvania, 1971).
Pipes, R. *Russia under the Old Regime*. Second edition (London, 1995).
Plagiarism in Early Modern England. Ed. P. Kewes (Basingstoke, 2003).
Planhol, X. de. 'Bandar-e-'Abbas.' In *Encyclopaedia Iranica*. Ed. E. Yarshalter (Available at http://www.iranica.com/newsite/. Accessed on 12 January 2007).

——. 'Caspian Sea.' In *Encyclopaedia Iranica*. Ed. E. Yarshalter (Available at http://www.iranica.com/newsite/. Accessed on 12 January 2007).

Po-chia Hsia, R. 'Introduction.' In *Calvinism and Religious Toleration in the Dutch Golden Age*. Eds. R. Po-chia Hsia and H. Van Nierop (Cambridge, 2002) 1–7.

Poe, M. 'Absolutism and the New Men.' In *Modernizing Muscovy*. Eds Kotilaine and Poe: 97–115.

——. *Foreign Descriptions of Muscovy: An Analytic Bibliography of Primary and Secondary Sources* (Columbus, OH, 1995).

——. *'A People Born to Slavery': Russia in Early Modern European Ethnography, 1476–1748* (Ithaca, NY, 2000).

——. *The Russian Elite in the Seventeenth Century*. 2 vols (Helsinki, 2004).

Pomeau, R., J. Erhard. *Littérature française*. Vol. 5 (Paris, 1984).

Polnoe sobranie zakonov' Rossiiskoi imperii. Vol. 1: *1649–76* (St Petersburg, 1830).

Polyevktov, M. 'Russian and Georgian Communications.' *Journal of Modern History* 3 (1930) 367–77.

Popov, A.N. 'Istoriia vozmuscheniia Sten'ki Razina.' *Russkaia beseda* 1 (1857) 47–104.

——. *O postroenii korablia Orla' v gosudarstvovanie tsaria Alekseia Mikhailovicha* (Moscow, 1858).

Porshnev, B.F. *Muscovy and Sweden in the Thirty Years' War, 1630–1635* (Cambridge, 1995).

Porter, R. *Enlightenment: Britain and the Creation of the Modern World* (London, 2000).

Posol'stvo Kunraada fan' Klenka k' tsariam' Alekseiu Mikhailovichu i Fedoru Alekseevichu. Trans. and ed. A.M. Loviagin (St Petersburg, 1900).

Postma, J. 'The Dimension of the Dutch Slave Trade from Western Africa.' *Journal of African History* 2 (1972) 237–48.

Prak, M. *Gezeten Burgers: De elite in een Hollandse stad, Leiden 1700–1800* (Leiden, 1985).

——. *The Dutch Republic in the Seventeenth Century: The Golden Age* (Cambridge, 2005).

Price, J.L. *Dutch Society, 1588–1713* (Harlow, 2000).

Prud'homme van Reine, R. *Rechterhand van Nederland* (Amsterdam, 1996).

——. 'De Republiek als grote en kleine mogendheid ter zee (1648–1763).' In *Met Man en Macht*. Eds. Bruijn and Wels: 105–39.

——. *Schittering en schandaal: Biografie van Maerten en Cornelis Tromp* (Amsterdam, 2001).

Pundani, V.V. 'Evropeizatsiia Rossii v XVII v.' In *Mezhdunarodnaia nauchnaia konferentsiia Rossiia i Zapadnaia Evropa: Dialog kul'tur* (Kurgan, 1993) 37–8.

'Puteshestvie Iana Iansena Streisa po Rossiiu v 1668 gody.' *Severnyi arkhiv'* 5 (1824) 275–90.

'Puteshestvye po Rossii gollandtsa Streisa.' *Russkii arkhiv* 1 (1880) 17–108.

(Rabus, P.) 'Reize van Avril.' *De Boekzaal van Europe* (July-August 1693) 106–15.

——. 'Berigt wegens Jan Jansz Struis.' *De Boekzaal van Europe* (May-June 1694) 562–5.

Randall, M. 'Appropriate(d) Discourse: Plagiarism and Decolonization.' *New Literary History* 3 (1991) 525–41.

Reger, W.M. 'In the Service of the Tsar: European Mercenary Officers and the Reception of Military Reform in Russia, 1654–1667.' Unpublished Ph.D. Dissertation (Urbana-Champaign: U. of Illinois, 1997).

Relaes van't gepasseerde voor ende op de Inkomste ende Receptie van den Heere van Klenck ('s-Gravenhage, 1676).

Relations veritables et curieuses de l'isle de Madagascar et du Brésil ... par François Cauche (de Rouen) et al. Ed. Claude-Barthélemy Morisot (Paris: A. Courbé, 1651).

Religion and Culture in Early Modern Russia and Ukraine. Eds S. Baron and N. Kollmann (De Kalb, IL, 1997).

Repertorium der Nederlandse vertegenwoordigers, residerende in het buitenland. Ed. O. Schutte (Den Haag, 1976).

Repin, N.N. 'Gollandskie kuptsy v Arkhangel'ske vo vtoroi polovine xviii v.' In *Niderlandy i Severnaia Rossiia.* Eds Besplatnikh et al.: 14–36.

(Anonymous). Review of 'Les Voyages de Jean Struys en Moscovie, Tartarie, Perse, & plusieurs autres Païs étrangers.' *Journal des Sçavans* (Lundi 21 Juillet 1681) 150–2.

Riasanovsky, N.V. *A History of Russia.* Sixth edn. (Oxford, 2000).

Richard, Francis. 'L'Apport des Missionnaires européens à la connaissance de l'Iran en Europe et de l'Europe en Iran.' In *Études Safavides.* Ed. Calmard: 251–66.

Richter, W.M. von. *Geschichte der Medicin in Russland.* Vols 2 and 3 (Moskva, 1815–1817).

Rietbergen, P. 'Zover de aarde reikt.' *De zeventiende eeuw* 1 (1986) 17–40.

——. 'Varieties of Asia? European perspectives, *c.*1600–*c.*1800.' *Itinerario* 3–4 (2001) 69–89.

——. *Europe: A Cultural History.* Second edn. (London, 2006).

Roberts, M. 'Introduction.' In *Swedish Diplomats at Cromwell's Court, 1655–1656* (London, 1988) 1–46.

Roodenburg, H. 'Social Control Viewed from Below.' In *Social Control in Europe.* Vol. 1. Eds H. Roodenburg and P. Spierenburg (Columbus, OH, 2004) 145–58.

Roorda, D.J. *Partij en factie* (Groningen, 1978).

Roosen, W. 'Early Modern Diplomatic Ceremonial: A Systems Approach.' *Journal of Modern History* 3 (1980) 452–76.

Rousseau, J.J. *Les Confessions.* 2 vols (Paris, 1973).

Rowlands, A. 'Conditions of the Life of the Masses.' In *Early Modern Europe.* Ed. E. Cameron (Oxford, 2001) 31–62.

Royen, P.C. van. *Zeevarenden op de koopvaardijvloot omstreeks 1700* (Amsterdam, 1987).

Russia under Western Eyes, 1517–1825. Ed. Anthony Cross (London, 1971).

Russkaia istoricheskaia biblioteka. Vol. 21: *Dela tainogo prikaza* (St Petersburg, 1907).

Saerdam, W. van. *Jan Jansen Struys* (Den Haag, 1929).

Safavid Persia: The History and Politics of an Islamic Society. Ed. C. Melville (London, 1996).

Said, E. 'Orientalism Reconsidered.' *Cultural Critique* 1 (1985) 89–107.

——. *Orientalism.* New edn. (London, 2003).

Sakharov, A.N. 'Demokratiia i volia v nashem otechestve.' *Svobodnaia mysl'* 17 (1991) 42–53.

Samuel Pepys and the Second Dutch War: Pepys's Navy White Book and Brooke House Papers. Ed. R. Latham (Aldershot, 1995).

Savory, R. *Iran under the Safavids* (Cambridge, 1980).
Scammell, G.V. *The First Imperial Age* (London, 1989).
——. 'European Exiles, Renegades and Outlaws and the Maritime Economy of Asia *c*.1500–1750.' *Modern Asia Studies* 4 (1992) 641–61.
Schama, S. *The Embarrassment of Riches* (Berkeley, CA, 1988).
——. *Patriots and Liberators* (New York, 1992).
Scheidegger, G. *Perverses Abendland-barbarisches Rußland* (Zürich, 1993).
Scheltema, J. *Peter de Groote, keizer van Rusland in Holland en te Zaandam in 1697 en 1717*. 2 vols (Amsterdam, 1814).
——. *Rusland en de Nederlanden*. 4 vols (Amsterdam, 1817).
Schierbeek, A. *Jan Swammerdam, 1637–1680: His Life and Works* (Amsterdam, 1967).
Schiphouwer, W.J. 'Jan Jansen Struys.' *De Zaende* 6 (1951) 199–205.
Schmidt, B. *Innocence Abroad: The Dutch Imagination and the New World, 1570–1670* (Cambridge, 2001).
Schöffer, I. 'De republiek der Verenigde Nederlanden, 1609–1702.' In *De Lage Landen van 1500 tot 1780*. Eds I. Schöffer et al. (Amsterdam, 1978) 167–267.
De schriklijke reis van Jan Jansz. Struys, 1668–1673. Ed. D.J. Douwes. Zaandijk, 1974.
Schulten, K. 'Ontstaan van de Republiek en het Staatse Leger.' In *Met Man en Macht*. Eds Bruijn and Wels: 13–43.
Schwartz, S.B. 'Introduction.' In *Implicit Understandings*. Ed. S.B. Schwarz (Cambridge, 1994) 1–19.
Scott, J. ' "Good Night Amsterdam": Sir George Downing and Anglo-Dutch Statebuilding.' *English Historical Review* 118 (2003) 334–56.
Sedov, N.V. 'Peremeny v odezhde praviashchikh verkhov Rossii v kontse xvii v.' In *Mesto Rossii v Evrazii-The Place of Russia in Eurasia*. Ed. Szvak: 173–81.
Selm, Bert van. ''... te bekomen voor een Civielen prijs'. De Nederlandse boekprijs in de zeventiende eeuw als onbekende grootheid.' *De zeventiende eeuw* 1 (1990) 98–116.
Shepherd, J.R. *Statecraft and Political Economy on the Taiwan Frontier, 1600–1800* (Stanford, CA, 1993).
'Sindbad the Seaman and Sindbad the Landsman.' In *The Arabian Nights*. Ed. and trans. R.F. Burton (New York, 2001) 331–89.
'Skazanie Adolfa Lizeka o posol'stve ot Imperatora Leopol'da k Tsariu Aleksiiu Mikhailovichu v 1675 godu.' *Zhurnal' ministerstva narodnago prosveshcheniia* (Noiabr' 1837) 327–94.
'Skazanie letopisi o grabe Astrakhani.' In *Materialy dlia istorii vozmushcheniia Sten'ki Razina* (Moskva, 1857) 241–61.
Smith, Adam. *An Inquiry into the Nature and Causes of the Wealth of Nations* (Oxford, 1976; origin.: 1776).
Smith, George Vinal. *The Dutch in Seventeenth-Century Thailand* (DeKalb, IL, 1977).
Smith, W.D. 'The Function of Commercial Contacts in the Modernization of European Capitalism: Amsterdam as Information Exchange in the Seventeenth Century.' *Journal of Economic History* 4 (1984) 985–1005.
Sobranie gosudarstvennykh gramot i dogovorov, khraniashchikhsia v gosudarstvennoi kollegii inostrannykh del. 5 vols (Moscow, 1813–94).
Solov'ev, V.M. *Anatomiia russkogo bunta: Stepan Razin-mify i real'nost'* (Moskva, 1994).

Soloviev, S.M. *Istoriia Rossii*. Vol. 3: Parts 11–15. Third edn. (St Petersburg, 1911).

——. *Istoriia Rossii s drevneishikh vremen*. Vol 6: Parts 11 and 12 (Moscow, 1961).

——. *A History of Russia from Earliest Times*. Vol. 21. Trans. and ed. T. Allan Smith (Gulf Breeze, FL, 2000).

——. *A History of Russia from Earliest Times*. Vol. 23. Trans. and ed. M.L. Lahana (Gulf Breeze, FL, 1998).

(Sparwenfeld, J.G.) *J.G. Sparwenfeld's Diary of a Journey to Russia 1684–87*. Ed. U. Birgegard (Stockholm, 2002).

Spierenburg, P.C. *Judicial Violence in the Dutch Republic* (Amsterdam, 1978).

Spisok russkikh voennykh sudov s 1668 po 1860 god (St Petersburg, 1872).

Spufford, M. 'Literacy, Trade and Religion in the Commercial Centres of Europe.' In *A Miracle Mirrored*. Eds Davids and Lucassen: 229–83.

Stagl, Justin. *A History of Curiosity: The Theory of Travel, 1550–1800* (Chur, 1995).

Starkov, V.F., V.L. Derzhavin. *Ekspeditsiia Villema Barentsa na Novoi Zemle, 1596–1597 gg.* (Moskva, 2003).

Stevens, C.B. *Soldiers on the Steppe: Army Reform and Social Change in Early Modern Russia* (De Kalb, IL, 1995).

Stiffe, A.W. 'Ancient Trading Centres of the Persian Gulf: VI. Bandar Abbas.' *Geographical Journal* 2 (1900) 211–15.

Struys, J.J., D. Butler. *Tri puteshestviia*. Ed. A.I. Tsepkov (Riazan' 2006).

Swart, K.W. 'Holland's Bourgeoisie and the Retarded Industrialization of the Netherlands.' In *Failed Transitions*. Eds Krantz and Hohenberg: 44–8.

Swift, Jonathan. *Gulliver's Travels*. Ed. Louis Landa (Boston, 1960; original: Lemuel Gulliver, *Travels into Several Remote Nations of the World*. In Four Parts. London, 1726).

Tagebuch des Generalen Patrick Gordon. Eds and trans. M.A. Obolenskii and M.C. Posselt (Moscow and St Petersburg, 1849–53).

Tavernier, J.-B. *Les Six Voyages en Turquie et en Perse et aux Indes*. 2 vols (Paris: G. Clouzier, 1676–7).

Tavernier, Roger. *Russia and the Low Countries: An International Bibliography, 1500–2000* (Groningen, 2006).

Temple, Sir William. *Observations upon the United Provinces of the Netherlands*. Ed. G. Clark. Oxford, 1972 (London: Gellibrand, 1673).

Thévenot, J. *Rélation d'un Voyage fait au Levant* (Paris: Louis Billaine, 1665).

Thomas, K. *Man and the Natural World* (Oxford, 1996).

Thomas, Nicholas. *Colonialism's Culture: Anthropology, Travel and Government* (Cambridge, 1994).

Tignor, R., et al. *Worlds Together, Worlds Apart* (New York, 2002).

Tikhomirov, B.N. *Razinshchina* (Moskva-Leningrad, 1930).

Tillyard, R.M.W. *The Elizabethan World Picture* (New York, 1944).

Tönnies, F. *Gemeinschaft und Gesellschaft* (Leipzig, 1887).

Torntoft, P. 'William III and Denmark–Norway, 1697–1702.' *The English Historical Review* 81 (January 1966) 1–25.

Toynbee, A. *Civilization on Trial* (New York, 1948).

Trap, J.P. *Statistisk-topographisk Beskrivelse of Hertugdømmet Slesvig*. Vol. 2. (Kjøbenhavn, 1864).

Tri puteshestviia Ia.Ia. Streis. Ed. A. Morozov. Trans. E. Borodina (Moskva, 1935).

Troost, W. *Stadhouder-Koning Willem III: Een Politieke Biografie* (Hilversum, 2001).

Tsvetaev, D. *Protestantstvo i protestanty v Rossiiu epokhi preobrazovanii* (Moskva, 1890).

Tushin, Iu.P. *Russkoe moreplavanie na Kaspiiskom, Azovskom i Chernom moriakh (XVII vek)* (Moscow, 1978).

Tverskaia, D. *Moskva vtoroi poloviny XVII veka* (Moskva, 1959).

De Twee Vermaarde Fortuyns Kinderen ofte Het Wonderlyk Leeven en zeldsame bedryven van Jonker Michiel van der Moesel, Heere van Toornvliet en Van Niclaes de Molembais (Delft: J. de Wilde, 1681).

Uchenie i khitrost' ratnogo stroeniia pekhotnykh liudei (Moscow: Moskovskii pechatnyi dvor, 1647).

Uhlenbeck, C.C. *Verslag aangaande een onderzoek van de archieven van Rusland ten bate der Nederlandsche geschiedenis* ('s-Gravenhage, 1891).

Ulenspiegel (Antwerp: van Hoochstraten, *c.*1530).

Unger, R.W. *Dutch Shipbuilding before 1800: Ships and Guilds* (Assen, 1978).

Uroff, Benjamin P. 'Grigory Karpovich Kotoshikhin, "On Russia in the Reign of Alexis Mikhailovich": An Annotated Translation.' Unpublished Ph.D. Dissertation (New York: Columbia University, 1970).

Valentijn, François. *Oud en Nieuw Oost-Indië, vervattende een naauwkeurige en uitvoerige verhandelinge van Nederlands mogentheyt in die gewesten* (Dordrecht and Amsterdam: J. van Braam, G. onder de Linden, 1724–6).

Valkenier, P. *'t Verwerd Europa* (Amsterdam: H. and D. Boom, 1675).

Valle, Pietro della. *Viaggi di Pietro della Valle, il Pellegrino*. 4 vols (Rome: Vitale Mascarde, 1650–3).

Veen, Ernst van. 'How the Dutch Ran a Seventeenth-Century Colony: The Occupation and Loss of Formosa, 1624–1662.' *Itinerario* 20 (1996): 59–77.

Veer, Gerrit de. *Waerachtige beschryvinghe van Willem Barents* (Den Haag, 1917; first edition: Amsterdam: Cornelis Claesz, 1598–9).

Velden, H. van der. 'Cambyses for Example.' *Simiolus* 1 (1995) 5–39.

——. 'Cambyses Reconsidered: Gerard David's *Exemplum Iustitiae* for Bruges Town Hall.' *Simiolus* 1 (1995) 40–62.

Veluvenkamp, Ia. V. 'Kompaniia 'de Vogelar i Klenk' v gollandsko-russkikh kommercheskikh otnosheniiakh xvii v.' In *Niderlandy i Severnaia Rossiia*. Ed. Besplatnikh: 37–73.

Veluwenkamp, J.W. *Archangel: Nederlandse ondernemers in Rusland, 1550–1785* (Amsterdam, 2000).

——. 'The Arkhangelsk Business Venture of the Amsterdam Merchant David Leeuw, 1714–1724.' In *Around Peter the Great*. Eds C. Horstmeier et al.: 92–102.

'Verantwoording.' In *Iournael ofte Gedenckwaedige beschrijvinghe*. Eds Verhoeven, Verkruijsse: 19–33.

Verhaal van drie voorname reizen naar Oostindien; te weten van Johan Jacobsz Saar, Volkert Evertsz, en Albrecht Herport. Trans. J. H. Glazemaker (Amsterdam: J. Rieuwertsz., P. Arentsz, 1671).

Verhoeven, G., P. Verkruijsse. 'De vele gezichten van Bontekoe.' In *Iournael ofte Gedenckwaedige beschrijvinghe*. Eds Verhoeven, Verkruijsse: 39–79.

Veselago, F. 'Botik' dedushka Russkago flota.' In *Svedeniia o botike dedushke russkago flota za 200 let' s' 1688 po 1888 god'* (Sankt Peterburg, 1888) 19–43.

——. *Kratkaia istoriia russkago flota*. 2 vols (Sankt Peterburg: 1893–5).

——. *Kratkaia istoriia russkogo flota*. Second edn. (Moskva, Leningrad, 1939).

Vink, M.P.M. 'Images and Ideologies of Dutch-South Asian Contact.' *Itinerario* 2 (1997) 82–123.

Viskovatov, A. *Kratkii istoricheskii obzor morskikh pokhodov russkikh i morekhodstva ikh voobshche do iskhoda XVII stoletiia* (Sankt Peterburg, 1994).

Vitsen, Nikolaas. *Puteshestvie v Moskoviiu 1664–1665. Dnevnik*. Trans. V.G. Grisman (Sankt Peterburg, 1996).

Vles, J. *Le roman picaresque hollandais des XVIIe et XVIIIe siècles et ses modèles espagnols et français* ('s-Gravenhage, 1926).

Vliet, A.P. van. 'De Staatse Vloot in de Tachtigjarige Oorlog.' In *Met Man en Macht*. Eds Bruijn and Wels: 44–62.

Vliet, Ieremias van. *Naukeurige Beschryvinge van het Koningryck Siam* (Leiden: F. Haaring, 1692).

Vliet, Jeremias van. *Historiael Verhael der Sieckte ende Doot van Pra Interra Tsia 22en Coninck in Siam & den Regherende Pra Onghsry 1640* (Tokyo, 1956).

Voltaire (François-Marie Arouet). *Candide, ou L'optimisme* (Geneve: Cramer, 1759).

——. *Dictionnaire philosophique*. Fifth edn. (London, Amsterdam: Rey, 1765). (Available at http://www.voltaire-integral.com. Accessed 6 August 2006).

Vondel, Joost van den. *Op den Zeetriomf der Heerschappije van Venetië*. In *De Werken van Vondel*. Ed. J. Sterck et al. Vol. 8 (1656–1660) (Np.: 1656). (Available at http://www.dbnl.org/tekst/vond001dewe08_01/vond001dewe08_01_0016.htm. Accessed 12 April 2007).

Voyages de Monsieur le Chevalier Chardin, en Perse, et autres Lieux de l'Orient. Vol. 3 (Amsterdam: Jean Louis de Lorme, 1711).

Vries, J. de. *The Economy of Europe in an Age of Crisis* (Cambridge, 1976).

——. 'Holland: Commentary.' In *Failed Transitions*. Eds Krantz and Hohenberg: 55–7.

——. and A. van der Woude. *The First Modern Economy: Success, Failure, and Perseverance of the Dutch Economy, 1500–1815* (Cambridge, 1997).

Waeyen, Johannes van der. *Ernstige betuiginge der gereformeerde kercke tot wederlegginge van den gronden van sr. Jean de Labadie* (Amsterdam: van Someren, 1670).

Wagner, P. 'Introduction.' In John Cleland, *Fanny Hill or Memoirs of a Woman of Pleasure* (Harmondsworth, 1985) 7–30.

Waller, F.G. *Biographisch Woordenboek van Noord Nederlandsche Graveurs* (Amsterdam, 1974).

Wallerstein, I. *The Modern World System*. 3 vols (London, 1974–1989).

Weber, M. 'Die protestantische Ethik und der Geist des Kapitalismus.' In M. Weber, *Gesammelte Aufsätze zur Religionssoziologie*. Vol. 1 (Tübingen, 1922) 17–206.

——. *Economy and Society: An Outline of Interpretive Sociology*. Eds G. Roth, C. Wittich. (Berkeley, CA, 1978).

Westermann, M. *A Worldly Art* (New Haven, CT, 2004).

Wheatcroft, Andrew. 'Orientalism: The Impact of the Image of the East on the West.' Paper presented at the *Annual Conference of the Society for the History of Authorship, Reading and Publishing* (The Hague, 12 July 2006).

White, H. 'Foucault Decoded: Notes from Underground.' *History and Theory* 1 (1973) 23–54.

White, R. *The Middle Ground: Indians, Empires, and Republics in the Great Lakes Region, 1650–1815* (Cambridge, 1991).

Wijnroks, E. *De Handel tussen Rusland en de Nederlanden 1560–1640* (Hilversum, 2003).
Williams, R. *Keywords*. Rev. edn. (London, 1983).
Wills, J. Review of *Colbert, Mercantilism, and the French Quest for Asian Trade* by Glenn Ames. *Annals of the American Academy of Political and Social Science* (March 1998) 206–7.
Witsen, N.C. *Aeloude en hedendaegsche scheeps-bouw en bestier* (Amsterdam: C. Commelijn et al., 1671).
——. *Architectura navalis et Regimen Nauticum. Ofte Aaloude en Hedendaagsche Scheeps-Bouw en Bestier* (Amsterdam: Pieter and Joan Blaeu, 1690).
——. *Noord en Oost Tartarye, ofte bondig ontwerp van enige dier landen en volken, welke voormaels bekent zijn geweest* (Amsterdam: F. Halma, 1705; first edn.: Amsterdam: Np., 1692).
——. *Moscovische reyse 1664–5. Journaal en aentekeningen*. 3 vols (Den Haag, 1966–67).
Wittram, R. *Peter I, Czar und Kaiser*. 2 vols (Göttingen, 1964).
Wladimiroff, I. 'Andries Winius and Nicolaas Witsen, Tsar Peter's Dutch Connection.' In *Around Peter the Great*. Eds C. Horstmeier et al. (Groningen, 1997) 5–23.
Wolff, L. *Inventing Eastern Europe: The Map of Civilization on the Mind of the Enlightenment* (Stanford, CA, 1994).
Die wonderlijcke reyse van Jan Mandevijl (Antwerpen: Wed. Guilleam van Parijs, 1592).
Woude, A.M. van der. *Het Noorderkwartier*. 3 vols (Wageningen, 1972).
Wrigley, E.A. 'The Divergence of England.' *Transactions of the Royal Historical Society*. Sixth series. Vol. 10 (2002) 117–41.
www.vocsite.nl/schepen/detail.html.
Zabelin, I. *Domashnii byt' Russkikh' tsarei*. 2 vols. Second edn. (Moskva, 1872).
Zagoskin, N.P. *Russkie vodnye puti i sudovoe delo v do-petrovskoi Rossii* (Kazan, 1909).
Zahedieh, N. 'Overseas Expansion and Trade.' In *The Origins of Empire*. Ed. Canny: 398–422.
Zapadnoevropeiskie kuptsy i ikh tovary v Rossii xvii veka. Eds A.V. Demkin, A.A. Preobrazhenskii (Moskva, 1992).
Zapiski Ia.K. Nomena o prebyvanii Petra Velikago v Niderlandakh v 1697/98 i 1716/17 gg. Ed. V. Kordt (Kiev, 1904).
Zapiski inostrantsev o vosstanii Stepana Razina. Vol. 1. Ed. A.G. Man'kov (Leningrad, 1968).
Zevakin, E.S. 'Persidskii vopros v russko-evropeiskikh otnosheniiakh XVII v.' *Istoricheskie zapiski* 8 (1940) 129–62.
Zhivov, V.M. 'The Emergence of the Individual.' In *Religion and Culture in Early Modern Russia and Ukraine*. Eds S.H. Baron and N. Shields Kollmann (DeKalb, IL, 1997) 184–98.
Zuber, R., M. Cuénin. *Littérature française*. Vol. 4 (Paris, 1984).
Zuiden, D.S. van. 'Nieuwe bijdrage tot de kennis van de Hollandsch-Russische relaties in de 16e-18e eeuw.' *Economisch-Historisch Jaarboek* 2 (1916) 258–95.

Index